T0203005

Lecture Notes in Computer Science 14635

The series Lecture Notes in Computer Science (LNCS), including its subseries Lecture Notes in Artificial Intelligence (LNAI) and Lecture Notes in Bioinformatics (LNBI), has established itself as a medium for the publication of new developments in computer science and information technology research, teaching, and education.

LNCS enjoys close cooperation with the computer science R & D community, the series counts many renowned academics among its volume editors and paper authors, and collaborates with prestigious societies. Its mission is to serve this international community by providing an invaluable service, mainly focused on the publication of conference and workshop proceedings and postproceedings. LNCS commenced publication in 1973.

Stephen Smith · João Correia · Christian Cintrano
Editors

Applications of Evolutionary Computation

27th European Conference, EvoApplications 2024
Held as Part of EvoStar 2024
Aberystwyth, UK, April 3–5, 2024
Proceedings, Part II

Springer

Editors
Stephen Smith 🆔
University of York
York, UK

João Correia 🆔
University of Coimbra
Coimbra, Portugal

Christian Cintrano 🆔
University of Málaga
Málaga, Spain

ISSN 0302-9743   ISSN 1611-3349 (electronic)
Lecture Notes in Computer Science
ISBN 978-3-031-56854-1   ISBN 978-3-031-56855-8 (eBook)
https://doi.org/10.1007/978-3-031-56855-8

This Springer imprint is published by the registered company Springer Nature Switzerland AG
The registered company address is: Gewerbestrasse 11, 6330 Cham, Switzerland

Paper in this product is recyclable.

Preface

This volume contains the proceedings of EvoApplications 2024, the International Conference on the Applications of Evolutionary Computation. The conference was part of Evo*, the leading event on bio-inspired computation in Europe, and was held in Aberystwyth, UK, as a hybrid event, between Wednesday, April 3, and Friday, April 5, 2023.

EvoApplications, formerly known as EvoWorkshops, aims to bring together high-quality research focusing on applied domains of bio-inspired computing. At the same time, under the Evo* umbrella, EuroGP focused on the technique of genetic programming, EvoCOP targeted evolutionary computation in combinatorial optimization, and EvoMUSART was dedicated to evolved and bio-inspired music, sound, art, and design. The proceedings for these co-located events are available in the LNCS series.

EvoApplications 2024 received 77 high-quality submissions distributed among the main session on Applications of Evolutionary Computation and 10 additional special sessions chaired by leading experts on the different areas: Analysis of Evolutionary Computation Methods: Theory, Empirics, and Real-World Applications (Thomas Bartz-Beielstein, Carola Doerr, and Christine Zarges); Applications of Bio-inspired Techniques on Social Networks (Giovanni Iacca and Doina Bucur); Computational Intelligence for Sustainability (Valentino Santucci, Fabio Caraffini, and Jamal Toutouh); Evolutionary Computation in Edge, Fog, and Cloud Computing (Diego Oliva, Seyed Jalaleddin Mousavirad, and Mahshid Helali Moghadam); Evolutionary Computation in Image Analysis, Signal Processing, and Pattern Recognition (Pablo Mesejo and Harith Al-Sahaf); Machine Learning and AI in Digital Healthcare and Personalized Medicine (Stephen Smith and Marta Vallejo); Problem Landscape Analysis for Efficient Optimisation (Bogdan Filipič and Pavel Krömer); Resilient Bio-inspired Algorithms (Carlos Cotta and Gustavo Olague); Soft Computing Applied to Games (Alberto P. Tonda, Antonio M. Mora, and Pablo García-Sánchez); and Surrogate-Assisted Evolutionary Optimisation (Tinkle Chugh, Alma Rahat, and George De Ath). We selected 24 of these papers for full oral presentation, while 9 works were presented in short oral presentations and as posters. Moreover, these proceedings also include contributions from the Evolutionary Machine Learning (EML) joint track, a combined effort of the International Conference on the Applications of Evolutionary Computation (EvoAPPS) and European Conference on Genetic Programming (EuroGP), organized by Penousal Machado and Mengjie Zhang. EML received 28 high-quality submissions. After careful review, eleven were selected for oral presentations and six for short oral presentations and posters. Since EML is a joint track, the "Evolutionary Machine Learning" part of these proceedings contains 16 of these papers. The remaining one is published in the EuroGP proceedings. All accepted contributions, regardless of the presentation format, appear as full papers in this volume.

An event of this kind would not be possible without the contribution of a large number of people:

- We express our gratitude to the authors for submitting their works and to the members of the Program Committee for devoting selfless effort to the review process.
- We would also like to thank Nuno Lourenço (University of Coimbra, Portugal) for his dedicated work as Submission System Coordinator.
- We thank Evo* Graphic Identity Team, Sérgio Rebelo, Jéssica Parente, and João Correia (University of Coimbra, Portugal), for their dedication and excellence in graphic design.
- We are grateful to Zakaria Abdelmoiz (University of Málaga, Spain) and João Correia (University of Coimbra, Portugal) for their impressive work managing and maintaining the Evo* website and handling the publicity, respectively.
- We credit the invited keynote speakers, Jon Timmis (Aberystwyth University, UK) and Sabine Hauert (University of Bristol, UK), for their fascinating and inspiring presentations.
- We would like to express our gratitude to the Steering Committee of EvoApplications for helping organize the conference.
- Special thanks to Christine Zarges (Aberystwyth University, UK) as local organizer and to Aberystwyth University, UK, for organizing and providing an enriching conference venue.
- We are grateful to the support provided by SPECIES, the Society for the Promotion of Evolutionary Computation in Europe and its Surroundings, for the coordination and financial administration.

Finally, we express our continued appreciation to Anna I. Esparcia-Alcázar, from SPECIES, Europe, whose considerable efforts in managing and coordinating Evo* helped build a unique, vibrant, and friendly atmosphere.

April 2024

Stephen Smith
João Correia
Christian Cintrano

Organization

EvoApplications Conference Chair

Stephen Smith   University of York, UK

EvoApplications Conference Co-chair

João Correia   University of Coimbra, Portugal

EvoApplications Publication Chair

Christian Cintrano   University of Málaga, Spain

Analysis of Evolutionary Computation Methods: Theory, Empirics, and Real-World Applications Chairs

Thomas Bartz-Beielstein   TH Köln, Germany
Carola Doerr   CNRS and Sorbonne Université, France
Christine Zarges   Aberystwyth University, UK

Applications of Bio-inspired Techniques on Social Networks Chairs

Giovanni Iacca   Università di Trento, Italy
Doina Bucur   University of Twente, The Netherlands

Computational Intelligence for Sustainability Chairs

Valentino Santucci   Università per Stranieri di Perugia, Italy
Fabio Caraffini   Swansea University, UK
Jamal Toutouh   University of Málaga, Spain

Evolutionary Computation in Edge, Fog, and Cloud Computing Chairs

Diego Oliva   Universidad de Guadalajara, México
Seyed Jalaleddin Mousavirad   Hakim Sabzevari University, Iran
Mahshid Helali Moghadam   RISE Research Institutes of Sweden, Sweden

Evolutionary Computation in Image Analysis, Signal Processing, and Pattern Recognition Chairs

Pablo Mesejo   Universidad de Granada, Spain
Harith Al-Sahaf   Victoria University of Wellington, New Zealand

Machine Learning and AI in Digital Healthcare and Personalized Medicine Chairs

Stephen Smith   University of York, UK
Marta Vallejo   Heriot-Watt University, UK

Problem Landscape Analysis for Efficient Optimisation Chairs

Bogdan Filipič   INRAE, Jožef Stefan Institute, Slovenia
Pavel Krömer   Technical University of Ostrava, Czech Republic

Resilient Bio-inspired Algorithms Chairs

Carlos Cotta   University of Málaga, Spain
Gustavo Olaguer   CICESE, Mexico

Soft Computing Applied to Games Chairs

Alberto P. Tonda   INRAE, France
Antonio M. Mora   Universidad de Granada, Spain
Pablo García-Sánchez   Universidad de Granada, Spain

Surrogate-Assisted Evolutionary Optimisation Chairs

Tinkle Chugh   University of Exeter, UK
Alma Rahat   Swansea University, UK
George De Ath   University of Exeter, UK

Evolutionary Machine Learning Chairs

Penousal Machado   University of Coimbra, Portugal
Mengjie Zhang   Victoria University of Wellington, New Zealand

EvoApplications Steering Committee

Stefano Cagnoni   University of Parma, Italy
Pedro A. Castillo   University of Granada, Spain
Anna I. Esparcia-Alcázar   Universitat Politècnica de València, Spain
Mario Giacobini   University of Torino, Italy
Paul Kaufmann   University of Mainz, Germany
Antonio Mora   University of Granada, Spain
Günther Raidl   Vienna University of Technology, Austria
Franz Rothlauf   Johannes Gutenberg University Mainz, Germany
Kevin Sim   Edinburgh Napier University, UK
Giovanni Squillero   Politecnico di Torino, Italy
Cecilia di Chio   King's College London, UK
   (Honorary Member)

Program Committee

Jacopo Aleotti   University of Parma, Italy
Mohamad Alissa   Edinburgh Napier University, UK
Anca Andreica   Babes-Bolyai University, Romania
Claus Aranha   University of Tsukuba, Japan
Aladdin Ayesh   De Montfort University, UK
Kehinde Babaagba   Edinburgh Napier University, UK
Jaume Bacardit   Newcastle University, UK
Marco Baioletti   Università degli Studi di Perugia, Italy
Illya Bakurov   Universidade NOVA de Lisboa, Portugal
Wolfgang Banzhaf   Michigan State University, USA
Tiago Baptista   University of Coimbra, Portugal
Thomas Bartz-Beielstein   TH Köln, Germany

Giulio Biondi	University of Florence, Italy
Philip Bontrager	New York University, USA
János Botzheim	Eötvös Loránd University, Hungary
Jörg Bremer	University of Oldenburg, Germany
Will Browne	Queensland University of Technology, Australia
Doina Bucur	University of Twente, The Netherlands
Maxim Buzdalov	ITMO University, Russia
Stefano Cagnoni	University of Parma, Italy
Fabio Caraffini	De Montfort University, UK
Oscar Castillo	Tijuana Institute of Technology, Mexico
Pedro Castillo	University of Granada, Spain
Josu Ceberio	University of the Basque Country, Spain
Ying-Ping Chen	National Yang Ming Chiao Tung University, Taiwan
Francisco Chicano	University of Málaga, Spain
Anders Christensen	University of Southern Denmark, Denmark
Tinkle Chugh	University of Exeter, UK
Christian Cintrano	University of Málaga, Spain
Anthony Clark	Pomona College, USA
José Manuel Colmenar	Universidad Rey Juan Carlos, Spain
Feijoo Colomine	Universidad Nacional Experimental del Táchira, Venezuela
Stefano Coniglio	University of Southampton, UK
Antonio Cordoba	University of Seville, Spain
Oscar Cordon	University of Granada, Spain
João Correia	University of Coimbra, Portugal
Carlos Cotta	Universidad de Málaga, Spain
Fabio D'Andreagiovanni	CNRS, Sorbonne University – UTC, France
Gregoire Danoy	University of Luxembourg, Luxembourg
George De Ath	University of Exeter, UK
Amir Dehsarvi	University of York, UK
Antonio Della Cioppa	University of Salerno, Italy
Bilel Derbel	CRIStAL (Univ. Lille), France
Travis Desell	Rochester Institute of Technology, USA
Laura Dipietro	HI
Federico Divina	Pablo de Olavide University, Spain
Carola Doerr	Sorbonne University, CNRS, France
Bernabe Dorronsoro	University of Cadiz, Spain
Tome Eftimov	Jožef Stefan Institute, Slovenia
Abdelrahman Elsaid	Rocheser Institute of Technology, USA
Ahmed Elsaid	University of Puerto Rico at Mayagüez, Puerto Rico

Contents – Part II

**Machine Learning and AI in Digital Healthcare and Personalized
Medicine**

Problem Landscape Analysis for Efficient Optimization

Soft Computing Applied to Games

Surrogate-Assisted Evolutionary Optimisation

Contents – Part I

Analysis of Evolutionary Computation Methods: Theory, Empirics, and Real-World Applications

Computational Intelligence for Sustainability

Evolutionary Computation in Edge, Fog, and Cloud Computing

Evolutionary Computation in Image Analysis, Signal Processing and Pattern Recognition

Evolutionary Machine Learning

Hindsight Experience Replay with Evolutionary Decision Trees for Curriculum Goal Generation

Erdi Sayar[1]([✉]), Vladislav Vintaykin[1], Giovanni Iacca[2], and Alois Knoll[1]

[1] Technical University of Munich, Munich, Germany
{erdi.sayar,vladislav.vintaykin}@tum.de, knoll@mytum.de
[2] University of Trento, Trento, Italy
giovanni.iacca@unitn.it

Abstract. Reinforcement learning (RL) algorithms often require a significant number of experiences to learn a policy capable of achieving desired goals in multi-goal robot manipulation tasks with sparse rewards. Hindsight Experience Replay (HER) is an existing method that improves learning efficiency by using failed trajectories and replacing the original goals with hindsight goals that are uniformly sampled from the visited states. However, HER has a limitation: the hindsight goals are mostly near the initial state, which hinders solving tasks efficiently if the desired goals are far from the initial state. To overcome this limitation, we introduce a curriculum learning method called HERDT (HER with Decision Trees). HERDT uses binary DTs to generate curriculum goals that guide a robotic agent progressively from an initial state toward a desired goal. During the warm-up stage, DTs are optimized using the Grammatical Evolution algorithm. In the training stage, curriculum goals are then sampled by DTs to help the agent navigate the environment. Since binary DTs generate discrete values, we fine-tune these curriculum points by incorporating a feedback value (i.e., the Q-value). This fine-tuning enables us to adjust the difficulty level of the generated curriculum points, ensuring that they are neither overly simplistic nor excessively challenging. In other words, these points are precisely tailored to match the robot's ongoing learning policy. We evaluate our proposed approach on different sparse reward robotic manipulation tasks and compare it with the state-of-the-art HER approach. Our results demonstrate that our method consistently outperforms or matches the existing approach in all the tested tasks.

Keywords: Decision Tree · Reinforcement Learning · Curriculum Learning · Sparse Reward · Multi-goal Tasks

1 Introduction

Reinforcement learning (RL) is a well-known computational paradigm for discovering optimal actions through trial and error, so as to maximize rewards without explicit guidance [1]. In recent years, deep RL, which combines RL with

© The Author(s), under exclusive license to Springer Nature Switzerland AG 2024
S. Smith et al. (Eds.): EvoApplications 2024, LNCS 14635, pp. 3–18, 2024.
https://doi.org/10.1007/978-3-031-56855-8_1

deep neural network (DNN)-based function approximators, has made remarkable advancements and achieved impressive outcomes, exceeding human-level performance e.g. in playing Atari games [2,3], beating Go champions [4], and solving robotic tasks [5–8]. In these scenarios, the design of an effective reward function [9] is one of the most challenging aspects. This is because the reward function must be carefully tailored to the specific task at hand, and it must be able to capture the desired behavior of the agent. However, in many cases, the admissible behavior of the agent is not known in advance, which makes it difficult to design an effective reward function. As a result, binary rewards are often used in RL algorithms. This is particularly true for robotic tasks, where binary rewards are used to simply indicate task success or failure, offering a comparatively simpler alternative to the intricate design of more complex reward functions. However, as binary rewards only provide information about whether the task is completed or not, without providing more detailed information about the actual agent's progress towards the goal, RL algorithms encounter difficulties at learning effectively [10]. To address the challenge of sparse rewards, Hindsight Experience Replay (HER) [11] offers a promising solution. HER substitutes the *desired goals* with the *achieved goals*, which are sampled uniformly from the visited states. In this way, HER can convert failed episodes into successful ones. However, a limitation of HER lies in its disregard for the importance of the states visited during the learning process, resulting in sample inefficiency. Different methods have been introduced to address the issue of sample inefficiency in RL, e.g. through prioritization of the replay buffer, such as Energy-Based Prioritization [12], which prioritizes experiences with higher energy, and Maximum Entropy-based Prioritization [13], which samples replay trajectories more frequently based on entropy.

Another challenge in deep RL is its explainability. In this regard, the growing interest in explainable RL [14] has been driven by the general lack of interpretability of deep RL, being built on top of opaque DNN [15]. Interpretability can be achieved by using interpretable models, such as Decision Trees (DTs), which can be highly interpretable, as long as they are shallow [16]. To interpret an RL agent with a DT, various approaches have been proposed. Coppens et al. [17] proposed distilling the output of a pre-trained deep RL policy network into a "soft" DT. Another method, called VIPER [18], uses a policy extraction technique to convert a complex, high-performing DNN-based policy into a simpler DT-based policy. Ding et al. [19] used Cascading DTs (CDTs), where the feature learning tree is cascaded with a decision-making tree. Roth et al. [20] proposed Conservative Q-Improvement (CQI) to produce a policy in the form of a DT, resulting in smaller trees compared to existing methods, without sacrificing policy performance. Hallawa et al. [21] proposed a methodology based on Genetic Programming (GP) to produce Behavior Trees (BTs) combined with various forms of RL such as Q-learning, DQN, and PPO. Other authors extended this approach by using Grammatical Evolution (GE) to produce DTs in combination with Q-learning performing online learning on the leaves of the tree, and testing the resulting agents on various RL tasks, including tasks with discrete action spaces [22–24], tasks with continuous action spaces [25], and multi-agent tasks [26]. However, most of these approaches focus on the use of DTs for defining the policy of the agent. To the best of our knowledge, no previous work has used DTs to generate curriculum goals during the training process, hence guiding the agent to the desired goal even in contexts with sparse rewards. This is precisely the main focus of the present work. Our hypothesis is that by using DTs, it is possible to generate curriculum waypoints in a straightforward and interpretable way, hence facilitating the agent's navigation and task solving, as opposed to directly generating control commands to the agent using DTs. Here, we propose a curriculum learning approach based on DTs, that we dub Hindsight Experience Replay with Decision Trees (HERDT). Our methodology works as follows. Initially, we transform a robotic environment into a straightforward grid representation, comprising an initial position and a desired goal. Subsequently, we employ GE to optimize the DT structures, which are then further optimized by means of Q-learning, as done in [22]. By doing so, we ensure that the DTs effectively capture the environment's characteristics and decision-making process. Then, we leverage the optimized DTs to generate curriculum goals that serve as navigational waypoints for guiding the agent through the environment. In the experimentation, we compare our method with the baseline HER on standard multi-goal RL benchmark tasks and perform thorough ablation studies. To summarize, the main contributions of this paper are the following:

- We propose HERDT, a curriculum learning approach composed of two stages, namely: (1) A *warm-up stage*, where we construct a grid environment with the same initial and desired positions as in the robotic environment. We then use the GE algorithm to optimize binary DTs with the grid environment. As a result, DTs acquire capabilities to provide guidance to an agent in the robotic environment. (2) A *training stage*, where we sample curriculum goals from those optimized binary DTs to guide the robotic agent toward the desired goal. Since binary DTs output discrete values given an input state, we further fine-tune these curriculum points to adjust their difficulty and ensure that they are neither too easy nor too hard for the robot's ongoing learning policy.
- We compare our approach against HER on a set of benchmark tasks with the 7-DOF Fetch Robotic-arm MuJoCo simulation environment [27].
- In the ablation studies, we investigate the impact of recursively generated curriculum goals and the effect of the fine-tuning of the curriculum points on the success rate of the tasks at hand.

The rest of the paper is structured as follows. The next section provides the background concepts. Then, Sect. 3 describes our proposed method. Section 4 presents the numerical results. Finally, Sect. 5 provides the conclusions.

2   Background

Hindsight Experience Replay (HER). As mentioned earlier, RL methods often struggle to explore the environment effectively, especially under sparse reward conditions. HER [11] addresses this sample efficiency problem by enabling agents to learn from their failures. When an agent fails to achieve its desired goal, HER replays the experience as if the agent had achieved a different and achievable goal. By doing so, the agent can receive a reward and learn at least how to accomplish a task from the achieved states. In this way, HER converts failed episodes into successful ones.

GE with Q-Learning. According to the GE with Q-learning approach proposed in [22], a population of genotypes, each one encoded as a fixed-length list of codons, is evolved. During the evaluation, each genotype is converted into a phenotype, which represents the policy of the agent based on a binary DT. Then, the agent acts accordingly in the environment and receives a reward signal. A Q-value is calculated from the reward signal, using the Q-learning approach [28], and the cumulative reward is used as a fitness value for selecting the individuals in GE (here, an "individual" refers to a DT). Then, a standard one-point crossover operator is applied, which simply sets a random cutting point and creates two individuals by mixing the two sub-strings of the genotype. This means that individuals whose genes are not expressed in the phenotype are not pruned. Then, a classical uniform mutation operator mutates each gene according to a given probability. In this way, DT policies can be optimized to perform optimal actions leading to receiving a high reward from the environment.

Multi-goal RL. Multi-goal RL, in which an agent learns to achieve multiple goals sampled from a goal distribution, can be modeled as a goal-directed Markov decision process with continuous state and action spaces $\langle \mathcal{S}, \mathcal{A}, \mathcal{G}, \mathcal{T}, r, p, \gamma \rangle$, where $\mathcal{S}$ is a continuous state space, $\mathcal{A}$ is a continuous action space, $\mathcal{G}$ is a goal distribution (indicating with g a desired goal sampled from $\mathcal{G}$), $\mathcal{T}(s'|s,a)$ is the transition function, $r(s, g)$ denotes the immediate reward obtained by an agent upon reaching state $s \in \mathcal{S}$ given goal g, $p(s_0, g)$ is a joint probability distribution over initial states and desired goals, and $\gamma \in [0, 1]$ is a discount factor. Specifically, the reward function is defined as:

$$r(s, g) = \mathbf{1}[\|\phi(s) - g\|_2 \leq \epsilon] - 1, \tag{1}$$

where $\mathbf{1}$ is the indicator function, ϕ is a predefined function that maps a state to the achieved goal[1], and ϵ is a fixed threshold. In this context, the learning task can be modeled as an RL problem that seeks a policy $\pi : \mathcal{S} \times \mathcal{G} \to \mathcal{A}$, with the primary objective of maximizing the expected discounted sum of rewards for any given goal.

While HER can be applied with various off-policy RL algorithms, in this work we opt for utilizing DDPG [29], in alignment with the HER setting presented in

---

[1] The physical interpretation of the achieved goal depends on the task at hand. For some robotic manipulation tasks, the robot needs to pick and place (Fig. 5b), push (Fig. 5c), or slide (Fig. 5d) an object. In this case, the achieved goal corresponds to the x-y-z position of the object. Conversely, if there is no object in the task (Fig. 5a), the achieved goal is defined as the position of the end-effector of the robot.

[30]. DDPG is an off-policy actor-critic algorithm that consists in a deterministic policy $\pi_\theta(s, g) : \mathcal{S} \times \mathcal{G} \to \mathcal{A}$, parameterized by θ, and a state-action value function $Q_\phi(s, a, g) : \mathcal{S} \times \mathcal{A} \times \mathcal{G} \to \mathbb{R}$, parameterized by ϕ. Gaussian noise with zero mean ($\mu = 0$) and constant std. dev. is added to the deterministic policy π_θ to improve exploration. The behavior policy, π, is then used for collecting the results over the episodes:

$$\pi(s, g) = \pi_\theta(s, g) + \mathcal{N}(\mu, \sigma^2). \tag{2}$$

The Q-value is trained by minimizing the Temporal Difference (TD) error, which is defined as the following loss function:

$$\mathcal{L}_{critic} = \mathbb{E}_{(s_t, a_t, r_t, s_{t+1}, g) \sim \mathcal{B}} \left[(y_t - Q_\phi(s_t, a_t, g))^2 \right] \tag{3}$$

where t is the timestep and $y_t = r_{t+1} + \gamma Q_\phi(s_{t+1}, \pi_\theta(s_{t+1}, g), g)$. Subsequently, the policy π is updated using policy gradient on the following loss function:

$$\mathcal{L}_{actor} = -\mathbb{E}_{(s_t, g) \sim B} \left[(Q(s_t, \pi_\theta(s_t, g), g)) \right]. \tag{4}$$

As Q-function, we use the universal value function (UVF), which is defined as $Q_\phi(s_t, a_t, g)$, where s_t is the current state, a_t is the current action, g_t is the goal, and π is the policy. The UVF is a generalization of the traditional value function that integrates the goal into the value estimation and it has been shown to be able to successfully generalize to unseen goals, which makes it a promising approach for multi-goal RL [31]. The UVF estimates the expected return from state s under policy π, given goal g.

3   Proposed Method

Our proposed HERDT method builds on some of the aforementioned related works and consists of two main parts:

- *Warm-up stage*: before starting to train the robotic agent, as in [22] we optimize binary DTs with GE while using Q-learning to learn the state-action value of the grid environment constructed starting from the robot environment, see Fig. 1. The first environment is a discrete representation of the latter, including the initial position and target position.
- *Training stage*: during this stage, the current state of the robot is provided to the DTs in each episode, and the next curriculum goals are sampled accordingly. Since the DTs produce discrete values, we refine these generated curriculum points by adjusting them based on the Q-value of the robotic agent. This refinement ensures that these points are appropriately calibrated, i.e., that they are neither overly simplistic nor excessively challenging for the robotic agent during the learning process.

The conversion to the grid environment is motivated by two primary factors. Firstly, evolving a DT in a continuous environment using GE would be too computationally intensive. Secondly, employing a binary DT to generate discrete curriculum goals is particularly suitable for grid environments.

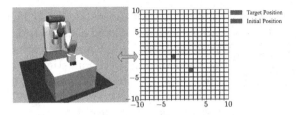

Fig. 1. Constructing a grid environment (right) from a robotic environment (left).

Table 1. Grammar rules used to evolve DTs. The symbol "|" indicates the option to select from various symbols. "comp_op" is a concise term for "comparison operator", with "lt" and "gt" representing "less than" and "greater than" operators, respectively. "input_var" indicates one of the potential inputs. It is important to emphasize that each input variable is associated with the same set of constants.

Rule	Production
dt	$< if >$
if	$if < condition > then < action > else < action >$
condition	$input\_var < comp\_op >< const >$
action	$leaf| < if >$
comp_op	$lt|gt$
$const$	$[0, 20)$ with step 1

3.1 Optimizing DTs

The grid environment is constructed using the same initial and target positions as the robotic task. As mentioned earlier, some robotic tasks, such as pushing and sliding, inherently require that robots move objects only on a flat table surface. These can be effectively modeled using a 2D grid environment, as curriculum points are generated only on the table. In contrast, tasks like reaching and pick-and-place, require the robot to navigate in a 3D space and as such they can be modeled using a 3D grid environment. We assume that we have an agent in the grid environment that can only move forward, backward, right, and left (in addition, up and down in the 3D case). For the grid environment, we define the following reward function:

$$r(s, g)_{grid} = \begin{cases} 1 & \text{if goal is reached} \\ \frac{-1}{10 \, s - g\|_2} & \text{otherwise} \end{cases} \tag{5}$$

where s and g are the agent state and target position, respectively. The DTs are evolved with the grammar rules given in Table 1, as described in [22,32]. The genotype of an individual is encoded as a fixed-length list of codons, which are represented as integers. The genotype is then translated to the corresponding phenotype, which is a DT. The policy $a_t = \pi_{DT}(\cdot)$ encoded by the tree is executed and the reward signals are obtained from the grid environment for each timestep t. Then, the average reward $R_t = \frac{1}{n} \sum_t^n r_t$ is used as fitness value for GE and at the same time the Q-values of the DT's leaves are updated using the Q-learning approach [28]:

$$Q(s_t, a_t) = Q(s_t, a_t) + \alpha \left(R_{t+1} + \gamma \max_a Q(s_{t+1}, a) - Q(s_t, a_t) \right). \qquad (6)$$

Then, the action for each leaf is selected as $\arg\max_a Q(s, a)$. Each action corresponds to a fixed-step movement on the grid environment. These actions are employed as curriculum goals during the robot's training and are scaled by the Q-value of the critic network to ensure they are appropriately challenging, without being overly difficult for the robot. In the selection process, a parent is replaced if its offspring has a better fitness value. The standard one-point crossover operator is used as a crossover mechanism, randomly selecting a cutting point and generating two individuals by mixing the two sub-strings of the genotype. A standard uniform mutation operator is utilized to randomly select a new value from within the variable's range of variation. This process is visually illustrated in Fig. 2.

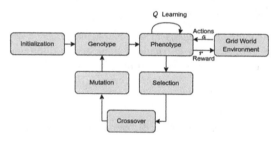

Fig. 2. Evolutionary process for the DTs.

Table 2. DT actions.

$a = \pi_{DT}$	Description
0	+x direction
1	−x direction
2	+y direction
3	−y direction
4	+z direction
5	−z direction

3.2   Generating Curriculum Goals

As seen, the actions generated by the DT policy in the grid environment serve as intermediate goals for guiding the robot toward the final goal. These actions are discrete values that describe the next step to move into the grid environment, thus determining the curriculum points for the robot.

After optimizing the DT policies, we provide the current state of the robot, s_t, to the state-to-goal mapping function $\phi(s)$, which is given to the DT policy, which in turn outputs a discrete value between 0 and 5 (0 and 3 in 2D) as the action. All the actions and their corresponding definitions are listed in Table 2. Each value corresponds to the next step in the grid environment. For example, the action 0 means that the robot should move an object to the next position in the positive x dimension in the grid environment.

The action (i.e., the next target) of the DT policy depends on the grid size. Specifically, if the grid size is small, then the next target will be closer to a task location that has already been reached by the robot. However, this can lead to a slower optimization process for the DTs. Consequently, the robot cannot efficiently improve its policy further as it has already explored the area around its current position. On the other hand, if the grid size is large, then the next target will be far from the robot's current position. This can lead to faster optimization of the DT, but it may not be optimal for the robot as it will not guide the robot efficiently.

Because of these reasons, we multiply the action output from the DT policy by a feedback value (i.e., the Q-value from the critic network) so that we can adjust the next target position to be neither too difficult nor too easy for the robot. An example of curriculum goals generated throughout the entire training process is shown in Fig. 3, with colors ranging from red to blue representing the curriculum goals across different episodes of the training.

Fig. 3. Visualization of the generated curriculum goals in the FetchPush task. The colors (ranging from red to light blue) represent the curriculum goals across different episodes of the training. The gray color represents the desired positions. (Color figure online)

The overall HERDT algorithm is given, in the form of pseudo-code, in Algorithm 1.

In the warm-up stage (lines 7–14), as explained in Sect. 3.1, the DT policy generates discrete actions, such as 'up' or 'down', based on the corresponding grid location of the given achieved goal. These actions are interpreted according to Table 2. By integrating the action from the DT policy into the current achieved goal, we can determine the subsequent goal for the robotic agent to pursue. Since the DT policy outputs discrete actions, directly adding these actions to the current achieved goal might lead to abrupt jumps from one goal to the next one, which may be hard for the robot to achieve. To address this, we introduce a refinement step in line 18, where κ (the curriculum goal) is multiplied by a factor proportionate to the Q-value. The Q-value represents the expected return for the robot when it starts from a specific state and acts according to the policy. A higher Q-value indicates a higher expected reward for that state, suggesting that the robot has effectively learned that particular state-action pair. As a result, in situations where the Q-value is high, we infer that the robot has learned the state well, and we adjust the next desired goal to be slightly farther away compared to scenarios with lower Q-values. We can achieve this by calculating $\nu = \kappa \cdot |c - Q'|$, where ν indicates the adjusted curriculum goal, c is a hyperparameter, and Q' is defined as follows:

$$Q' = \begin{cases} -1 & \text{if } Q < -1 \\ 0 & \text{if } Q > 0 \\ Q & \text{otherwise.} \end{cases} \tag{7}$$

For each timestep, the current policy π of the robotic agent (the actor neural network) outputs the action a_t, given the current state s_t and the desired goal ν (line 20), which is then executed in the robotic environment, and the next state and reward are received (line 21). Subsequently, the experiences $(s_t, a_t, r_t, s_{t+1}, \nu)$ are collected and added to the replay buffer (line 22). If the achieved state is regarded as a hindsight goal (HER idea), then the reward is recalculated and added to the replay buffer (lines 23–25). A minibatch is sampled from the replay buffer. The Q-value function and the policy π are updated using minibatch in the loss functions Eq. (3) and Eq. (4), respectively (lines 29–30). In the test rollout, we calculate the success rate by setting the target goal and

---

Algorithm 1. Hindsight Experience Replay with Decision Trees (HERDT)

1: **Input:** no. of iterations K, no. of epochs L, no. of episodes M, no. of timesteps T, no. of test rollouts n_{test}, ϵ
2: Select an off-policy algorithm $\mathbb{A}$ ▷ In our case $\mathbb{A}$ is DDPG
3: Initialize replay buffer $\mathcal{B} \leftarrow \emptyset$
4: Sample initial state s_0 and target goal g from robotic environment
5: Construct grid environment
6: Initialize DTs with the grammar rules in Table 1 and randomly assign values to leaf nodes
7: **for** $t = 1 \ldots K$ **do**   ▷ Warm-up stage
8:       Encode genotype of a DT
9:       Translate the genotype into a phenotype $a_{DT} = \pi_{DT}(s)$
10:      Execute the action a_{DT} in the grid environment
11:      Obtain the next state s_{t+1} and reward r_t
12:      Update Q-value in Eq. (6) and use it as fitness value for GE
13:      Select individuals, apply crossover and mutation operators
14: **end for**
15: **for** epoch $= 1 \ldots L$ **do**   ▷ Training stage
16:      **for** episode $= 1 \ldots M$ **do**
17:          Sample state s and curriculum goal $\kappa = \phi(s) + \pi_{DT}(\phi(s))$
18:          $\nu = \kappa \cdot |c - Q'|$ ▷ Eq. (7)
19:          **for** $t = 1 \ldots T$ **do**   ▷ Rollout episode
20:              $a_t = \pi(s_t, \nu)$
21:              Execute the action a_t, obtain the next state s_{t+1} and reward r_t
22:              Store transition $(s_t, a_t, r_t, s_{t+1}, \nu)$ in replay buffer $\mathcal{B}$
23:              Sample additional goals from achieved states for replay $G := \mathcal{S}(\text{episode})$ ▷ Hindsight goal
24:              **for** $g' \in G$ **do**
25:                  Recompute reward r'_t
26:                  Store transition $(s_t, a_t, r'_t, s_{t+1}, g')$ in replay buffer $\mathcal{B}$
27:              **end for**
28:          **end for**
29:          Sample a minibatch b from replay buffer $\mathcal{B}$
30:          Update Q and π with b to minimize $\mathcal{L}_{critic}$ in Eq. (3) and $\mathcal{L}_{actor}$ in Eq. (4)
31:      **end for**
32:      $success\_rate \leftarrow 0$
33:      **for** $t = 1 \ldots n_{test}$ **do**   ▷ Test rollouts
34:          $a_t = \pi(s_t, g)$
35:          Execute the action a_t, obtain a next state s_{t+1} and reward r_t
36:          **if** $\|\phi(s_{t+1}) - g\|_2^2 \leq \epsilon$ **then**   ▷ Eq. (1)
37:              $success\_rate \leftarrow success\_rate + 1/n_{test-rollouts}$
38:          **end if**
39:      **end for**
40: **end for**

---

evaluating how many rollouts out of n_{test} successfully achieve the target goal (lines 32–39).

4   Experimental Results

In this section, we assess the proposed HERDT approach by experimenting with a 7-degree-of-freedom robotic arm [33] within the MuJoCo simulation environment [27]. Robotic manipulation tasks are indeed well-established benchmark tasks for multi-goal RL [30]. In this study, four standard manipulation tasks (FetchReach, FetchPickAndPlace, FetchPush, and FetchSlide) are selected, as shown in Fig. 5 and the selected tasks are briefly described below:

- *FetchReach*: This task consists of controlling the robotic arm to move its gripper to a target position in a 3D space (shown as a red dot in Fig. 5a).
- *FetchPickAndPlace*: This task consists of controlling the robotic arm to grasp the black object with its gripper and place it at the target position in a 3D space (shown as a red dot in Fig. 5b).
- *FetchPush*: This task consists of controlling the robotic arm to push the black object with its clamped gripper to the target position in a 2D space (shown as a red dot in Fig. 5c).
- *FetchSlide*: This task consists of controlling the robotic arm to make the black object slide along the sliding table from its initial position to the target position in a 2D space (shown as a red dot in Fig. 5d).

In the warm-up stage, we optimize the DTs using the GE algorithm. These optimized DTs are employed to generate curriculum goals during the training stage for the robotic tasks. For example, we illustrate an optimized DT and its corresponding grid word environment for the FetchPush task in Fig. 4. The mechanism behind how DTs generate curriculum goals during training-stage is explained below: As an example, the initial and target positions are indicated as $(x = 8, y = 10)$ and $(x = 9, y = 16)$ in Fig. 4, respectively. For each episode, the initial position $(x = 8, y = 10)$ is given to the DT, and node 0 checks whether x is lower than 9 or not. In our case, $x = 8$, so it is lower than 9, then the DT goes to node 1. This node checks if y is greater than 11 (in our case, it is not), and then it selects node 4, which checks y. Since y in our case is lower than 13, the DT selects node 7, which again checks y. In our case y is greater than 7, then node 9 is selected, which outputs action 0, i.e., movement in the +x direction. So, our next curriculum goal will be in $(x = 9, y = 10)$. If this position is given to the DT, then node 0 will check again x. In this case, x is not lower than 9. Then, the DT selects node 2, which outputs action 2, corresponding to movement in the +y direction. Our next curriculum goal will be in $(x = 9, y = 11)$ and after that position, the DT will always select action 0, which results in arriving at the target position after 6 iterations. The other action (2) is the movement in the −y direction as specified in Table 2.

After every epoch, the performance is evaluated by running 100 deterministic test rollouts. Then, the success rate of the tests is calculated by averaging the results, as detailed between lines 32 and 39. In the selected tasks, the state vector incorporates multiple components, including the target position and the current position, orientation, linear velocity, and angular velocity of both the end-effector and the object. The goal represents the desired position for an object. If an object reaches the target within a threshold distance, as defined in Eq. (1), then the task is considered successfully completed and the agent obtains a 0 reward value. If the object falls outside the defined range of the goal, the agent is penalized with a negative reward of -1. We test the proposed approach with 10 different seeds for all the experimental settings.

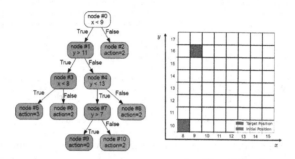

Fig. 4. An example DT generated for the FetchPush task (left) and its convergence graph (right).

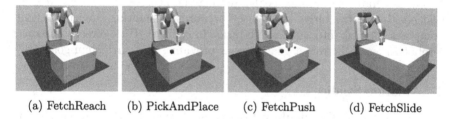

(a) FetchReach   (b) PickAndPlace   (c) FetchPush   (d) FetchSlide

Fig. 5. Overview of the robotic manipulation benchmark tasks.

The trend of the test success rate, shown for all environments in Fig. 6, clearly shows the impact of generating curriculum goals on the success rate. From the figures, it can be evidently inferred that HERDT outperforms HER (in terms of success rate and convergence) across all the considered robotic manipulation environments, except for the FetchSlide. It should be noted in fact that, for the FetchSlide task, HERDT converges faster and has a lower standard deviation than HER. However, the success rate of HER becomes slightly better than our method in the final episodes of the FetchSlide task. HERDT converges to the near-optimal policy after 5, 75, 35, and 50 epochs, while HER requires 6, 400, 65 and 300 epochs to converge for FetchReach, FetchPickAndPlace, FetchPush, and FetchSlide, respectively.

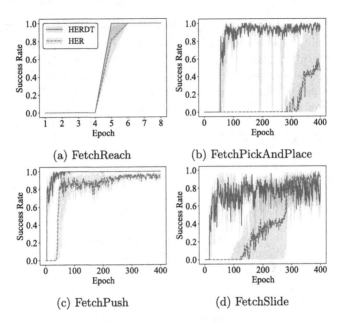

Fig. 6. Test success rate for the tasks. The mean success rate (line) and inter-quartile range (shaded area) are shown for 10 random seeds.

This can be attributed to the fact that our approach, HERDT, employs a DT that generates curriculum goals, guiding the agent step by step from its initial position to the target goal. In other words, generating curriculum goals decomposes the task into simpler sub-tasks. In contrast, HER directly instructs the robot to reach the target goal without providing any curriculum goals.

4.1 Ablation Studies

Experiments with Curriculum Goals. As discussed earlier, in our method the DT policy takes the achieved goals as input by converting the state s_t using the state-to-goal mapping function $\phi(s_t)$. In the first ablation study, we then aim to answer the following question: What if we recursively feed the generated goal g_t into the DT policy back again, such that $g_{t+1} = g_t + \pi_{DT}(g_t)$, and repeat this process n times such that $g_{t+n} = \frac{1}{n}\sum_{i=1}^{n} g_i$? How would this affect the performance of the agent, compared to generating the curriculum goal g_t and using it as the next desired goal during the next episode?

The idea behind this experiment is that we can generate the next curriculum goals g_{t+1} given the previous curriculum goal g_t, recursively. Such curriculum goals represent positions in a 3D space. We can repeat this process n times and obtain different future curriculum goals $g_{t+1}, g_{t+2}, \ldots, g_n$. However, these goals might be very far away from g_t, making the task excessively challenging for the robot. Additionally, some generated goals might lie outside the designated task environment. To mitigate this issue, we scale the recursively generated curriculum goals by multiplying their sum by $1/n$, where n is the number of curriculum goals.

Figure 7 illustrates the impact of the parameter n on the success rate. It can be inferred that when $n = 1$, the performance dwindles in the FetchPickAndPlace, FetchPush, and FetchSlide tasks. This is because curriculum goals approach the desired goals slowly and may not cover the entire distribution of sampled desired goals within the given number of training epochs. On the other hand, as n increases, the curriculum goals approach the desired goals more quickly. Yet, these curriculum goals might be too challenging for the robot to achieve, resulting in a lower success rate at the beginning of the training in all tasks except for the FetchPush task, where the parameter n does not significantly affect the outcomes.

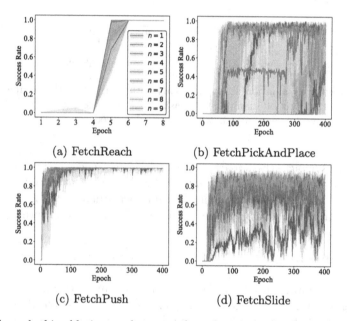

(a) FetchReach

(b) FetchPickAndPlace

(c) FetchPush

(d) FetchSlide

Fig. 7. Through this ablation study, we analyze the impact of recursively generated curriculum goals on the success rate of robotic manipulation tasks. The mean success rate (line) and inter-quartile range (shaded) are shown for 10 random seeds.

Experiments Without Feedback from Q-Value. In the second ablation study, we aim to answer the following question: How does the feedback from the Q-value (from the critic network) affect the success rate?

As discussed earlier, after optimizing the binary DTs, we modify the curriculum goals generated during the training process by multiplying them by the Q-value. This adjustment is made to ensure that their positions are optimized accordingly, making them neither too challenging nor too easy for the robot to

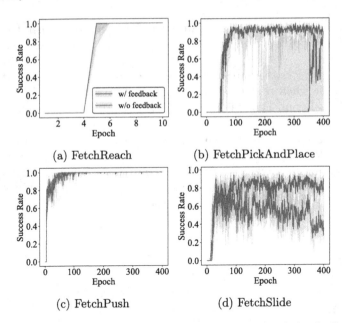

Fig. 8. Through this ablation study, we analyze the impact of the feedback on the success rate for the robot manipulation tasks. The average success rate (line) and inter-quartile range (shaded) are shown for 10 random seeds.

reach. Figure 8 depicts the impact of feedback on the success rate in benchmark tasks. Based on the simulation results, it becomes evident that the feedback value plays a crucial role in enhancing learning performance and fine-tuning the generated curriculum points to align with the robot's ongoing learning policy.

5   Conclusion

In this study, we introduced a novel curriculum learning RL method, HERDT, which leverages binary Decision Trees (DTs) to generate curriculum goals for the robotic agent. Prior to training the robotic agent, a *warm-up* phase is performed, during which DTs are optimized using the Grammatical Evolution (GE) algorithm. In the training stage, the optimized DTs generate curriculum goals to guide the agent toward the target goal. Since DTs output discrete values, we further refine these curriculum points based on feedback from the robotic agent. This fine-tuning helps the robotic agent find the right balance in the difficulty of the generated curriculum points. As a limitation, our method, which adds an additional DT-based curriculum generation to HER, obviously takes a longer time to compute than the original HER. However, as seen in our experiments, HERDT is more sample-efficient than HER. Another notable constraint of our implementation is that the DTs only accepts integer numbers. To address this limitation, we multiply floating-point numbers (position parameters) by a hyper-parameter and then truncate the fractional part. However, this truncation process reduces the algorithm's accuracy, as it treats positions with different fractional parts as the same position. To address this limitation and as part of future work, we will explore the use of multi-branch DTs as well as regression trees to handle continuous positions.

References

1. Sutton, R.S., Barto, A.G.: Reinforcement Learning: An Introduction. MIT press, Cambridge (2018)
2. Mnih, V., et al.: Playing Atari with deep reinforcement learning. arXiv:1312.5602 (2013)
3. Mnih, V., et al.: Human-level control through deep reinforcement learning. Nature **518**, 529–533 (2015)
4. Silver, D., et al.: Mastering the game of Go with deep neural networks and tree search. Nature **529**, 484–489 (2016)
5. Rajeswaran, A., Lowrey, K., Todorov, E.V., Kakade, S.M.: Towards generalization and simplicity in continuous control. Adv. Neural Inf. Process. Syst. **30** (2017)
6. Ng, A.Y., Coates, A., Diel, M., Ganapathi, V., Schulte, J., Tse, B., Berger, E., Liang, E.: Autonomous inverted helicopter flight via reinforcement learning. In: Ang, M.H., Khatib, O. (eds.) Experimental Robotics IX. STAR, vol. 21, pp. 363–372. Springer, Heidelberg (2006). https://doi.org/10.1007/11552246_35
7. Lillicrap, T.P., et al.: Continuous control with deep reinforcement learning. arXiv:1509.02971 (2019)
8. Zakka, K., et al.: RoboPianist: a benchmark for high-dimensional robot control. arXiv:2304.04150 (2023)
9. Ng, A.Y., Harada, D., Russell, S.: Policy invariance under reward transformations: theory and application to reward shaping. In: International Conference on Machine Learning. (1999)
10. Rengarajan, D., Vaidya, G., Sarvesh, A., Kalathil, D., Shakkottai, S.: Reinforcement learning with sparse rewards using guidance from offline demonstration. In: International Conference on Learning Representations (2022)
11. Andrychowicz, M., et al.: Hindsight experience replay. Adv. Neural Inf. Process. Syst. **30** (2017)
12. Zhao, R., Tresp, V.: Energy-based hindsight experience prioritization. In: Conference on Robot Learning, pp. 113–122. PMLR (2018)
13. Zhao, R., Sun, X., Tresp, V.: Maximum entropy-regularized multi-goal reinforcement learning. In: International Conference on Machine Learning, pp. 7553–7562. PMLR (2019)
14. Puiutta, E., Veith, E.M.S.P.: Explainable reinforcement learning: a survey. In: Holzinger, A., Kieseberg, P., Tjoa, A.M., Weippl, E. (eds.) CD-MAKE 2020. LNCS, vol. 12279, pp. 77–95. Springer, Cham (2020). https://doi.org/10.1007/978-3-030-57321-8_5
15. Lipton, Z.C.: The mythos of model interpretability: in machine learning, the concept of interpretability is both important and slippery. Queue **16**(3), 31–57 (2018)
16. Molnar, C.: Interpretable machine learning. Lulu. com (2020)
17. Coppens, Y., et al.: Distilling deep reinforcement learning policies in soft decision trees. In: IJCAI Workshop on Explainable Artificial Intelligence, pp. 1–6 (2019)

18. Bastani, O., Pu, Y., Solar-Lezama, A.: Verifiable reinforcement learning via policy extraction. Adv. Neural Inf. Process. Syst. **31** (2018)
19. Ding, Z., Hernandez-Leal, P., Ding, G.W., Li, C., Huang, R.: CDT: cascading decision trees for explainable reinforcement learning. arXiv:2011.07553 (2020)
20. Roth, A.M., Topin, N., Jamshidi, P., Veloso, M.: Conservative Q-improvement: reinforcement learning for an interpretable decision-tree policy. arXiv:1907.01180 (2019)
21. Hallawa, A., et al.: Evo-RL: evolutionary-driven reinforcement learning. In: Genetic and Evolutionary Computation Conference Companion, pp. 153–154 (2021)
22. Custode, L.L., Iacca, G.: Evolutionary learning of interpretable decision trees. IEEE Access **11**, 6169–6184 (2023)
23. Ferigo, A., Custode, L.L., Iacca, G.: Quality diversity evolutionary learning of decision trees. In: Symposium on Applied Computing, pp. 425–432. ACM/SIGAPP (2023)
24. Custode, L.L., Iacca, G.: Interpretable pipelines with evolutionary optimized modules for reinforcement learning tasks with visual inputs. In: Genetic and Evolutionary Computation Conference Companion, pp. 224–227 (2022)
25. Custode, L.L., Iacca, G.: A co-evolutionary approach to interpretable reinforcement learning in environments with continuous action spaces. In: IEEE Symposium Series on Computational Intelligence, pp. 1–8. IEEE (2021)
26. Crespi, M., Ferigo, A., Custode, L.L., Iacca, G.: A population-based approach for multi-agent interpretable reinforcement learning. Appl. Soft Comput. **147**, 110758 (2023)
27. Todorov, E., Erez, T., Tassa, Y.: MuJoCo: A physics engine for model-based control. In: IEEE/RSJ International Conference on Intelligent Robots and Systems, pp. 5026–5033. IEEE (2012)
28. Watkins, C.J., Dayan, P.: Q-learning. Mach. Learn. **8**, 279–292 (1992)
29. Lillicrap, T.P., et al.: Continuous control with deep reinforcement learning. arXiv:1509.02971 (2015)
30. Plappert, M., et al.: Multi-goal reinforcement learning: challenging robotics environments and request for research. arXiv:1802.09464 (2018)
31. Schaul, T., Horgan, D., Gregor, K., Silver, D.: Universal value function approximators. In: International Conference on Machine Learning. (2015)
32. Ryan, C., Collins, J.J., Neill, M.O.: Grammatical evolution: evolving programs for an arbitrary language. In: Banzhaf, W., Poli, R., Schoenauer, M., Fogarty, T.C. (eds.) EuroGP 1998. LNCS, vol. 1391, pp. 83–96. Springer, Heidelberg (1998). https://doi.org/10.1007/BFb0055930
33. Brockman, G., et al.: OpenAI Gym. arXiv:1606.01540 (2016)

Cultivating Diversity: A Comparison of Diversity Objectives in Neuroevolution

Didrik Spanne Reilstad and Kai Olav Ellefsen$^{(\boxtimes)}$

Department of Informatics, University of Oslo, Oslo, Norway
kaiolae@ifi.uio.no

Abstract. Inspired by biological evolution's ability to produce complex and intelligent beings, *neuroevolution* utilizes evolutionary algorithms for optimizing the connection weights and structure of artificial neural networks. With evolutionary algorithms often failing to produce the same level of diversity as biological evolution, explicitly *encouraging diversity* with additional optimization objectives has emerged as a successful approach. However, there is a lack of knowledge regarding the performance of different types of diversity objectives on problems with different characteristics. In this paper, we perform a systematic comparison between objectives related to *structural diversity*, *behavioral diversity*, and our newly proposed *representational diversity*. We explore these objectives' effects on problems with different levels of *modularity*, *regularity*, *deceptiveness* and *discreteness* and find clear relationships between problem characteristics and the effect of different diversity objectives – suggesting that there is much to be gained from adapting diversity objectives to the specific problem being solved.

Keywords: Neuroevolution · Diversity · Evolutionary Algorithm · Neural Networks

1 Introduction

Traditionally, artificial neural networks are trained using gradient-based learning through a combination of backpropagation and stochastic gradient descent [13]. An alternative to gradient-based learning, partly inspired by the ability of biological evolution to produce the complexity that is natural brains in animals and humans, is *neuroevolution* (NE). Neuroevolution harnesses the capabilities of an evolutionary algorithm to optimize the hyperparameters, but also the topology and activation function of neural networks, which are capabilities typically unavailable to gradient-based approaches [26]. Additionally, instead of optimizing a single neural network, neuroevolution employs and maintains a population of neural networks, enabling extreme exploration and parallelization of solutions.

Evolutionary algorithms attempt to emulate the diversity of complex organisms produced by biological evolution but are typically far from achieving the

---

Source code: https://github.com/dreilstad/Neuroevolution.

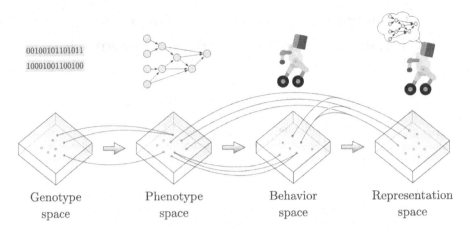

Fig. 1. The mapping between genotype-, phenotype-, and behavior space, including our proposed representation space. The mappings (i.e. grey lines) are non-injective. Thus, different genotypes can correspond to the same phenotype, structurally different neural networks can produce identical behavior, and unique neural networks producing different behaviors can learn the same representation. Figure adapted from [23]. (Color figure online).

same level of diversity that we see in nature. This lack of diversity often results in sub-optimal solutions as the population has converged to a local optimum. A way to explicitly increase diversity in evolutionary algorithms while maintaining good performance is to use a multi-objective evolutionary algorithm to simultaneously optimize solutions according to a performance objective and a diversity objective.

Using an additional diversity objective has demonstrated the ability to discover novel solutions and ultimately lead to better performance more efficiently [26]. A common type of diversity objective is *behavioral diversity*, which encourages different behaviors [20,23]. *Genetic diversity* objectives can be used to encourage genetically different solutions [9,28], whereas *structural diversity* objectives encourages structurally different solutions [1,6,9] (Fig. 1).

Encouraging diversity in neuroevolution with diversity objectives has demonstrated impressive results. Still, there is a lack of knowledge regarding the relationship between the type of diversity objective and the characteristics of the problem. For instance, is there a type of diversity objective that is more effective for problems with certain characteristics? Behavioral diversity objectives can be quite effective but are limited by usually being domain-dependent and having to be adapted to each problem. In contrast, structural diversity objectives are domain-independent but can be expensive to compute. A research gap in diversity-driven neuroevolution is combining the advantages of both behavioral and structural diversity into a new type of diversity.

Our main contributions are 1) A systematic evaluation of the effects of different types of diversity objectives on problems with different characteristics, revealing relationships between the performance of diversity objectives and task characteristics, and 2) Introducing a new objective called *representational diversity* that is designed to encourage a diversity of *learned neural network representations* (Fig. 1).

2   Related Work

2.1   Diversity-Driven Neuroevolution

One of the most remarkable feats of biological evolution is the ability to produce a diversity of complex organisms that are all high-performing in their niche. The biological diversity seen in nature today is the result of an evolutionary process that is millions of years in the making. In recent years, research on neuroevolution (and evolutionary computing in general) has focused more and more on *diversity* [26]. Population-based evolutionary algorithms should, in theory, create diversity by themselves through evolutionary operators. In practice, they often converge too early and lack the diversity needed to avoid local optima.

Explicitly encouraging diversity will drive exploration and avoid early convergence to local optima. Novelty Search [15] does this by rewarding novel behaviors instead of performance, which was seen to be effective in deceptive domains. The success of Novelty Search has inspired research into *Quality-Diversity* optimization methods [16,21], where the goal is to generate a collection of diverse yet high-performing solutions rather than a single solution.

2.2   Diversity Objectives in Neuroevolution

Another successful approach has been to employ a *multi-objective evolutionary algorithm* (MOEA) with both a performance objective and an additional diversity objective (i.e. multiobjectivization [20]). The effect of encouraging diversity will depend on the space in which one tries to encourage diversity. As illustrated in Fig. 1, diversity in one space does not guarantee diversity in the others, due to the non-injective mapping between them.

Behavioral Diversity. Encouraging behavioral diversity alone or in addition to performance has been successfully applied to a variety of tasks [15,20,22,23]. To encourage behavioral diversity, one first needs to specify a way to measure the *behavioral distance* between two individuals. Commonly used distances are divided into two categories: (1) domain-dependent *ad hoc* distances, and (2) domain-independent *generic* distances.

Ad hoc behavioral distances, by being designed for each domain, are often quite effective compared to *generic* behavioral distances [23], but defining the behavioral distance can be difficult when applied to complex real-world problems. *Generic* behavioral distances can be applied directly across domains with minimal adjustments (e.g. using the Hamming distance on input-output history vectors of networks [23]). This generality can be less informative and lead to a performance trade-off, but *generic* distances can perform on par with *ad hoc* distances on certain problems [5]. In a comparison of multiple *generic* behavioral distances, the Hamming distance was found to perform well, with a relatively low computational cost [7].

Structural Diversity. A more problem-independent alternative to behavioral diversity is to encourage structurally different individuals with *structural diversity*. However, computing the distance between two neural networks (i.e. graphs) is an NP-hard problem, making a complete one-to-one connectivity distance metric not feasible due to the algorithmic complexity involved [22]. One solution is to use *approximate* structural distance metrics, which are less computationally expensive at the cost of accuracy.

Recent research suggests benefits to encouraging certain structural properties inspired by biology instead. Inspired by the adaptability, flexibility, and robustness of natural brains, it is believed that the structure of a neural network should express traits such as *modularity*, *regularity*, and *hierarchy* in order to exhibit the same behavioral complexity as humans and animals [27]. As a result, encouraging *modularity* [1,9] in the neural network structure has been shown to improve the performance. Encouraging *modularity diversity* [6] (that is, a diverse set of modular neural network decompositions in the population) has also been suggested as a way to diversify the population at an abstraction level above individual connections.

Representational Diversity. As described above, there are advantages and limitations to behavioral and structural diversity objectives. Therefore, we explore a new type of diversity called *representational diversity* which attempts to encompass both the structure and behavior of neural networks, including the generality of structural diversity and the effectiveness of behavioral diversity.

Inspired by work in neuroscience, researchers proposed using similarity metrics such as *centered kernel alignment* (CKA) and *canonical correlation analysis* (CCA) for measuring the similarities between learned representations of deep neural networks using neuron activations [11]. The purpose was to study the similarity between the learned representations of two neural networks that were initialized differently, and the proposed similarity index was found to reliably identify correspondences between learned representations.

In neuroevolution, encouraging a diversity of neural network representations has not been extensively explored except in the *Creative Thinking Approach* (CTA) [18] by using the neuron activations of hidden layers to characterize the neural network representation. An important difference between CTA and our approach is that the neuron activation values are not binarized and a more informative distance measure is used (e.g. an activation of 0.51 and 1.0 is interpreted the same with CTA when binarizing and using the Hamming distance).

2.3   Problem Characteristics of Interest

A number of interesting problem characteristics are studied to determine which diversity objectives are more suitable for different problems. The following characteristics have previously been studied individually and have been shown to significantly impact the evolutionary search.

Modularity. Modularity is understood to be the degree to which the problem structure can be decomposed into independent sub-tasks and the potential for a modular neural network structure to evolve. Modularity been studied extensively in neuroevolution [1,6,10], and is an important *organizing principle* in biological neural networks of human and animal brains [19] and a key driver of *evolvability* – the ability to rapidly adapt to novel environments [1].

Regularity. Regularity is expressed if there is a repeating or oscillatory nature to the problem and its structure. Examples of problems exhibiting regularity are problems where solving a repeating sub-task is required, a sequence of different sub-tasks is required to be solved, or discovering an oscillatory pattern is useful for solving the task. Discovering oscillatory patterns is an important aspect of robot locomotion tasks that are often targeted in neuroevolution (e.g. [6,15,17]).

Deceptiveness. Deceptiveness as a problem characteristic has been the subject of several studies on evolutionary algorithms (EAs), especially with a focus on diversity maintenance techniques [12,15,17]. The deceptiveness of a problem is often linked to the objective function and is the degree to which the problem contains obvious sub-optimal solutions (i.e. local optima) that are easily converged to and normally hard to avoid.

Environment Representation. Environment representation as a characteristic is understood to be whether the problem structure and representation is *discrete* or *continuous*. Defining an appropriate representation of a problem is an important aspect of evolutionary computation (and optimization in general) that can greatly affect performance. Different studies may use vastly different representations for the same problem, which changes the problem difficulty and with it the reported performance of evolutionary algorithms [25].

3   Targeted Problems

For the comparison of diversity objectives, the following problems are targeted: The Retina problem, The Tartarus problem, Maze navigation, and Robot locomotion. The main reasons for choosing the targeted problems are: 1) They have previously been used in related research to study diversity-driven neuroevolution; 2) They are easy to set up and reproduce from scratch or with existing simulator frameworks available for a few of the problems; and 3) They have distinct characteristics, thus giving more valuable insight into the relationship between the type of problem and the effectiveness of the diversity objectives. Only a brief description of each problem is presented due to space constraints. The targeted problems are described in more detail in [24], including implementation details.

To measure *generic* behavioral diversity, we follow previous work in applying the Hamming distance on the binarized input-output history of networks [6,23]. The *ad hoc* behavioral diversity of individuals is defined as the distance to its k-nearest neighbors of an archive of previously novel individuals, as in [15] – where this distance is a problem-dependent measurement, described below for each problem.

3.1 The Retina Problem

The `Retina` problem is a *pattern-recognition* problem (see Fig. 2). The problem has been used in several previous studies focused on the evolution of modular structures in neural networks. In particular, to study if modular structures appear in neural networks when applied to modular problems [1,10] and to study structural diversity objectives in neuroevolution [6,9].

(a) The retina problem.   (b) Target patterns for the left and right side of the retina.

Fig. 2. The 3×3 Retina problem, a classification problem which can benefit from neural networks with modular structure.

Early testing of the original `Retina` problem with 2×2 patterns as used in [1,6], indicated similar performance across treatments and was not informative. Therefore, a harder version of the `Retina` problem with larger and more target patterns is implemented (see Fig. 2). For the problem-specific *ad hoc* behavioral distance, we apply the euclidean distance between the outputs of two neural networks for a set of user-defined test patterns as follows:

$$d_{\text{ad-hoc}}(x, y) = \|\beta_x - \beta_y\| \tag{1}$$

where the behavior vector β_x of individual x consists of the set of *raw* output o of the neural network for k test patterns.

$$\beta_x = \{o^{(1)}, \ldots, o^{(k)}\} \tag{2}$$

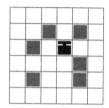 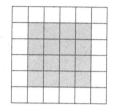 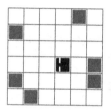

| (a) Example of an initial board configuration. | (b) Valid start positions for agent and blocks. | (c) Example of a final board configuration. |

Fig. 3. In the Tartarus problem, the task is for an agent (black) to move blocks (brown) to the edge of the environment within 80 moves. The agent and the blocks are initialized randomly away from the edge (gray). 1 point is awarded for blocks at edges and 2 points for blocks in corners. Figures adapted from [8]. (Color figure online).

3.2 The Tartarus Problem

The Tartarus problem [2,7,8] is a grid-based optimization problem. The goal of the agent is to move blocks to the edges and corners of the environment (see Fig. 3). The setup for the Tartarus problem follows the setup in [7]. In addition, there exists an explicitly *deceptive* version of the Tartarus problem used in [2]. In the Deceptive-Tartarus problem, blocks in the corners are still awarded 2 points but blocks at the edges are awarded −1 point.

The *ad hoc* behavioral distance, first described in [2], rewards individuals for solving board configurations differently from others. The distance between individuals x and y is defined as the Manhattan distance between the corresponding blocks in the two sets of final board configurations:

$$d_{\text{ad-hoc}}(x, y) = \text{Manhattan}(\beta_x(i), \beta_y(i)) \tag{3}$$

where the behavior vector β_x of individual x contains the final positions $p^{(c)}$ of corresponding blocks for all k board configurations:

$$p^{(c)} = \{(x_1, y_1), \ldots, (x_6, y_6)\} \tag{4}$$

$$\beta_x = \left\{ p^{(c)}, c \in [1, k] \right\} \tag{5}$$

3.3 Deceptive Maze Navigation Problem

A deceptive Maze Navigation problem was used to demonstrate the efficiency of Novelty Search [15] and has since been used in various other studies [14,17,20, 23]. The task is for a mobile robot to navigate a maze domain from a starting point and find the goal within a fixed time limit using sensor inputs (see Fig. 4).

The maze domains are designed to be deceptive to a reasonable fitness function and contain multiple dead-ends (i.e. local optima) that prevent a direct route to the goal and require the robot to navigate the environment properly.

The setup of the maze domains, mobile robot, and *ad hoc* behavioral distance follows the setup in [15]. The *ad hoc* behavioral distance between two individuals,

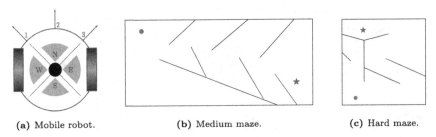

(a) Mobile robot.   **(b)** Medium maze.   **(c)** Hard maze.

Fig. 4. The mobile robot in **(a)**, equipped with a 4-pie slice goal radar (blue) and 3 range sensors (red), is tasked with navigating the mazes in **(b)** and **(c)** from a starting position (blue circle) and find the goal (red star). Maze domains are from [15]. (Color figure online)

first described in [15] for Novelty Search, is defined as the euclidean distance between their end locations p_x and p_y in the maze at the end of simulation:

$$d_{\text{ad-hoc}}(x, y) = \|p_x - p_y\| \tag{6}$$

3.4   Robot Locomotion Problem

The Robot Locomotion problem is, as the name implies, a *locomotion* problem where a robot has to transport itself from one location to another. This type of problem has previously been targeted in neuroevolution studies [6,9,15,17] with various robot configurations (e.g. bipedal robot or hexapod). A `Bipedal-Walker` is simulated (see Fig. 5), where the objective is to reach the end of the environment within a set time limit without falling. The task is difficult because the neural network must learn a gait that is efficient and fast enough to reach the end within the time limit. Learning such a gait requires coordination of the two legs, balancing the hull, and discovering an *oscillatory* pattern [15].

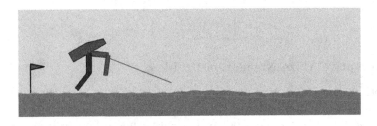

Fig. 5. Simulated bipedal walker for the Robot locomotion problem. Its goal is to reach the end of the environment (not pictured) within a time limit.

The *ad hoc* behavioral distance previously defined in [15,17] is used in this work. The offset of the bipedal robot's center of mass is sampled during the simulation and concatenated in a behavior vector β_i to define its behavior:

$$\beta_i = \{(x'_1, y'_1, \ldots, x'_m, y'_m)\} \tag{7}$$

where (x'_k, y'_k) is the offset of the center of gravity at the kth sample during the simulation of m samples. The *ad hoc* behavioral distance between two individuals, x and y, is the euclidean distance between behavior vectors:

$$d_{\text{ad-hoc}}(x, y) = \|\beta_x - \beta_y\| \tag{8}$$

This way of measuring novel behaviors rewards individuals with unique gaits.

3.5 Characterizing the Targeted Problems

A systematic characterization of commonly targeted problems is lacking in previous work. This work attempts to characterize the targeted problems with the purpose of potentially gaining more valuable insight into the various diversity objectives: Are some diversity objectives more suitable for problems with certain characteristics? A summary is given in Table 1, where the targeted problems are classified according to which degree it expresses the characteristics introduced in Sect. 2.3. It should be noted that the characterization performed in this work is subjective.

Table 1. Characterization of the targeted problems.

Targeted problem	Modularity			Regularity			Deceptiveness			Environment	
	None	Partially	A lot	None	Partially	A lot	None	Partially	A lot	Discrete	Continuous
Retina (2x2 and 3x3)			✓	✓				✓		✓	
Tartarus	✓				✓			✓		✓	
Deceptive-Tartarus	✓				✓				✓	✓	
Medium-Maze	✓			✓				✓			✓
Hard-Maze	✓			✓					✓		✓
Bipedal-Walker		✓			✓			✓			✓

4 Comparison of Diversity Objectives

4.1 Neuroevolution Setup

Neural Network Encoding. A graph-based direct neural network encoding is used to represent and evolve neural networks. The encoding is a simplified version of the NEAT encoding, where neural networks are represented as a list of nodes and a list of edges with weight and bias values. This simple and lightweight encoding has been employed in many previous studies (e.g. [5,6,20,22,23]).

Evolving Neural Networks. Neural networks are initialized as fully connected with no hidden neurons (i.e. every input neuron is connected to every output neuron), with randomly sampled weights and biases. Following previous studies (e.g. [6,20,23]), only mutations are employed to evolve networks with no crossover. Two types of mutations are implemented: (1) *Structural mutation* – (add/remove neuron, add/remove connection), and (2) *Parametric mutation* – (weight/bias mutation). As in previous neuroevolution studies [6,23], the *polynomial mutation* [4] scheme is used in this work for parametric mutation.

Multi–objective Evolutionary Algorithm. *Non-dominated Sorting Genetic Algorithm–II* (NSGA–II) is used as the evolutionary algorithm for optimizing multiple objectives. The NSGA–II algorithm is well-established and has previously been used in various neuroevolution studies, such as [5,6,17,20,23] to name a few. In the case of a single objective, NSGA–II is equivalent to an elitist evolutionary algorithm with tournament-based selection [3].

4.2 Introducing Representational Diversity for Neuroevolution

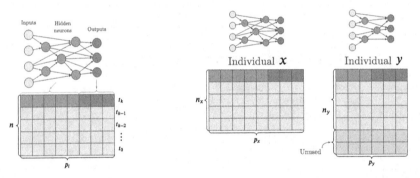

(a) Representation matrix.   (b) Computing the similarity between representations.

Fig. 6. (a) The learned representation of a neural network is characterized by a matrix of the neuron activations of p_i neurons for n samples (or simulation steps). n for two individuals can be different if either solves the task before the other. (b) To perform matrix multiplication when computing representation similarity (CKA or CCA), either the number of samples or neurons has to be equal. To allow for neural networks to evolve different topologies with different numbers of neurons, only the corresponding samples of the representation matrix with the smallest number of samples are considered when computing the similarity.

With representational diversity measures (Sect. 2.2), we aim to capture differences in how neural networks represent their input – by looking at their activation patterns (Fig. 6). The similarity metrics *centered kernel alignment* (CKA) and *canonical correlation analysis* (CCA) [11] can be applied to measure the similarity between learned representations of two individuals using the *representation matrix* (see Fig. 6a). Originally, these metrics were used on neural networks with identical topologies. Using neuroevolution in this work, the metrics are adapted to evolved neural networks with different topologies (see Fig. 6b). In short, the similarity metrics are adapted to allow neural networks to be evolved with a different number of neurons but only an equal number of samples are considered when computing the similarity between representations.

4.3   Treatments, Parameter Settings, and Statistical Testing

A comparison of diversity objectives is performed on four targeted problems. In total, seven experimental treatments are applied to every problem (see Table 2). Experimental parameters specific to each targeted problem are listed in [24]. To ensure differences between treatments are only due to different diversity objectives, the same parameters are used for treatments applied to the same problem. Parameter setting is done by using similar values used in previous studies and further adjusting them based on initial testing. Comprehensive parameter tuning has not been performed due to computational constraints, time, and the large number of possible combinations. Experiments were performed on a HPC cluster[1] with 128 CPUs and 512 GiB RAM. All experiments are repeated 50 times with random seeds (i.e. different stochastic events). All statistical significance testing of results presented in Sect. 5 apply the Mann-Whitney U test.

Table 2. Summary of all experimental treatments and their abbreviations.

Abbreviation	Objectives	Treatment
PA	Performance Alone	Maximize $F(x)$
NOVELTY	Performance + Novelty	Maximize $\begin{cases} F(x) \\ D(x) = \frac{1}{k}\sum_{i=1}^{k} d_{\text{ad-hoc}}(x, \mu_i) \end{cases}$
HAMMING	Performance + Hamming distance	Maximize $\begin{cases} F(x) \\ D(x) = \frac{1}{|P|}\sum_{j \in P} d_{\text{ham}}(x, j) \end{cases}$
MOD	Performance + Modularity (Q-score)	Maximize $\begin{cases} F(x) \\ D(x) = Q(x) \end{cases}$
MODDIV	Performance + Modularity Diversity	Maximize $\begin{cases} F(x) \\ D(x) = \frac{1}{|P|}\sum_{j \in P} \Delta_{decomp}(x, j) \end{cases}$
CKA	Performance + CKA	Maximize $\begin{cases} F(x) \\ D(x) = \frac{1}{|P|}\sum_{j \in P} 1 - \text{CKA}(x, j) \end{cases}$
CCA	Performance + CCA	Maximize $\begin{cases} F(x) \\ D(x) = \frac{1}{|P|}\sum_{j \in P} 1 - R^2_{\text{CCA}}(x, j) \end{cases}$

---

[1] https://www.uio.no/english/services/it/research/hpc/fox/.

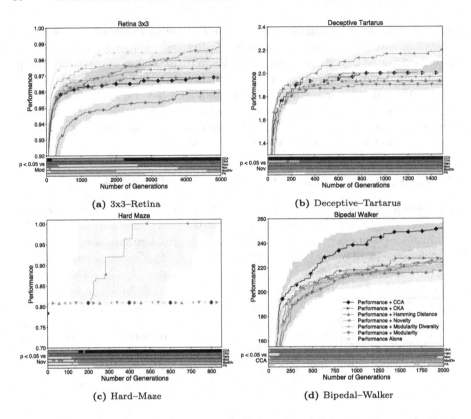

Fig. 7. The median best performance of all experimental treatments on the 3×3–Retina, Deceptive–Tartarus, Hard–Maze, and Bipedal–Walker problems with bootstrapped 95% confidence interval for 50 runs per treatment. Only figures with the most significant correlations are shown, but all figures can be found in [24].

5   Results

The results for the 2x2-Retina and 3x3-Retina problems show that the structural diversity treatments, MOD and MODDIV, outperform the rest (see Fig. 7a). The results for 2x2-Retina were consistent with previous work [6], and the same trends match closely the results for the harder 3x3-Retina introduced in this work. Interestingly, the MOD treatment was slightly more efficient compared to in [6]. The reason for this is likely due to setup differences where MOD potentially benefits from no restriction on the number of layers and the number of neurons per layer in our setup. Unexpectedly, PA performed surprisingly well unlike in [6], which could be explained by structural diversity becoming less important to the search when no neural network restrictions are used.

For the Tartarus problem, the NOVELTY and PA treatments outperformed the rest. None of the treatments reach the maximum performance score and converge to a score between 4.0–6.0, which is in line with previous work [7].

However, the HAMMING treatment performed slightly worse in this work. A possible explanation is the use of a recurrent neural network architecture as a memory element in [7] compared to only using the previous action as input.

As for the Deceptive-Tartarus problem, NOVELTY significantly outperformed the rest of the treatments (see Fig. 7b). Achieving a score of over 2.0 means that the agent is able to move more than a single box to the corners on average for all 30 board configurations (or alternatively more than one box in the corners but negative scores from boxes at edges) with the remaining treatments converge to the local optimum of 2.0. The results for NOVELTY and PA are similar to the results in previous work [2].

The results for the Medium-Maze showed NOVELTY significantly outperforming all other treatments. The treatments MODDIV, PA, and HAMMING were all able to consistently reach the goal, albeit slower than NOVELTY. For the Hard-Maze problem, NOVELTY is the only treatment able to reach the goal (see Fig. 7c). The rest of the treatments all converge to the attractive local optima near the goal with a performance score of 0.81. Only 850 of 1500 simulated generations are shown due to no further change in performance. The results for the Medium-Maze and Hard-Maze problems are consistent with previous work, wherein only the NOVELTY treatment reliably reaches the goal [15, 20].

Analyzing the results in Fig. 7d for the Bipedal-Walker problem, CCA significantly outperformed the rest of the treatments. CKA performs on par with PA but significantly worse than CCA. The structural and behavioral diversity treatments all converged to approximately the same performance.

5.1 Summary of Experiments

Figure 8 shows the median best performance score for all treatments across all targeted problems. A notable observation is that NOVELTY generally performs well on deceptive problems with the exception of the Bipedal-Walker problem. Furthermore, MOD performs well on the 2x2-Retina and 3x3-Retina problems but performs much worse on the other targeted problems. CCA resulted in much higher performance compared to other treatments on the Bipedal-Walker problem. Both CKA and HAMMING appear to perform poorly or at least result in mediocre performance on all problems. Overall, the control treatment PA performs well and is never the worst treatment.

A clear correlation between the performance of diversity objectives and the problem characteristics was found. *Ad hoc* behavioral diversity was found to outperform all other types of diversity on deceptive problems, as domain knowledge may be required to solve the task in contrast to domain-independent diversity objectives. The results for NOVELTY are not surprising as it has been shown to perform well on deceptive tasks in previous studies [12, 15, 20, 23], but there was a lack of knowledge about how other types of diversity objectives performed on deceptive problems such as the ones targeted in this work. Structural diversity objectives performed best on problems expressing a high degree of modularity, consistent with the findings of [6] where structural diversity outperformed behavioral diversity. Representational diversity showed promising results on a problem

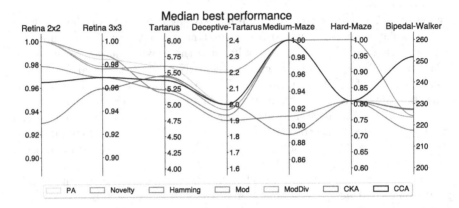

Fig. 8. Median best performance for 50 runs of treatments across all targeted problems. The axis for each problem is scaled to show the differences between treatments if there is one, and does not always show the maximum possible performance score.

with the characteristic of regularity, consistent with previous work indicating a diversity of representations can be beneficial for tasks with regularities [18].

A significant performance difference between diversity objectives of the same type was found for many of the targeted problems. (a) Of the behavioral diversity objectives, the *ad hoc* behavioral diversity objective outperformed the *generic* behavioral diversity objective on all targeted problems, (b) For the structural diversity objectives, encouraging a diversity of modular decomposition of the *input neurons* was more effective in the early phase of the evolutionary search than encouraging a diversity of modular decomposition of the *network* as a whole but converged earlier on certain problems, and (c) The difference in the performance of representational diversity objectives with different similarity metrics was significant. Results suggest that similarity metrics with invariance to invertible linear transformations (i.e. CCA) are more suitable for neuroevolution (i.e. neural networks with different topologies) than those without (i.e. CKA).

6   Conclusion

Encouraging diversity in neuroevolution with diversity objectives has seen increased interest in recent years, and demonstrated impressive results in terms of efficiency and performance of the evolutionary search. What is not yet clear is the relationship between the type of diversity objective and the characteristics of the problem, and if there is one, how to determine which types of diversity objectives are more suitable for which problems. Additionally, a type of diversity objective that is able to encompass both the structure and behavior of neural networks has not been extensively explored yet.

This work has presented initial indications of a correlation between the performance of different diversity objectives and the characteristics of the problem being solved. These correlations reinforce findings of previous work and reveal new relationships between diversity objectives and characteristics. More specifically, (1) Behavioral diversity is important for deceptive problems, (2) Structural diversity is effective on modular problems, and (3) Representational diversity works well on problems with regularities. Additionally, clear differences between diversity objectives of the same type were found (e.g. *ad hoc* vs *generic*). Furthermore, representational diversity was proposed as its own diversity type with early promising results on a robot locomotion problem.

Future work should validate the findings of this work on other problems to see if the correlation still holds. More parameter tuning, a different network initialization scheme, and an alternative neural network encoding or architecture (e.g. NEAT, CPPN, RNN) should be explored. Further study of representational diversity is needed, especially to validate that representational diversity is more suitable for problems with a high degree of regularity. In particular, combining the ability of CPPNs to produce complex patterns and representational diversity could potentially be effective on problems with a high degree of regularity.

References

1. Clune, J., Mouret, J.B., Lipson, H.: The evolutionary origins of modularity. Proc. Biol. Sci./Royal Soc. **280**, 20122863 (2013). https://doi.org/10.1098/rspb.2012. 2863
2. Cuccu, G., Gomez, F.: When novelty is not enough. In: Di Chio, C., Cagnoni, S., Cotta, C., Ebner, M., Ekárt, A., Esparcia-Alcázar, A.I., Merelo, J.J., Neri, F., Preuss, M., Richter, H., Togelius, J., Yannakakis, G.N. (eds.) EvoApplications 2011. LNCS, vol. 6624, pp. 234–243. Springer, Heidelberg (2011). https://doi.org/ 10.1007/978-3-642-20525-5_24
3. Deb, K., Pratap, A., Agarwal, S., Meyarivan, T.: A fast and elitist multiobjective genetic algorithm: NSGA-II. IEEE Trans. Evol. Comput. **6**(2), 182–197 (2002). https://doi.org/10.1109/4235.996017
4. Deb, K.: Multi-objective optimization using evolutionary algorithms (2008)
5. Doncieux, S., Mouret, J.B.: Behavioral diversity measures for evolutionary robotics. In: IEEE Congress on Evolutionary Computation, pp. 1–8. IEEE (2010)
6. Ellefsen, K.O., Huizinga, J., Torresen, J.: Guiding neuroevolution with structural objectives. Evol. Comput. **28**(1), 115–140 (2020). https://doi.org/10.1162/ evco_a_00250
7. Gomez, F.: Sustaining diversity using behavioral information distance, pp. 113–120 (2009). https://doi.org/10.1145/1569901.1569918
8. Griffiths, T.D., Ekárt, A.: Improving the tartarus problem as a benchmark in genetic programming. In: McDermott, J., Castelli, M., Sekanina, L., Haasdijk, E., García-Sánchez, P. (eds.) EuroGP 2017. LNCS, vol. 10196, pp. 278–293. Springer, Cham (2017). https://doi.org/10.1007/978-3-319-55696-3_18
9. Huizinga, J., Mouret, J.B., Clune, J.: Does aligning phenotypic and genotypic modularity improve the evolution of neural networks? In: Proceedings of the 25th Genetic and Evolutionary Computation Conference (GECCO), pp. 125–132. ACM, Denver (2016). https://doi.org/10.1145/2908812.2908836

10. Kashtan, N., Alon, U.: Spontaneous evolution of modularity and network motifs. Proc. Natl. Acad. Sci. **102**(39), 13773–13778 (2005). https://doi.org/10.1073/pnas. 0503610102
11. Kornblith, S., Norouzi, M., Lee, H., Hinton, G.: Similarity of neural network representations revisited. In: International Conference on Machine Learning, pp. 3519–3529. PMLR (2019)
12. Krčah, P.: Solving deceptive tasks in robot body-brain co-evolution by searching for behavioral novelty. In: 2010 10th International Conference on Intelligent Systems Design and Applications, pp. 284–289 (2010). https://doi.org/10.1109/ISDA.2010. 5687250
13. LeCun, Y., Bengio, Y., Hinton, G.: Deep learning. Nature **521**(7553), 436–444 (2015). https://doi.org/10.1038/nature14539
14. Lehman, J., Chen, J., Clune, J., Stanley, K.O.: Safe mutations for deep and recurrent neural networks through output gradients. CoRR arxiv:1712.06563 (2017)
15. Lehman, J., Stanley, K.: Abandoning objectives: evolution through the search for novelty alone. Evol. Comput. **19**, 189–223 (2011). https://doi.org/10.1162/ EVCO_a_00025
16. Lehman, J., Stanley, K.O.: Evolving a diversity of virtual creatures through novelty search and local competition. In: Proceedings of the 13th Annual Conference on Genetic and Evolutionary Computation, GECCO 2011, pp. 211–218. Association for Computing Machinery, New York (2011). https://doi.org/10.1145/2001576. 2001606
17. Lehman, J., Stanley, K.O., Miikkulainen, R.: Effective diversity maintenance in deceptive domains. In: Proceedings of the 15th Annual Conference on Genetic and Evolutionary Computation, GECCO 213, p. 215–222. Association for Computing Machinery, New York (2013). https://doi.org/10.1145/2463372.2463393
18. Li, J., Storie, J., Clune, J.: Encouraging creative thinking in robots improves their ability to solve challenging problems. In: Proceedings of the 2014 Annual Conference on Genetic and Evolutionary Computation, GECCO 2014, pp. 193–200. Association for Computing Machinery, New York (2014). https://doi.org/10.1145/ 2576768.2598222
19. Mountcastle, V.B.: The columnar organization of the neocortex. Brain J. Neurol. **120**(4), 701–722 (1997)
20. Mouret, J.B.: Novelty-based multiobjectivization, vol. 341, pp. 139–154 (2011). https://doi.org/10.1007/978-3-642-18272-3_10
21. Mouret, J.B., Clune, J.: Illuminating search spaces by mapping elites. ArXiv arxiv:1504.04909 (2015)
22. Mouret, J.B., Doncieux, S.: Using behavioral exploration objectives to solve deceptive problems in neuro-evolution. In: The 11th Annual Conference on Genetic and Evolutionary Computation (GECCO 2009), pp. 627–634. ACM, Montréal (2009). https://doi.org/10.1145/1569901.1569988
23. Mouret, J.B., Doncieux, S.: Encouraging behavioral diversity in evolutionary robotics: an empirical study. Evol. Comput. **20**(1), 91–133 (2012). https://doi. org/10.1162/EVCO_a_00048
24. Reilstad, D.S.: Cultivating Diversity: a Comparison of Diversity Objectives in Neuroevolution. Master's thesis, University of Oslo (2023). https://www.duo.uio.no/ handle/10852/103916
25. Rothlauf, F., Rothlauf, F.: Representations for Genetic and Evolutionary Algorithms. Springer, Heidelberg (2006). https://doi.org/10.1007/3-540-32444-5_2

26. Stanley, K., Clune, J., Lehman, J., Miikkulainen, R.: Designing neural networks through neuroevolution. Nat. Mach. Intell. **1** (2019). https://doi.org/10.1038/s42256-018-0006-z
27. Striedter, G.F.: Principles of Brain Evolution. Sinauer associates (2005)
28. Toffolo, A., Benini, E.: Genetic diversity as an objective in multi-objective evolutionary algorithms. Evol. Comput. **11**(2), 151–167 (2003). https://doi.org/10.1162/106365603766646816

Evolving Reservoirs for Meta Reinforcement Learning

Corentin Léger[1,2]([✉]), Gautier Hamon[1], Eleni Nisioti[1], Xavier Hinaut[2],
and Clément Moulin-Frier[1]

[1] Flowers Team, Inria and Ensta ParisTech, Bordeaux, France
corentin.leger@inria.fr
[2] Mnemosyne Team, Inria; LaBRI, Univ. Bordeaux, Bordeaux INP, CNRS UMR
5800; Univ. Bordeaux, CNRS, IMN, UMR 5293, Bordeaux, France

Abstract. Animals often demonstrate a remarkable ability to adapt to
their environments during their lifetime. They do so partly due to the
evolution of morphological and neural structures. These structures capture features of environments shared between generations to bias and
speed up lifetime learning. In this work, we propose a computational
model for studying a mechanism that can enable such a process. We
adopt a computational framework based on meta reinforcement learning as a model of the interplay between evolution and development. At
the evolutionary scale, we evolve reservoirs, a family of recurrent neural
networks that differ from conventional networks in that one optimizes
not the synaptic weights, but hyperparameters controlling macro-level
properties of the resulting network architecture. At the developmental
scale, we employ these evolved reservoirs to facilitate the learning of a
behavioral policy through Reinforcement Learning (RL). Within an RL
agent, a reservoir encodes the environment state before providing it to
an action policy. We evaluate our approach on several 2D and 3D simulated environments. Our results show that the evolution of reservoirs can
improve the learning of diverse challenging tasks. We study in particular three hypotheses: the use of an architecture combining reservoirs and
reinforcement learning could enable (1) solving tasks with partial observability, (2) generating oscillatory dynamics that facilitate the learning of
locomotion tasks, and (3) facilitating the generalization of learned behaviors to new tasks unknown during the evolution phase.

Keywords: Meta Reinforcement Learning · Reservoir Computing ·
Evolutionary Computation

1 Introduction

Animals demonstrate remarkable adaptability to their environments, a trait
honed through the evolution of their morphological and neural structures [30,46].

C. Léger and G. Hamon—Equal first authors.
X. Hinaut and C. Moulin-Frier—Equal last authors.
C. Léger—Work done as intern at Flowers and Mnemosyne.

© The Author(s), under exclusive license to Springer Nature Switzerland AG 2024
S. Smith et al. (Eds.): EvoApplications 2024, LNCS 14635, pp. 36–60, 2024.
https://doi.org/10.1007/978-3-031-56855-8_3

They are born equipped with both hard-wired behavioral routines (e.g. breathing, motor babbling) and learning capabilities for adapting based on their own experience. The costs and benefits of evolving hard-wired behaviors vs. learning capabilities depend on different factors, a central one being the level of unpredictability of environmental conditions across generations [17,42]. Environmental challenges that are shared across many generations favor the evolution of hard-wired behavior (e.g. breathing). On the other hand, traits whose utility can hardly be predicted from its utility in previous generations are likely to be learned through individual development (e.g. learning a specific language). Some brain regions might have evolved to generically facilitate the learning of diverse behaviors. For example, central pattern generators (CPGs) enable limb bambling, which may facilitate locomotion, pointing and vocalizations in humans [24]. Another example is the prefrontal cortex (PFC), a brain region that maps inputs within a high-dimensional non-linear space from which they can be decoded by other brain regions, acting as a reservoir for computations [14,23].

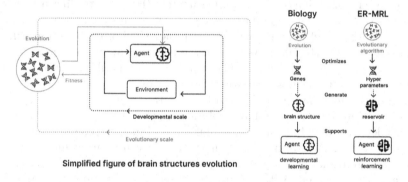

Fig. 1. (left) A simplified view of the evolution of brain structures. The generating parameters of neural structures are modified at an evolutionary loop. In the developmental loop, agents equipped with these neural structures learn to interact with their environment (right) Parallel to our computational approach. We propose a computational framework where an evolutionary algorithm optimizes hyperparameters that generate neural structures called reservoirs. These reservoirs are then integrated into RL agents that learn an action policy to maximize their reward in an environment

This prompts an intriguing question: How can neural structures, optimized at an evolutionary scale, enhance the capabilities of agents to learn complex tasks at a developmental scale? To address this question, we propose to model the interplay between evolution and development as two nested adaptive loops: neural structures are optimized through natural selection over generations (i.e. at an evolutionary scale), while learning specific behaviors occurs during an agent's lifetime (i.e. at a developmental scale). Figure 1 illustrates the interactions between evolutionary-scale and developmental-scale optimization. This model agrees with recent views on evolution that emphasize the importance of both scales for evolving complex skills [19, 20]. It is also compatible with the biological principle of a *genomic bottleneck*, i.e. the fact that the information contained in the genome of most organisms is not sufficient to fully describe their morphology [52]. In consequence, genomes must instead encode macro-level properties of morphological features such as synaptic connection patterns.

In line with these biological principles, we propose a novel computational approach, called Evolving Reservoirs for Meta Reinforcement Learning (ER-MRL), integrating mechanisms from Reservoir Computing (RC), Meta Reinforcement Learning (Meta-RL) and Evolutionary Algorithms (EAs). We use RL as a model of learning at a developmental scale [9, 29]. In RL, an agent interacts with a simulated environment through actions and observations, receiving rewards according to the task at hand. The objective is to learn an action policy from experience, mapping the observations perceived by the agent to actions in order to maximize cumulative reward over time. The policy is usually modeled as a deep neural network which is iteratively optimized through gradient descent. We use RC as a model of how a genome can encode macro properties of the agent's neural structure. In RC, the connection weights of a recurrent neural network (RNN) are generated from a handful of hyperparameters (HPs) controlling macro-level properties of the network related to connectivity, memory and sensitivity. Our choice of using RC relies on its parallels with biological brain structures such as CPGs and the PFC [15, 50], as well as on the fact that its indirect encoding of a neural network in global hyperparameters makes it compatible with the genomic bottleneck principle mentioned above. Being a cheap and versatile computational paradigm, RCs may have been favored by evolution [39].

We use Meta-RL to model how evolution shapes development [8, 32]. Meta-RL considers an outer loop, akin to evolution, optimizing HPs of an inner loop, akin to development. At the evolutionary scale, we use an evolutionary algorithm to optimize a genome specifying HPs of reservoirs. At a developmental scale, an agent equipped with a generated reservoir learns an action policy to maximize cumulative reward in a simulated environment. Thus, the objective of the outer evolutionary loop is to optimize hyperparameters of reservoirs in order to facilitate the learning of an action policy in the inner developmental loop.

Using this computational model, we run experiments in diverse simulated environments, e.g. 2D environments where the agent learns how to balance a pendulum and 3D environments where the agent learns how to control complex morphologies. These experiments provide support to three main hypotheses for how evolved reservoirs can affect development. First, they can facilitate solving partially-observable tasks, where the agent lacks access to all the information necessary to solve the task. In this case, we test the hypothesis that the recurrent nature of the reservoir will enable inferring the unobservable information. Second, it can generate oscillatory dynamics useful for solving locomotion tasks. In this case, the reservoir acts as a meta-learned CPG. Third, it can facilitate the generalization of learned behaviors to new tasks unknown during the evolution phase, a core hypothesis in meta-learning. In our case, our expectation is that HPs of reservoirs evolved across different environments will capture some abstract properties useful for adaptation.

In Sect. 2, we detail the methods underlying our proposed model, including RL (Sect. 2.1), Meta-RL (Sect. 2.2), RC (Sect. 2.3) and EAs (Sect. 2.4). We then explain their integration into our ER-MRL architecture (Sect. 3). Our results, aligned with the three hypotheses, are presented in Sect. 4. Computational specifics and supplementary experiments can be found in the appendix. The source code and videos are accessible at this link.

2 Background

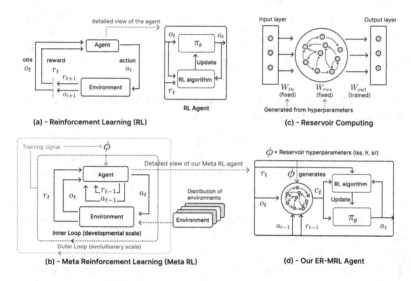

Fig. 2. Our proposed architecture, called ER-MRL, integrates several ML paradigms. We consider an RL agent learning an action policy (a), having access to a reservoir (c). We consider two nested adaptive loops in the spirit of Meta-RL (b). Our proposed architecture (d) consists in evolving HPs ϕ for the generation of reservoirs in an outer loop. In an inner loop, the agent learns an action policy, that takes as input the neural activation of the reservoir. The policy is trained using RL in order to maximize episodic return. Section 2 provides the computational details of each ML paradigm.

2.1 Reinforcement Learning as a Model of Development

Reinforcement Learning (RL) involves an agent that interacts with an environment by taking actions, receiving rewards, and learning an action policy in order to maximize its accumulated rewards (Fig. 2a). This interaction is formalized as a Markov Decision Process (MDP) [33]. An MDP is represented as a tuple (S, A, P, p_0, R), where S is the space of possible states of the environment, A is the space of available actions to the agent, $P(s_{t+1}|s_t, a_t)$ is the transition function specifying how the state at time $t+1$ is determined by the current state and action at time t, p_0 represents the initial state distribution, and $R(s_t, a_t)$ defines the reward received by the agent for a specific state-action pair. At each time step of an episode lasting T time steps, the agent observes the environment's state s_t, takes an action a_t, and receives a reward r_t. The environment then transitions to the next step according to $P(s_{t+1}|s_t, a_t)$. The objective of RL is to learn a policy $\pi_\theta(a|s)$ that maps observed states to actions in order to maximize the cumulative discounted reward G over time, where $G = \sum_{t=0}^{T} \gamma^t r_t$ [44]. The parameter $\gamma < 1$ discounts future rewards during decision making.

In Deep RL [21], the policy is implemented as an artificial neural network, whose connection weights are iteratively updated as the agent interacts with the environment. In all conducted experiments, we employ the Proximal Policy Optimization (PPO) RL algorithm [38] (see details in Sect. 6.1).

2.2 Meta Reinforcement Learning as a Model of the Interplay Between Evolution and Development

While RL has led to impressive applications [4,25,40], it suffers from several limitations: the learned policy is specific to the task at hand and does not necessarily generalize well to variations of the environment while requiring a large amount of data to converge. To address these issues Meta Reinforcement Learning (Meta-RL) [3] aims at training agents that *learn how to learn*, i.e. agents that can quickly adapt to new tasks or environments unknown during training. It is based on two nested adaptive loops: an outer loop, analogous to evolution, optimizes the HPs of an inner loop, analogous to development (Fig. 2b) [31,32]. The objective of the outer loop is to maximize the average performance of the inner loop on a distribution of environments. Formally, a set of HPs Φ are meta-optimized in the outer loop, with the objective of maximizing the average performance of a population of RL agents conditioned by Φ. In this paper, we leverage the RC framework where Φ corresponds to HPs encoding macro-level properties of a RNN, as explained in the next subsection.

2.3 Reservoir Computing as a Model of Neural Structure Generation

Meta-RL algorithms often directly optimize the weights of a RNN through backpropagation in the outer loop [10,11]. While this technique has demonstrated remarkable efficacy, it is ill-suited for addressing the research question outlined in the introduction. This is due to its lack of biological plausibility in two main aspects: (1) evolutionary-scale adaptation cannot rely on backpropagation mechanisms [43] and (2) the notion that evolution directly fine-tunes neural network weights contradicts the genomic bottleneck principle mentioned in the introduction [52]. Instead our method evolves RNNs based on the Reservoir Computing

(RC) paradigm. Instead of directly optimizing the neural network weights at the evolutionary scale, it optimizes HPs encoding macro-level properties of randomly generated recurrent networks.

The fundamental idea behind RC is to create a dynamic 'reservoir' of computation, where inputs are nonlinearly and recurrently recombined over time [22]. This provides a set of dynamic features from which a linear 'readout' can be easily trained: such training equivalent to selecting and combining interesting features to solve the given task (Fig. 2c).

A reservoir is generated from a few HPs which play a crucial role in shaping the efficiency of the reservoir dynamics. This includes the number of neurons in the reservoir, the spectral radius sr (controlling the level of recurrence in the generated network), input scaling iss (controlling the strength of the network's inputs), and leak rate lr (controlling how much the neurons retain past information); we explain reservoir HPs in more details in Appendix 6.1. In this paper, we propose to meta-optimize reservoir's HPs $\Phi = (sr, iss, lr)$ in a Meta-RL outer loop, using evolutionary algorithms explained in the next subsection. We will then explain how we propose to integrate RC with RL in Sect. 3.

2.4 Evolutionary Algorithms as a Model of Evolution

Evolutionary Algorithms (EAs) draw inspiration from the fundamental principles of biological evolution [2, 36], where species improve their fitness through the selection and variation of their genomes. EAs iteratively enhance a population of candidate parameterized solutions to a given optimization problem, iteratively selecting those with higher fitness levels (i.e. higher performance of the solution) and mutating their parameters for the next generation.

In our approach, we utilize the Covariance Matrix Adaptation Evolution Strategy (CMA-ES) [13] as our designated evolutionary algorithm in order to meta-optimize HPs Φ of reservoirs. In CMA-ES, a population of HPs candidates is sampled from a multivariate Gaussian distribution, with mean μ and covariance matrix V. The fitness of each sample Φ_i of the population is evaluated (see Sect. 3 for how we do it in our proposed method). The Gaussian distribution is then updated by weighting each sample proportionally to its fitness; resulting in a new mean and covariance matrix that are biased toward solutions with higher fitness. This process continues iteratively until either convergence towards sufficiently high fitness values of the generated HPs is achieved, or until a predefined threshold of candidates is reached.

3 Evolving Reservoirs for Meta Reinforcement Learning (ER-MRL)

General Approach. Our objective is to devise a computational framework to address a fundamental question: How can neural structures adapt at an evolutionary scale, enabling agents to better adapt to their environment at a developmental scale? For this aim, we aim to integrate the Machine Learning paradigms presented above. The architecture is illustrated in Fig. 2d and the optimization procedure in Fig. 3. We call our method ER-MRL, for "Evolving Reservoirs for Meta Reinforcement Learning".

The ER-MRL method encompasses two nested optimization loops (as in Meta-RL, Sect. 2.2). In the outer loop, operating at an evolutionary scale, HPs Φ for generating a reservoir (Sect. 2.3) are optimized using an evolutionary algorithm (Sect. 2.4). In the inner loop, focused on a developmental scale, a RL algorithm (Sect. 2.1) learns an action policy π_θ using the reservoir state as inputs. In other words, the outer loop meta-learns HPs able to generate reservoirs resulting in maximal averages performance on multiple inner loops. The whole process is illustrated in Fig. 3 and detailed below.

Inner Loop. To represent the development of an agent, we consider a RL agent (Sect. 2.1) that interacts with an environment through observation o_t, actions a_t and rewards r_t at each time step t (Fig. 2a). In our proposed ER-MRL method, this agent is composed of three main parts: a reservoir generated by HPs $\Phi = \{iss, lr, sr\}$ (see Sect. 6.1 for more details), a feed forward action policy network π_θ and a RL algorithm. At each time step, we feed the reservoir with the current o_t, and the previous action and reward a_t and r_t (Fig. 2d). Contrarily to standard RL, the policy π_θ does not directly access the observation of the environment's state o_t, but the context c_t of the reservoir instead (i.e. the vector of all reservoir's neurons activations at time t). Because reservoirs are recurrent neural networks, c_t not only encompasses information about the current time step, but also integrates information over previous time steps. In some experiments, we also use ER-MRL with multiple reservoirs. In this case, we still generate the reservoirs from a set of HPs Φ, and the context c_t given to the policy is the concatenation of hidden states of all reservoirs. We then train our policy $\pi_\theta(a|c_t)$ using RL.

Outer Loop. The outer loop employs the Covariance Matrix Adaptation Evolutionary Strategy (CMA-ES) (Sect. 2.4) to optimize reservoir HPs Φ. The objective is to generate reservoirs which, on average over multiple agents, improve learning abilities. For each set of HPs, we assess the performance of our agents in multiple inner loops (we utilize 3 in our experiments), each one with a different random seed. Using different random seeds implies that, while using the same HPs set, each agent will be initialized with different connection weights of both their reservoirs, their policies and the initial environment state. Note that while the generated reservoirs have different connection weights, they share the same macro-properties in terms of spectral radius sr, input scaling iss and leak rate lr (since they are generated from the same HPs set). In assessing an agent's fitness within its RL environment, we compute the mean episodic reward over the final 10 episodes of its training. To obtain the fitness of a reservoir HPs, we calculate the mean fitness of three agents across three different versions of the same environment. These steps are iterated until we reach a predetermined threshold of CMA-ES iterations (set at 900 in our experiments).

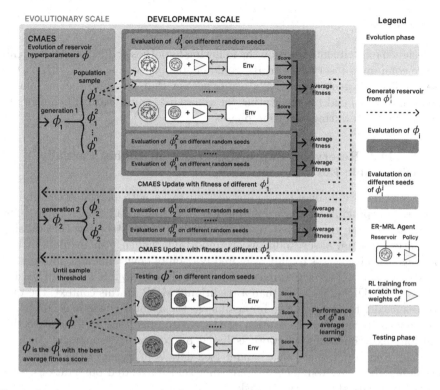

Fig. 3. In the evolution phase (top), CMA-ES refines Reservoir HPs Φ. At each generation i of the evolution loop (left), a population $\Phi_i : \{\Phi_i^1, \ldots, \Phi_i^n\}$ of HPs is sampled from the CMA-ES Gaussian distribution. Each Φ_i^j undergoes evaluation on multiple random seeds, generating multiple reservoirs. An ER-MRL agent is created for each reservoir, with its action policy being trained from the states of that reservoir (lighter grey frames). The fitness of a sampled Φ_i^j is determined by the average score of all ER-MRL agents generated from it (mid-grey frames). The fitness values are used to update the CMA-ES distribution for the next generation (dotted arrow). This process iterates until a predetermined threshold is reached. In the Testing phase (bottom), the best set of HPs Φ^* from all CMA-ES samples is employed. Multiple reservoirs are generated within ER-MRL agents, and their performance is evaluated.

Evaluation. To evaluate our method, we select the HPs Φ^* that generated the best fitness function during the whole outer loop optimization with CMA-ES (see bottom of Fig. 3). We then generate 10 ER-MRL agents with different random seeds (with a different reservoir sampled from Φ^* for each seed, together with random initial policy weights θ) and train the action policy π_θ of each agent using RL. We report our results in the next section, comparing the performance of these agents against vanilla RL agents using a feedfoward policy.

4   Results

We designed experiments to study the following hypotheses: The ER-MRL architecture combining reservoirs and RL could enable (1) solving tasks with partial observability, (2) generating oscillatory dynamics that facilitate the learning of locomotion tasks, and (3) facilitating the generalization of learned behaviors to new tasks unseen during evolution phase.

4.1   Evolved Reservoirs Improve Learning in Highly Partially Observable Environments

In this section, we evaluate our approach on tasks with partial observability, where we purposefully remove information from the agent observations. Our hypothesis is that the evolved reservoir can help reconstructing this missing information. Partial observability is an important challenge in the field of RL, where agents have access to only a limited portion of environmental information to make decisions. This is referred to as a Partially Observable Markov Decision Process (POMDP) [26] rather than a traditional MDP. In this context, the task becomes harder to learn, or even impossible, as the agent needs to make decisions based on an incomplete observation of the environment state. To explore this issue, our experimental framework is based on control environments, such as CartPole, Pendulum, and LunarLander (see details in Fig. 9 of the appendix). We modify these environments by removing velocity-related observations, thus simulating a partially-observable task.

Let's illustrate this issue with the first environment (CartPole), where the agent's goal is to keep the pole upright on the cart while it moves laterally. If we remove velocity-related observations (both for the cart and the pole's angle), a standard feedfoward RL agent cannot effectively solve the task. The reason is straightforward: without this information, the agent doesn't know the cart's movement direction or whether the pole is falling or rising. We apply the same process to the other two environments, removing all velocity-related observations for our agents. Can the ER-MRL architecture address this challenge? To find out, we independently evolve reservoirs using ER-MRL for each task. We search for effective HPs tailored to the partial observability of each environment. To evaluate our approach, we compare the learning curves of ER-MRL agents (from the test phase, see bottom of Fig. 3) on these three partially observable environments against an agent with a feedforward policy.

Figure 4 presents the results for the three selected partially observable tasks. We observe, as expected, that vanilla RL agents cannot learn how to solve the task under partial observability (for the reasons mentioned above). In comparison, our approach leads to performance scores close to those obtained by a RL algorithm with full observability. This indicates that the evolved reservoir is able to reconstruct missing information related to velocities from its own internal recurrent dynamics. This confirms the hypothesis that an agent with a reservoir can solve partially observable tasks by using the internal reservoir state to reconstruct missing information. We explain with more details why this method

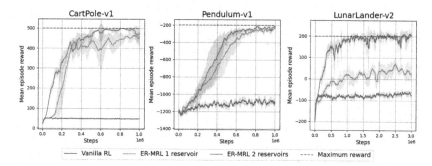

Fig. 4. Learning curves for partially observable tasks. The x-axis represents the number of timesteps during the training and the y-axis the mean episodic reward. Learning curves of our ER-MRL methods correspond to the testing phase described in the bottom of Fig. 3. Vanilla RL corresponds to a feedforward policy RL agent. The curves and the shaded areas represent the mean and the standard deviation of the reward for 10 random seeds. See Sect. 6.3 for a comparison with another method.

could work in Sect. 6.3 of the appendix. The difference in results between the model with 2 reservoirs on LunarLander environment suggests that solving it requires encoding at least two different timescales dynamics. Our interpretation here is that solving LunarLander requires to deal with both an "approaching" and "landing" phase, unlike the two other environments.

4.2   Evolved Reservoirs Could Generate Oscillatory Dynamics that Facilitate the Learning of Locomotion Tasks

In this section, we evaluate our approach on agents with 3D morphology having to learn locomotion tasks shown in Fig. 10. We postulate that the integration of an evolved reservoir can engender oscillatory patterns that aid in coordinating body movements, akin to Central Pattern Generators (CPGs). CPGs, rooted in neurobiology, denote an interconnected network of neurons responsible for generating intricate and repetitive rhythmic patterns that govern movements or behaviors [24] such as walking, swimming, or other cyclical movements. Existing scientific literature hypothesizes that reservoirs, possessing significant rhythmic components, share direct connections with CPGs [37]. We propose to study this hypothesis using motor tasks involving rhythmic movements.

We employed 3D MuJoCo environments (detailed in Fig. 10 of the appendix), where the goal is to exert forces on various rotors of creatures to propel them forward. Notably, while the ultimate goal across these tasks remains constant (forward movement), the creatures exhibit diverse morphologies, including humanoids, insects, worms, bipeds, and more. Furthermore, the action and observation spaces vary for each morphology. We individually evaluate our ER-MRL architecture on each of these tasks.

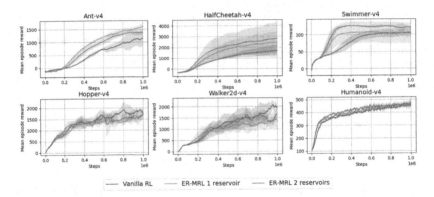

Fig. 5. Learning curves for locomotion tasks. Same conventions as Fig. 4

Our approach demonstrates improved performance in some tasks (Ant, HalfCheetah, and Swimmer) compared to a standard RL baseline, particularly noticeable in the early stages of learning, as illustrated in Fig. 5. This suggests that the evolved reservoir may generate beneficial oscillatory patterns, facilitating the learning of locomotion tasks, in line with the notion that reservoirs could potentially function as CPGs, aiding in solving motor tasks. Although carefully testing this hypotheses would require more analysis, we present in Sect. 6.4 in the appendix preliminary data suggesting that the evolved reservoir is able to generate oscillatory dynamics that could facilitate learning in the Swimmer environment. However, as shown in Fig. 5, performance enhancement was not observed in the Walker and Hopper environments compared to the RL baseline. Locomotion in both environments demands precise closed-loop control strategies to maintain an agent's equilibrium. In such cases, generated oscillatory patterns may not be as beneficial.

4.3   Evolved Reservoirs Improve Generalization on New Tasks Unseen During Evolution Phase

In this section, we address a key aspect of our study: the ability of evolved reservoirs to facilitate adaptation to novel environments. This inquiry is crucial in assessing the potential of evolved neural structures to generalize and enhance an agent's adaptability beyond the evolution phase. Building on the promising results of ER-MRL with two reservoirs in previous experiments, we focus exclusively on this configuration for comparison with the RL baseline.

Generalizing Across Different Morphologies with Similar Tasks. In prior experiments, ER-MRL demonstrated effectiveness in environments like Ant, HalfCheetah, and Swimmer. This success led us to explore whether reservoirs evolved for two of these tasks could be adaptable to the third, indicating potential generalization across different morphologies. However, due to variations in environments, including differences in morphology, observation and action spaces, and reward functions, generalization from one set of tasks to another presents a complex challenge. To ensure fair task representation of each environment in the final fitness, we employ the normalization formula detailed in Sect. 6.6. Subsequently, we select the reservoir HPs Φ^* that yielded the highest fitness and evaluate them in a distinct environment. For instance, if we evolve reservoirs on Ant and HalfCheetah, we test them in the Swimmer task.

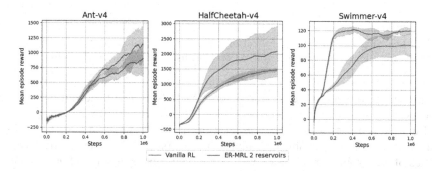

Fig. 6. Learning curves for generalization on similar locomotion tasks with different morphologies. The curves evaluate the performance of ER-MRL on an environment that was unseen during the evolution phase. For instance, the left plot shows performance of an agent on Ant, using reservoirs evolved on only HalfCheetah and Swimmer.

In Fig. 6, we observed a notable improvement in the performance of ER-MRL agents with reservoirs evolved for different tasks, particularly in HalfCheetah and Swimmer environments. This substantiates the capacity of evolved reservoirs to generalize to new tasks and encode diverse dynamics from environments with distinct morphologies. However, it's worth noting that this improvement wasn't replicated in the Ant task. This could be attributed to the unique characteristics of the Ant environment, with its stable four legged structure, in contrast to the simpler anatomies of Swimmer and HalfCheetah. For a detailed analysis, please refer to Sect. 6.7 in the appendix.

Generalizing Across Different Tasks with Similar Morphologies. We have seen how reservoirs facilitated ER-MRL agent's ability to generalize across locomotion tasks with different morphologies. Now, we shift our focus to tasks with consistent morphologies but distinct objectives. To delve into this, we turn to the Humanoid and HumanoidStandup environments (shown in Fig. 12 of the appendix), both presenting tasks within the realm of humanoid movement. One task involves learning to walk as far as possible, while the other centers around the challenge of standing up from the ground. As in our previous study, we follow

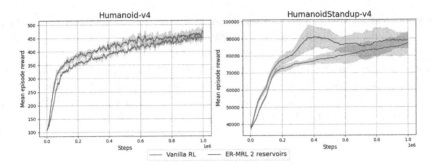

Fig. 7. Learning curves for generalization on different locomotion tasks with similar morphologies. The reservoirs are evolved on one task and tested on the other one.

the procedure of evolving reservoir-generating HPs on one task and evaluating their performance on the other.

Figure 7 provides a visual representation of our findings. While the performance improvement may not be dramatic, it underscores the generalization capabilities of reservoirs across tasks with similar morphologies but differing objectives. This observation, though promising, invites further investigation, given the limited number of experiments conducted in this context. This aspect represents an avenue for future research.

5   Discussion

In this paper, we have addressed the compelling question of whether reservoir-like neural structures can be evolved at an evolutionary time scale, to facilitate the learning of agents on a multitude of sensorimotor tasks at a developmental scale. Our results demonstrate the effectiveness of employing evolutionary algorithms to optimize these reservoirs, especially on Reinforcement Learning tasks involving partial observability, locomotion, and generalization of evolved reservoirs to unseen tasks.

Our ER-MRL approach has parallels to previous algorithms in RL that employ an indirect encoding for mapping a genome to a particular neural network architecture [12,28,41]. Our choice of employing reservoirs comes with the benefit of a very small genomic size (reservoirs are parameterised by a handful of parameters that we show in Appendix 6.1) without reducing the complexity of the phenotype (the number of weights of the reservoir policy is independent of the number of hyper-parameters). Moreover, our approach clearly distinguish neural structures optimized at the evolutionary scale (the reservoirs) vs. at the developmental scale (the RL action policy).

Nonetheless, some limitations persist within our methodology. The combination of reservoir computing and reinforcement learning remains underexplored in the existing literature [6,7], leaving substantial room for refining the algorithmic framework for improved performance. Moreover, our generalization experiments and quantitative analyses warrant further extensive testing to gain deeper insights. Notably, our approach does incur a computational cost due to the time required to train a new policy with RL for each generated reservoir. Future studies could devise more efficient evolutionary strategies or employ alternative optimization techniques.

However, because our method remains agnostic to specific environment and agent's characteristics (a reservoir architecture being independent of the shape of its inputs and outputs), we could in theory evolve reservoirs across a very wide range of environments and agent's morphologies. Such evolved generalist reservoirs could then result in highly reduced computational cost at the developmental scale, as our results suggest, compared to training recurrent architectures from scratch.

Moving forward, there are several promising avenues for exploration. Firstly, a more comprehensive understanding of the interaction between RL and RC could significantly improve the performance of such methods on developmental learning tasks. Secondly, integrating our approach with more sophisticated Meta-RL algorithms could offer a mean to initialize RL policy weights with purposefully selected values rather than random ones. Additionally, a broader framework allowing for the evolution of neural structures with greater flexibility, such as varying HPs and neuron counts, could yield more intricate patterns during the evolution phase, potentially resulting in substantial improvements in agent performance across developmental tasks [28, 41].

Our research bridges the gap between evolutionary algorithms, reservoir computing and meta-reinforcement learning, creating a robust framework for modelling neural architecture evolution. We believe that this integrative approach opens up exciting perspectives for future research in RC and Meta-RL to propose new paradigms of computations. It also provides a computational framework to study the complex interplay between evolution and development, a central issue in modern biology [16, 18, 27, 49].

Acknowledgments. Financial support was received from: the University of Bordeaux's France 2030 program/RRI PHDS framework, French National Research Agency (ANR) grants: ECOCURL ANR-20-CE23-0006 and DEEPPOOL ANR-21-CE23-0009-01. We benefited HPC resources: IDRIS under the allocation A0091011996 made by GENCI, using the Jean Zay supercomputer, and Curta from the University of Bordeaux.

6 Appendix

In this appendix, we provide comprehensive insights and clarifications on the methodologies employed in our study. Specifically, we elaborate on aspects such as the parameters governing our experiments, including the RL (PPO), the RC and the evolutionary (CMA-ES) algorithms we used. Furthermore, we furnish a detailed exposition of the environments utilized in our research. Lastly, we conduct supplementary analyses aimed at enhancing our understanding of some observed phenomena in the obtained results. In addition, we present results from experiments that were not featured in the main text to offer a more comprehensive view of our findings.

6.1 Methods

Proximal Policy Optimization (PPO). PPO, categorized as a policy gradient technique [45], undertakes exploration of diverse policies through stochastic gradient ascent. This process involves assigning elevated probabilities to actions correlated with high rewards, subsequently adjusting the policy to aim for higher expected returns. The adoption of PPO stems from its well-established reputation as a highly efficient and stable algorithm in the scientific literature, although its use does not have major theoretical implications for this particular project.

Reservoir Hyperparameters. In Reservoir Computing, the spectral radius controls the trade-off between stability and chaoticity of reservoir dynamics: in general "edge of chaos" dynamics are often desired [5]. Input scaling determines the strength of input signals, and the leak rate governs the memory capacity of reservoir neurons over time. These HPs specify the generation of the reservoir weights. Once the reservoir is generated, its weights are kept fixed and only a readout layer, mapping the states of the reservoir neurons to the desired output of the network are learned. Other HPs exist to initialize a reservoir, but they have not been studied in the experiments of the paper (as it has been tested that they have much less influence on the results).

6.2 Experiment Parameters

General Parameters. In our experiments, we adapted the number of timesteps during the training phase of our ER-MRL agent in the inner loop, based on whether we were evolving the reservoir HPs or testing the best HPs set discovered during the CMA-ES evolution. For the evolution phase, which was computationally intensive, we utilized 300,000 timesteps per training. Conversely, when evaluating our agents against standard RL agents, we employed 1,000,000 timesteps. Notably, in the case of the LunarLander environment, we extended the testing to 3,000,000 timesteps, as the learning curve had not yet converged at 1,000,000 timesteps.

PPO Hyperparameters. Regarding the parameters of our RL algorithm, PPO, we used the default settings provided by the Stable Baselines3 library [35]. For tasks involving partial observability, we made a slight adjustment by setting the learning rate to 0.0001, as opposed to the standard 0.0003. This modification notably enhanced performance, potentially indicating that reservoirs contained a degree of noise, warranting a lower learning rate to stabilize RL training.

CMA-ES Hyperparameters. For the parameters of our evolutionary algorithm, CMA-ES, we adopted the default settings of the CMA-ES sampler from the Optuna library [1].

Reservoirs Hyperparameters. For the reservoirs, we only modified the parameters mentioned in Sect. 6.1 and the number of neurons. We consistently used 100 neurons per reservoirs during all experiments. All the other HPs were

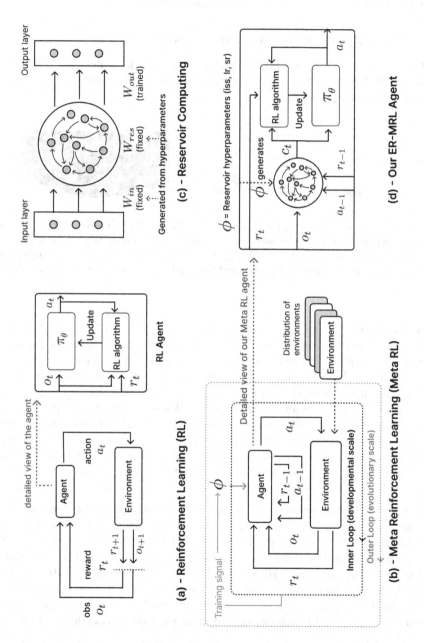

Fig. 8. Rotated view of Fig. 2 presenting the background methods used, and how our ER-MRL agents incorporate them

kept the same and are the default reservoir parameters used in ReservoirPy [48]. We conducted additional analyses and observed that they exerted a relatively modest influence on tasks of this nature. However, a more refined analysis of the importance of these HPs could be interesting in future works.

6.3 Partially Observable Environments

In the following section, we present the different Reinforcement Learning environments from the Gymnasium library [47], used during our experiments on partial observability.

Fig. 9. Partially observable environments used, The goal of CartPole (left) is to learn how to balance the pole on the cart. The goal of Pendulum (middle) is to learn how to maintain the pendulum straight up by applying forces on it. The goal of LunarLander (right) is to learn how to land between the two flags by generating forces on the different spaceship reactors.

Results Analysis. To better understand the reservoir's capabilities on these tasks, we conducted several tests on supervised learning problems where a sequence of actions, rewards, and observations (without velocity) was provided to a reservoir with a linear readout. In one case, the model had to reconstruct full observation information (position, angle, velocity, angular velocity), and in the other, it had to reconstruct positions and angles over several time steps (doing this only for the last 2 time steps allows a PPO to achieve maximum reward later on). In both cases, this model successfully solved the tasks with very high performance. Moreover, it was also capable of predicting future observations, which can be extremely valuable to find an optimal action policy.

Benchmark Comparison. Regarding benchmarks, our approach compares favorably with the results reported in the blog post from Raffin [34] where he used another model combining a RNN (LSTM [51]) with a RL algorithm (PPO, the one we also used) on the same partially observable tasks. The performance on each environment are pretty similar, but it is the training timesteps needed to reach the maximum performance that varies the most between the methods. Indeed for the LunarLander environment, our method is able to learn in less timesteps after evolving reservoirs, but it is the contrary with CartPole and Pendulum tasks.

It is worth noting that even if both approaches have similarities, ER-MRL consists in optimizing the HPs of reservoirs at an evolutionary scale, whereas the method presented in the blog post trains a recurrent architecture from scratch. This divergence complicates direct comparisons between both methods. Indeed, our results are derived after an extensive phase of computation in a Meta-RL outer loop, but the subsequent evaluation with the final reservoir configuration is comparatively swift. as only the RL policy (linear readout) requires training. In contrast, the LSTM-PPO method does not incorporate a computationally intensive meta-learning phase, but their training process takes more time per timestep update. Indeed, each training step of the this demands more computation, due to having to train the LSTM from scratch in addition to PPO, compared to our method where only the linear readout is trained at the developmental scale.

However to ensure a fair and comprehensive comparison with other baselines, especially in tracking the time required to achieve presented results, more experiments are necessary.

6.4   MuJoCo Forward Locomotion Environments

Fig. 10. MuJoCo environments, the goal of these tasks is to apply force to the rotors of the creatures to make them move forward. On the top row, we have from left to right the Ant, HalfCheetah and Swimmer environments, and on the bottom row, the Hopper, Walker and Humanoid environments. The environment observations comprise positional data of distinct body parts of the creatures, followed by the velocities of those individual components, while actions entail the torques applied to the hinge joints.

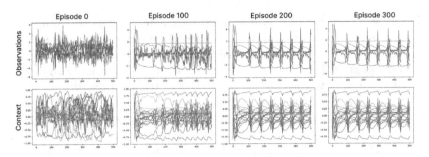

Fig. 11. Differences between the observations of a RL agent (top) with the context of an ER-MRL agent (bottom) at the same stage of training. Each episode lasts 1000 timesteps in the environment. The curves of the RL agent represent the real observation values from the environment, and the curves of the ER-MRL one part of the context given to the agent's policy: the activation values of 20 reservoir neurons (out of 100).

Results Analysis. In this section, we present how reservoirs could act as Central Pattern Generators within agents learning a locomotion task in these 3D environments.

It can be observed that the separation between the two models seems to occur starting from 100,000 timesteps at the top-right of Fig. 5. Therefore, we recorded videos of the RL and ER-MRL agents to better understand the performance difference between the two models. Furthermore, we conducted a study at the level of the input vector in the agent's policy (o_t for RL agent, and c_t for ER-MRL agent). As seen in Fig. 11, it is noticeable that very early in the learning process, the reservoir exhibits much more rhythmic dynamics than the sole observation provided by the environment. This could be due to the link between the reservoir and CPGs, potentially facilitating the acquisition and learning of motor control in these tasks.

Expanding on this, it's notable that CPGs, shared across various species, have evolved to embody common structures. Drawing parallels from nature, our investigation delves into whether generalization (results in Sect. 4.3) across a spectrum of motor tasks may mirror the principles found in biological systems.

However, further experiments, accompanied by robust quantitative analysis, are necessary to gain valuable insights into whether reservoirs can function as CPG-like structures.

6.5   MuJoCo Humanoid Environments

Interesting Reservoir Results. As seen in Sect. 2.3, one of the basic principles of RC is to project input data into a higher-dimensional space. In the case of the Humanoid tasks, where our results are displayed in Fig. 5 and Fig. 7, the initial observation and action space is larger (400 dimensions) compared to the context dimension for one or two reservoirs of 100 neurons (the dimension is equal to

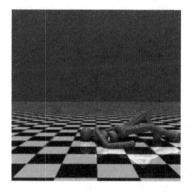 

Fig. 12. MuJoCo environments with humanoid morphologies. On the left figure, the goal is to learn how to stand up, and on the right the goal is to walk forward as far as possible

the number of neurons). This means that even by reducing the input dimension in the RL policy network, the reservoir improves the quality of the data. For other morphologies, the dimension of input data is inferior to the dimension our reservoir context.

6.6   Normalized Scores for Generalization

To prevent any particular task from disproportionately influencing the fitness score due to variations in reward scales, we use a fitness function for CMA-ES that aggregates the normalized score, denoted as $nScore$, across both environments. The normalization process is defined as :

$$nScore = \frac{score - randomScore}{baselineScore - randomScore}$$

Where $randomScore$ and $baselineScore$ represent the performances of a random and of a standard PPO agent, respectively.

6.7   Reservoir Hyperparameters Analysis

In preceding sections, we observed how HPs play a pivotal role in enabling ER-MRL agents to generalize across tasks. Now, we delve deeper into understanding why some reservoirs aid in generalization for specific tasks while others do not. To gain this insight, we constructed a hyperparameter map to visualize the regions of HPs associated with the best fitness in each environment. We selected the best 30 sets of HPs, comprising the spectral radius and leak rate values of the reservoirs, out of a pool of 900 for all MuJoCo locomotion tasks (refer to Fig. 10) and plotted them on a 2D plane.

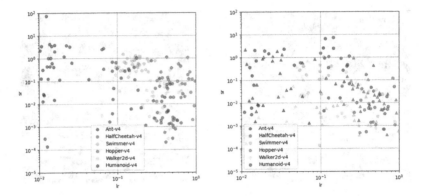

Fig. 13. The left figure represents parameters obtained with a single reservoir, while the right figure corresponds to configurations with two reservoirs (depicted as either circles or triangles).

In Fig. 13, we observe that the HPs for most environments are clustered closely together. Conversely, those for the Ant environment form a distinct cluster, characterized by notably lower leak rates. The leak rate reflects how much information a neuron retains in the reservoir, influencing its responsiveness to input data and connections with other neurons. A lower leak rate implies a more extended memory, possibly instrumental in capturing long-term dynamics. This observation aligns with the stable morphology of the Ant, potentially allowing the agent to prioritize long-term dynamics for efficient locomotion. This would partially explain why generalization wasn't successful on this environment in Sect. 4.3, when reservoirs were evolved on other types of morphologies.

7 Additional experiments

We also led other experiments that we didn't mention in the main text. As mentioned above in Sect. 6.1, we consistently employed reservoirs with a size of 100 neurons to ensure a standardized basis for result comparison. This configuration equates one reservoir to 100 neurons, two reservoirs to 200 neurons, and so forth. We conducted additional experiments to investigate the impact of varying the number of reservoirs and neurons within them. We observed that altering the number of neurons within a reservoir had a limited effect. For example, reducing the number of neurons to as low as 25 did not significantly affect performance on the partially observable environments. Increasing the size of the reservoirs didn't seem to improve the performance a lot either, except for the Humanoid environments (with a large observation space) where reservoirs equipped with a lot of neurons (1000) performed slightly better than others. While we opted for 100 neurons per reservoir in our experiments, there is surely potential for further optimization.

Furthermore, we explored experiments involving partially observable reservoirs, in which only a subset of the observation was provided to the policy. The results demonstrated that it is not always necessary to fully observe the contextual information within the reservoir to successfully accomplish tasks. On the CartPole environment, we tested 3 type of models with a reservoir of 100 fully observable neurons (the policy has access to 100 out of the 100 neurons), a reservoir of 1000 fully observable neurons, and another reservoir with only 100 partially observable neurons out of 1000. We observed that the model with 1000 fully observable neurons performed worse than the two other, who had similar results.

Regarding generalization experiments, we investigated the impact of varying the number of reservoirs. Although experiments with three reservoirs yielded intriguing insights, such as distinct memory types characterized by leak rate in the different reservoirs, the overall performance was notably lower compared to configurations with two reservoirs. This observation can likely be attributed to the increased complexity of learning due to the larger observation space, despite the potential for richer dynamics. We also noted instances where several reservoirs maintained very similar hyperparameters for specific tasks, potentially indicating the importance of capturing particular dynamics.

Additionally, we considered the possibility of employing smaller reservoirs in greater numbers. This approach could capture a diverse range of interesting features, such as different dynamics, while keeping the total number of neurons low. This strategy would be particularly advantageous for tasks characterized by small observation and action spaces, but would also imply a wider space of reservoirs HPs search in return.

References

1. Akiba, T., Sano, S., Yanase, T., Ohta, T., Koyama, M.: Optuna: a next-generation hyperparameter optimization framework. In: Proceedings of the 25th ACM SIGKDD International Conference on Knowledge Discovery & Data Mining, pp. 2623–2631 (2019)
2. Bäck, T., Schwefel, H.P.: An overview of evolutionary algorithms for parameter optimization. Evol. Comput. **1**(1), 1–23 (1993)
3. Beck, J., et al.: A survey of meta-reinforcement learning. arXiv preprint arXiv:2301.08028 (2023)
4. Berner, C., et al.: Dota 2 with large scale deep reinforcement learning. arXiv preprint arXiv:1912.06680 (2019)
5. Bertschinger, N., Natschläger, T.: Real-time computation at the edge of chaos in recurrent neural networks. Neural Comput. **16**(7), 1413–1436 (2004)
6. Chang, H., Futagami, K.: Reinforcement learning with convolutional reservoir computing. Appl. Intell. **50**, 2400–2410 (2020)
7. Chang, H.H., Song, H., Yi, Y., Zhang, J., He, H., Liu, L.: Distributive dynamic spectrum access through deep reinforcement learning: a reservoir computing-based approach. IEEE Internet Things J. **6**(2), 1938–1948 (2018)
8. Clune, J.: Ai-gas: Ai-generating algorithms, an alternate paradigm for producing general artificial intelligence. arXiv preprint arXiv:1905.10985 (2019)
9. Doya, K.: Reinforcement learning: computational theory and biological mechanisms. HFSP J. **1**(1), 30 (2007)

10. Duan, Y., Schulman, J., Chen, X., Bartlett, P.L., Sutskever, I., Abbeel, P.: Rl squared: fast reinforcement learning via slow reinforcement learning. arXiv preprint arXiv:1611.02779 (2016)
11. Finn, C., Abbeel, P., Levine, S.: Model-agnostic meta-learning for fast adaptation of deep networks. In: International Conference on Machine Learning, pp. 1126–1135. PMLR (2017)
12. Ha, D., Dai, A., Le, Q.V.: HyperNetworks (2016). http://arxiv.org/abs/1609.09106. arXiv:1609.09106 [cs]
13. Hansen, N.: The CMA evolution strategy: a tutorial. arXiv preprint arXiv:1604.00772 (2016)
14. Hinaut, X., Dominey, P.F.: A three-layered model of primate prefrontal cortex encodes identity and abstract categorical structure of behavioral sequences. J. Physiol.-Paris 105(1–3), 16–24 (2011)
15. Hinaut, X., Dominey, P.F.: Real-time parallel processing of grammatical structure in the fronto-striatal system: a recurrent network simulation study using reservoir computing. PLoS ONE 8(2), e52946 (2013)
16. Hougen, D.F., Shah, S.N.H.: The evolution of reinforcement learning. In: 2019 IEEE Symposium Series on Computational Intelligence (SSCI), pp. 1457–1464. IEEE (2019)
17. Johnston, T.D.: Selective costs and benefits in the evolution of learning. In: Rosenblatt, J.S., Hinde, R.A., Beer, C., Busnel, M.C. (eds.) Advances in the Study of Behavior, vol. 12, pp. 65–106. Academic Press (1982). https://doi.org/10.1016/S0065-3454(08)60046-7. http://www.sciencedirect.com/science/article/pii/S0065345408600467
18. Johnston, T.D.: Selective costs and benefits in the evolution of learning. In: Advances in the Study of Behavior, vol. 12, pp. 65–106. Elsevier (1982)
19. Kauffman, S.A.: The Origins of Order: Self Organization and Selection in Evolution. Oxford University Press, Oxford (1993)
20. Laland, K.N., et al.: The extended evolutionary synthesis: its structure, assumptions and predictions. Proc. Royal Soc. B: Biol. Sci. 282(1813), 20151019 (2015). https://doi.org/10.1098/rspb.2015.1019. https://royalsocietypublishing.org/doi/10.1098/rspb.2015.1019
21. Li, Y.: Deep reinforcement learning: an overview. arXiv preprint arXiv:1701.07274 (2017)
22. Lukoševičius, M., Jaeger, H.: Reservoir computing approaches to recurrent neural network training. Comput. Sci. Rev. 3(3), 127–149 (2009)
23. Mante, V., Sussillo, D., Shenoy, K.V., Newsome, W.T.: Context-dependent computation by recurrent dynamics in prefrontal cortex. Nature 503(7474), 78–84 (2013)
24. Marder, E., Bucher, D.: Central pattern generators and the control of rhythmic movements. Curr. Biol. 11(23), R986–R996 (2001)
25. Mnih, V., et al.: Playing atari with deep reinforcement learning. arXiv preprint arXiv:1312.5602 (2013)
26. Monahan, G.E.: State of the art-a survey of partially observable Markov decision processes: theory, models, and algorithms. Manag. Sci. 28(1), 1–16 (1982)
27. Moulin-Frier, C.: The ecology of open-ended skill acquisition. Ph.D. thesis, Université de Bordeaux (UB) (2022)
28. Najarro, E., Sudhakaran, S., Risi, S.: Towards self-assembling artificial neural networks through neural developmental programs. In: Artificial Life Conference Proceedings, vol. 35, p. 80. MIT Press, Cambridge (2023)

29. Nussenbaum, K., Hartley, C.A.: Reinforcement learning across development: what insights can we draw from a decade of research? Dev. Cogn. Neurosci. **40**, 100733 (2019)
30. Pearson, K.: Neural adaptation in the generation of rhythmic behavior. Ann. Rev. Physiol. **62**(1), 723–753 (2000)
31. Pedersen, J., Risi, S.: Learning to act through evolution of neural diversity in random neural networks. In: Proceedings of the Genetic and Evolutionary Computation Conference, pp. 1248–1256 (2023)
32. Pedersen, J.W., Risi, S.: Evolving and merging hebbian learning rules: increasing generalization by decreasing the number of rules. In: Proceedings of the Genetic and Evolutionary Computation Conference, pp. 892–900 (2021)
33. Puterman, M.L.: Markov decision processes. Handb. Oper. Res. Manag. Sci. **2**, 331–434 (1990)
34. Raffin, A.: Ppo vs recurrentppo (aka ppo lstm) on environments with masked velocity (sb3 contrib). https://wandb.ai/sb3/no-vel-envs/reports/PPO-vs-RecurrentPPO-aka-PPO-LSTM-on-environments-with-masked-velocity-VmlldzoxOTI4NjE4. Accessed Nov 2023
35. Raffin, A., Hill, A., Gleave, A., Kanervisto, A., Ernestus, M., Dormann, N.: Stable-baselines3: reliable reinforcement learning implementations. J. Mach. Learn. Res. **22**(1), 12348–12355 (2021)
36. Reddy, M.J., Kumar, D.N.: Computational algorithms inspired by biological processes and evolution. Curr. Sci. 370–380 (2012)
37. Ren, G., Chen, W., Dasgupta, S., Kolodziejski, C., Wörgötter, F., Manoonpong, P.: Multiple chaotic central pattern generators with learning for legged locomotion and malfunction compensation. Inf. Sci. **294**, 666–682 (2015)
38. Schulman, J., Wolski, F., Dhariwal, P., Radford, A., Klimov, O.: Proximal policy optimization algorithms. arXiv preprint arXiv:1707.06347 (2017)
39. Seoane, L.F.: Evolutionary aspects of reservoir computing. Phil. Trans. R. Soc. B **374**(1774), 20180377 (2019)
40. Silver, D., et al.: Mastering the game of go without human knowledge. Nature **550**(7676), 354–359 (2017)
41. Stanley, K.O., D'Ambrosio, D.B., Gauci, J.: A hypercube-based encoding for evolving large-scale neural networks. Artif. Life **15**(2), 185–212 (2009). https://doi.org/10.1162/artl.2009.15.2.15202
42. Stephens, D.W.: Change, regularity, and value in the evolution of animal learning. Behav. Ecol. **2**(1), 77–89 (1991). https://doi.org/10.1093/beheco/2.1.77
43. Stork: Is backpropagation biologically plausible? In: International 1989 Joint Conference on Neural Networks, pp. 241–246. IEEE (1989)
44. Sutton, R.S., Barto, A.G.: Reinforcement Learning: An Introduction. MIT press, Cambridge (2018)
45. Sutton, R.S., McAllester, D., Singh, S., Mansour, Y.: Policy gradient methods for reinforcement learning with function approximation. Adv. Neural Inf. Process. Syst. **12** (1999)
46. Tierney, A.: Evolutionary implications of neural circuit structure and function. Behav. Proc. **35**(1–3), 173–182 (1995)
47. Towers, M., et al.: Gymnasium (2023). https://doi.org/10.5281/zenodo.8127026. https://zenodo.org/record/8127025
48. Trouvain, N., Pedrelli, L., Dinh, T.T., Hinaut, X.: *ReservoirPy*: an efficient and user-friendly library to design echo state networks. In: Farkaš, I., Masulli, P., Wermter, S. (eds.) ICANN 2020. LNCS, vol. 12397, pp. 494–505. Springer, Cham (2020). https://doi.org/10.1007/978-3-030-61616-8_40

49. Watson, R.A., Szathmáry, E.: How can evolution learn? Trends Ecol. Evol. **31**(2), 147–157 (2016)
50. Wyffels, F., Schrauwen, B.: Design of a central pattern generator using reservoir computing for learning human motion. In: 2009 Advanced Technologies for Enhanced Quality of Life, pp. 118–122. IEEE (2009)
51. Yu, Y., Si, X., Hu, C., Zhang, J.: A review of recurrent neural networks: LSTM cells and network architectures. Neural Comput. **31**(7), 1235–1270 (2019)
52. Zador, A.M.: A critique of pure learning and what artificial neural networks can learn from animal brains. Nat. Commun. **10**(1), 3770 (2019)

Hybrid Surrogate Assisted Evolutionary Multiobjective Reinforcement Learning for Continuous Robot Control

Atanu Mazumdar[(✉)] and Ville Kyrki

Department of Electrical Engineering and Automation (EEA), Aalto University,
Espoo, Finland
{atanu.mazumdar,ville.kyrki}@aalto.fi

Abstract. Many real world reinforcement learning (RL) problems consist of multiple conflicting objective functions that need to be optimized simultaneously. Finding these optimal policies (known as Pareto optimal policies) for different preferences of objectives requires extensive state space exploration. Thus, obtaining a dense set of Pareto optimal policies is challenging and often reduces the sample efficiency. In this paper, we propose a hybrid multiobjective policy optimization approach for solving multiobjective reinforcement learning (MORL) problems with continuous actions. Our approach combines the faster convergence of multiobjective policy gradient (MOPG) and a surrogate assisted multiobjective evolutionary algorithm (MOEA) to produce a dense set of Pareto optimal policies. The solutions found by the MOPG algorithm are utilized to build computationally inexpensive surrogate models in the parameter space of the policies that approximate the return of policies. An MOEA is executed that utilizes the surrogates' mean prediction and uncertainty in the prediction to find approximate optimal policies. The final solution policies are later evaluated using the simulator and stored in an archive. Tests on multiobjective continuous action RL benchmarks show that a hybrid surrogate assisted multiobjective evolutionary optimizer with robust selection criterion produces a dense set of Pareto optimal policies without extensively exploring the state space. We also apply the proposed approach to train Pareto optimal agents for autonomous driving, where the hybrid approach produced superior results compared to a state-of-the-art MOPG algorithm.

Keywords: Multiobjective reinforcement learning · multiobjective evolutionary optimization · multiobjective policy gradient · Pareto front · surrogate assisted optimization

1 Introduction

Deep reinforcement learning (RL) algorithms have been applied to solve complex decision making problems in the field of robot control. However, many real world control problems involve multiple conflicting objectives that are to be optimized

S. Smith et al. (Eds.): EvoApplications 2024, LNCS 14635, pp. 61–75, 2024.
https://doi.org/10.1007/978-3-031-56855-8_4

simultaneously. For example, in the autonomous driving of a vehicle, we need to find optimal policies that minimize travel time and CO_2 emissions for the given preferences of objectives (generally in the form of weights). Prior works such as [11,21] optimize a single policy using gradient based methods for a given preference. However, in certain cases, we need to find the set of Pareto optimal policies that represent a diverse set of preferences between objectives (also known as Pareto optimal policies). Finding a dense set of Pareto optimal policies requires an extensive search of the state space, which reduces sample efficiency. Multi-policy approaches [12] such as [19,23] used population based policy optimization for different preferences to find a set of Pareto optimal policies. Most studies on multi-policy approaches have utilized a scalarized multiobjective policy gradient (MOPG) algorithm to find the Pareto optimal policies. These approaches converge quickly but are generally unable to achieve well distributed Pareto optimal policies. Furthermore, achieving a dense Pareto front requires executing the MOPG algorithm for various weight preferences.

Multiobjective evolutionary algorithms (MOEAs) can effectively generate a uniformly distributed Pareto front. However, optimizing deep neural networks for solving MORL problems with MOEAs is still challenging and sample inefficient. A two-stage policy optimization approach was proposed in [4] that used multiobjective soft actor-critic and a multiobjective covariance matrix adaptation evolution strategy (MO-CMA-ES) to harness the advantages of both algorithms. The evolutionary stage first vectorizes the policy-independent parameters as chromosomes of individuals. It utilizes CMA-ES to generate offspring individuals that are evaluated through simulations. Later, the offspring individuals go through a multiobjective selection process, and after multiple generations, we have a uniformly distributed Pareto front. The disadvantage of such an evolutionary stage is that it utilizes expensive simulation experiments in the evolutionary stage that reduce sample efficiency.

In this paper, we propose a hybrid approach that utilizes the faster convergence of MOPG and a surrogate (or meta-model) assisted MOEA to obtain a uniformly distributed set of Pareto optimal policies. Computationally inexpensive random forest surrogate models are trained in the policy parameter space using the optimal policies found by the MOPG algorithm. The MOEA uses a robust selection method that finds the optimal policies using the surrogates' prediction, reducing the requirement of expensive simulation runs. Tests on benchmark MORL problems show that the hybrid approach produces solutions with better hypervolume and sparsity. The proposed hybrid approach was later applied to solve an end-to-end multiobjective autonomous driving problem, maximizing travel speed and fuel efficiency. Overall, the results show that a hybrid MOEA and MOPG approach can reduce the extensive state space exploration and provide a dense set of Pareto optimal policies.

The rest of the paper is arranged as follows. Section 2 provides a brief background on multiobjective Markov decision processes and surrogate assisted MOEA. In Sect. 3, we demonstrate the working of our proposed hybrid approach. Experimental results on benchmark MORL problems and solutions to the autonomous driving problem are shown in Sect. 4. We conclude our paper and provide remarks on future research directions in Sect. 5.

2 Background

2.1 Multiobjective Markov Decision Process

A multiobjective Markov decision process (MOMDP) for continuous control reinforcement learning problem is defined by the tuple $\langle \mathcal{S}, \mathcal{A}, \mathcal{P}, \mathbf{R}, \gamma, \mathcal{D} \rangle$, where $\mathcal{S}$ and $\mathcal{A}$ are the state and action spaces, respectively. The state transition probability, $\mathcal{P}(s'|s, a)$ and the reward vector $\mathbf{R} = [\mathbf{r}_1, \ldots, \mathbf{r}_m]^\top$ where m is the number of objectives. The initial state distribution is $\mathcal{D}$, and the vector discount factor is $\gamma = [\gamma_1, \ldots, \gamma_m] \in [0, 1]^m$. The expected vector return is $\mathbf{J}^\pi = [J_1{}^\pi, \ldots, J_m{}^\pi]$ and the expected return of the ith objective for policy $\pi_\theta : \mathcal{S} \to \mathcal{A}$ is:

$$J_i^\pi = \mathbb{E}\left[\sum_{t=0}^{T} \gamma_i^t \mathbf{r}_i\left(s_t, a_t\right) \mid s_0 \sim \mathcal{D}, a_t \sim \pi_\theta\left(s_t\right)\right], \tag{1}$$

where θ represents the policy parameters. A state transition from s_t to s_{t+1} occurs by action a_t with T being the horizon.

The overall multiobjective optimization problem can be formulated as $\max_\pi \mathbf{J}^\pi$ that simultaneously maximizes the expected return for all the objectives. A solution $\mathbf{J}^{\pi'}$ is dominated by policy $\mathbf{J}^\pi$ if $J_i^\pi \geq J_i^{\pi'}$ for all $i = 1, \ldots, m$ and $J_j^\pi > J_j^{\pi'}$ for at least one index j. If any other policies do not dominate a policy, it is called non-dominated. The set of not-dominated policies in the objective and policy parameter space is referred to as the Pareto front and Pareto set, respectively. However, obtaining the true Pareto front of complex MORL problems is generally not possible using computational methods and the Pareto optimal policies are an approximation.

2.2 Prediction Guided MORL

The prediction guided MORL (PGMORL) algorithm proposed in [23] is a state-of-the-art population based MORL algorithm. The algorithm optimizes a set of policies with various weight vectors at each generation using multiobjective policy gradient (MOPG) workers. For a given policy π_θ and a given weight vector (or preference) ω where $\sum_i \omega_i = 1$ the goal of an MOPG is to train a multiobjective policy that maximizes the expected weighted sum return $\mathcal{J}(\theta, \omega) = \omega^\top \mathbf{J}^\pi$. PGMORL is an evolutionary learning algorithm that selects policies and their respective weights in each generation of the learning process. It uses a vectorized value function $\mathbf{V}_\pi(s)$ to estimate the expected return vector of a policy π for state s. The parameters of the value function are updated by the squared loss between the estimated and target value vectors. Therefore, the same policy can be used with different weights during the training process without retraining the value function from scratch. The policy gradient used to update the policy is:

$$\nabla_\theta \mathcal{J}(\theta, \omega) = \sum_{i=1}^{m} \omega_i \nabla_\theta J_i(\theta)$$

$$= \mathbb{E} \left[\sum_{t=0}^{T} \omega^\top \mathbf{A}_\omega^\pi(s_t, a_t) \log \pi_\theta(a_t, s_t) \right]$$

$$= \mathbb{E} \left[\sum_{t=0}^{T} A_\omega^\pi(s_t, a_t) \log \pi_\theta(a_t, s_t) \right],$$

where $A_\omega^\pi(s_t, a_t) = \omega^\top \mathbf{A}_\omega^\pi(s_t, a_t)$ is the weighted sum scalarized advantage. PGMORL uses weighted sum Proximal Policy Optimization as the policy gradient algorithm that uses clipped surrogate objectives.

PGMORL initiates with a warmup phase with n uniformly distributed weights and randomly assigned policies that are trained in parallel for a predefined number of timesteps. Each parallel training (or task) is done by the previously defined MOPG algorithm here referred to as *MOPG workers*. The improvement in the objectives for respective weights is stored, and a monotonic hyperbolic model is fitted to estimate the improvement in the objective values. Next, a mixed integer optimization (that is discretized to reduce computation cost) is performed to find the best weight and policy combinations that improve the objectives compared to the current Pareto archive EP. Finally, the Pareto optimal policies are clustered by k-means clustering and a continuous Pareto front is achieved by linear interpolation of the policy parameters in each cluster.

2.3 Surrogate Assisted Multiobjective Evolutionary Optimization

Many real-world multiobjective optimization problems consist of computationally expensive objective functions that require extensive computational resources or time. Solving these problems using MOEAs is usually impractical as MOEAs require a high number of function evaluations. An alternative way to use MOEAs is to use surrogate models that approximate (or emulate) the expensive objectives. The proposed hybrid approach starts with a few samples of data (consisting of parameters and their respective objective values) acquired through expensive evaluations, and surrogate models are trained using the data. Choosing a suitable model that utilizes the most out of the provided data and with superior predicting accuracy is often challenging. The MOEA later utilizes the prediction of the surrogate models as objective functions to find the optimal solutions. Certain surrogate models, such as Gaussian process regression, have been widely used in surrogate assisted optimization due to their ability to provide uncertainty in the prediction. The uncertainty information has been widely used in the field of Bayesian optimization [10] aims to maximize an acquisition function that determines the next expensive function evaluation. Generally, these acquisition functions (such as expected improvement, probability of improvement, expected hypervolume improvement) [6,24] are designed to find the location that maximizes the objective values. The newly evaluated values are used to update the surrogates, which iteratively improves the approximation accuracy of the surrogates and converges the solutions towards the true Pareto front.

Decomposition-Based MOEAs. Decomposition-based MOEAs were designed to solve multiobjective optimization problems (MOPs) with a high number of objectives (generally $m > 3$). In general, these algorithms decompose the given MOP into subproblems by scalarizing the objectives or multiple MOPs. In this paper, we use RVEA [5], an MOEA that has shown superior performance in solving problems with many objectives. New offspring individuals are generated by simulated binary crossover and polynomial mutation [9] and added to the population. The individuals in the population are divided into sub-populations and assigned to uniformly distributed reference/weight vectors. Next, one individual from each subpopulation is selected using angle penalized distance (APD) as the selection criterion to improve both convergence and diversity of the solutions in the objective space. The selected individuals are used as parents for the next generation and the steps are repeated until the specified stopping criteria are met.

2.4 Probabilistic Selection in Decomposition-Based Multiobjective Evolutionary Algorithms

The probabilistic selection approaches proposed in [18] were originally designed to solve offline data-driven multiobjective optimization problems where further expensive function evaluations are not possible. The probabilistic selection approach optimizes the mean predicted objective values by the surrogate and simultaneously minimizes the uncertainty in the predicted objective values. The solutions produced by the probabilistic selection have a higher accuracy compared to the generic selection operator of an MOEA that uses the mean prediction from surrogates.

The algorithm initializes with the parent population that was used to build the surrogates. For probabilistic RVEA, one individual from each sub-population is selected using a probabilistic ranking of their APD. For each individual, Monte Carlo samples are drawn using the predicted distribution of objective values of the surrogates. The APD values of all the samples for each individual are computed. The ith individual in the jth subpopulation $\bar{P}_j$ is assigned a rank $R'_{i,j}$ using the distribution of APD, $d_{i,j}$:

$$R'_{i,j} = \sum_{n=1}^{|\bar{P}_j|} Pr_{\text{wrong}}(d_{n,j} > d_{i,j}) - 0.5. \tag{2}$$

The probability of wrongly selecting an individual with a higher APD value $Pr_{\text{wrong}}(d_{n,j} > d_{i,j})$ is calculated by utilizing the Monte Carlo samples. The zth individual I_z is selected from subpopulation $\bar{P}_j$, where $j = 1, \ldots, N$, for population of the next generation $P_{\text{next gen}}$ by:

$$P_{\text{next gen}} = \left\{ I_z | z = \underset{i \in \left\{ 1,...,|\bar{P}_j| \right\}}{\text{argmin}} \ R'_{i,j} \right\}. \tag{3}$$

We used probabilistic RVEA in this paper and lower confidence bound (LCB) [15] selection to find Pareto optimal policy parameters with high confidence.

3   Hybrid Surrogate Assisted Evolutionary Multiobjective Policy Gradient

Obtaining a dense set of optimal policies in a multiobjective reinforcement learning problem requires extensive exploration of the state space that reduces the sample efficiency. MOPG algorithms converge to the Pareto front quickly but generally cannot provide a dense set of solutions. Furthermore, there is no mechanism to exchange the learned parameters between policies that lead us to intermediate Pareto optimal policies. Operators in MOEAs such as crossover and mutation are a way to generate intermediate policies as they directly alter the parameter space. However, MOEAs generally require numerous function evaluations and thus are sample inefficient. A possible alternative to utilize evolutionary methods is to train a surrogate model in the policy parameter space predicting the returns. In what follows, we demonstrate the proposed hybrid surrogate assisted MOEA and MOPG approach for solving MORL problems.

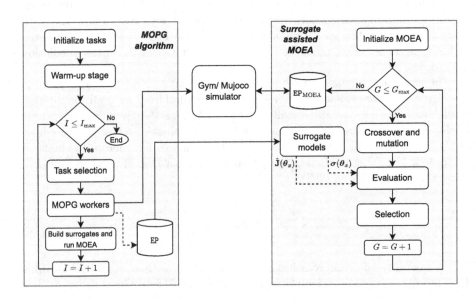

Fig. 1. Overview of the hybrid surrogate assisted MOEA and MOPG approach.

The overview of the proposed approach is shown in Fig. 1. Overall, there are two algorithm blocks: an MOPG algorithm and a surrogate assisted MOEA. The MOPG algorithm starts by initializing n random worker policies for the MOPG task and n uniformly distributed weight vectors. In the warm-up stage, each of the initialized policies is assigned to a weight vector. The policies are optimized by the MOPG algorithm for a certain number of iterations to take the policies out of low performance regions. Later, a task selection method decides the policies and their respective weights for the MOPG workers and optimizes the policies for I_{max} iterations. The policies and their episodic returns are stored in an external Pareto archive EP. At the end of each iteration in the MOPG loop, the data in EP is transferred to the surrogate assisted MOEA block. Surrogate models are trained to approximate the returns for each objective on the parameters of the stored policies. The predicted vector objective values by the surrogates is $\hat{\mathbf{J}}(\boldsymbol{\theta}_x)$ with uncertainty in the prediction, $\boldsymbol{\sigma}(\boldsymbol{\theta}_x)$ for parameter $\boldsymbol{\theta}_x$. The MOEA is initialized with a population of policy parameters and weight vectors (for decomposition based MOEAs) [5,8,26]. The offspring individuals are created by random crossover and mutation and added to the parent population. The population of parameters is evaluated using the surrogates. A selection criterion is used to select certain individuals from the population and these individuals become the new parent population. The evolution process is executed for G_{max} generations, and the final solutions are evaluated with the expensive simulator. An external Pareto archive EP_{MOEA} stored the non-dominated policy parameters and their respective evaluated objective values. The control is moved back to the MOPG block and the final Pareto optimal policies are combined non-dominated policies of EP and EP_{MOEA}.

It should be noted that the MOPG algorithm used in this paper is PGMORL [23], which is population based and referred to as evolutionary in the literature. However, since our approach already has an MOEA, we use the term iterations for the MOPG algorithm instead of generations as used in the [23].

3.1   Dataset for Surrogates

The quality of the solutions produced by surrogate assisted MOEA depends on the approximation accuracy of the surrogates and the dataset used. The first step in the proposed approach is to find a set of policies with good objective values and low noise. Therefore, we utilize the state-of-art prediction guided multiobjective policy gradient reinforcement learning (PGMORL) [23] to find a set of optimal policies. At each iteration, PGMORL updates an external Pareto archive (EP) that stores the non-dominated intermediate policies. Each element of EP consists of the policy independent parameters of the deep network and the corresponding objective vector as a tuple of two matrices: $(\boldsymbol{\Theta}, \mathcal{J})$, where $\boldsymbol{\Theta} = [\boldsymbol{\theta}^1, \ldots, \boldsymbol{\theta}^N]$ and $\mathcal{J} = [\mathbf{J}^1, \ldots, \mathbf{J}^N]$, and N is the number of policies in EP. The data in EP is passed to the surrogate assisted MOEA block to train the surrogates at each iteration's end. For our approach, we do not use all the intermediate policies for training the surrogate models, which will become expensive, thereby slowing down the optimization process. In addition, we are interested in finding the Pareto optimal policies and not the sub-optimal coverage set.

3.2 Surrogate Models

The data consists of the policy parameters and their respective objective vector. The number of parameters in deep network policies is generally in the order of 10^3, which is quite high. The hypervolume of the solutions found by a surrogate assisted MOEA depends on the approximation accuracy of the surrogates [6, 18]. The challenges in approximating the objective vector of the policies are as follows:

- The parameter size (or dimension) is huge. Thus surrogates such as Gaussian process regression as surrogates will become impractical. [3]
- The Pareto optimal policies are generally discontinuous in the policy space. [23]
- The surrogates should be able to predict the uncertainty in the prediction in addition to the mean prediction that the MOEA can utilize in the later stage. [13, 14, 25]

In this paper, we used random forest regression as surrogates due to their ability to handle high dimensional datasets [2, 20]. They can also handle discontinuities and are computationally inexpensive to train. Random forest surrogates have also proven to have better accuracy than state-of-art deep neural networks [1] and can be trained with less amount data. We train a random forest model consisting of B bootstrapped regression tree for each objective. The prediction of the ith objective's random forest is $\hat{J}_i(\boldsymbol{\theta}_x) = 1/B \sum_{b=1}^{B} f_b(\boldsymbol{\theta}_x)$ and the estimated uncertainty (standard deviation) in the prediction is $\sigma(\boldsymbol{\theta}_x) = (\sum_{b=1}^{B} (f_b(\boldsymbol{\theta}_x) - \hat{J}_i(\boldsymbol{\theta}_x))/(B-1))^{1/2}$. The unseen policy parameters is $\boldsymbol{\theta}_x$ and $f_b(\boldsymbol{\theta}_x)$ is the prediction of the bth regression tree. The MOEA utilizes the predictions of the random forest surrogates that maximize the approximated objective values.

3.3 Optimization Method

Using Bayesian optimization techniques that aim to improve the optimal solutions can be applied to improve the optimal solutions. However, a high number of surrogate updates are required to sufficiently increase the approximation accuracy since the dimension of the policy space is large (with $> 10^3$ parameters) [13, 22]. In addition, we are interested in finding uniformly distributed policies closer to the policies in EP. The MOEA should therefore optimize the objectives and simultaneously try to minimize the uncertainty in the solutions. One such approach is to modify the objective function and maximize the lower confidence bound (LCB) [15] of the surrogates' prediction. The modified objective function for LCB is $\hat{J}_i^{\text{LCB}}(\boldsymbol{\theta}_x) = \hat{J}_i(\boldsymbol{\theta}_x) - \sigma(\boldsymbol{\theta}_x)$. Another approach that can be used is probabilistic selection proposed for MOEA (PMOEA) [18] that modifies the selection criterion to select individuals with a higher probability of being close to the true objective function value. In this paper, both LCB and PMOEA were integrated with RVEA, and the predictions from random forest surrogates were used for optimization.

The random forest surrogates are retrained with the updated EP, and the MOEA is executed at the end of each iteration of PGMORL. The solutions found by the MOEA are evaluated by running the simulation. The normalizing parameters (of the observation) used for running the simulation were chosen from EP that was closest to the solution in the objective space found by the MOEA. An external Pareto archive (EP_{MOEA}) consisting of the non-dominated policies found by the MOEA stores the evaluated objective values of the solutions and their respective policy parameters.

4   Results

4.1   Benchmark Experiments

An ablation study was conducted on four bi-objective continuous control Mujoco environments previously proposed by [23] to test the performance of the proposed hybrid MORL approach. The environments chosen were MO-Ant, MO-HalfCheetah, MO-Swimmer and MO-Walker2d. The MOPG algorithm used was PGMORL and we used the same experiment settings as provided in [23] for the respective environment. One random forest surrogate model with 500 trees was trained for each objective using the dataset. We ran RVEA for 1000 generations after every iteration of PGMORL. Other hyper-parameters for RVEA were kept the same as provided in [5]. All codes were implemented in Python and are available for reproducibility.[1]

Table 1. Mean and standard deviation of hypervolume (higher is better) and sparsity (lower is better) for three different approaches for various benchmark environments. The best performing approach has been shown in bold.

Environments		Mean		LCB		PMOEA	
		Hypervolume	Sparsity	Hypervolume	Sparsity	Hypervolume	Sparsity
MO-HalfCheetah-v2	Mean	5.44E+06	6.08E+04	**5.76E+06**	**1.27E+03**	5.73E+06	2.76E+03
	Std	3.81E+04	2.54E+04	9.73E+03	7.56E+02	1.77E+04	2.06E+03
MO-Swimmer-v2	Mean	2.06E+04	2.94E+01	2.07E+04	**9.74E+00**	**2.11E+04**	2.25E+01
	Std	6.81E+03	1.60E+01	6.82E+03	6.77E+00	6.59E+03	1.68E+01
MO-Walker2d-v2	Mean	4.25E+06	2.25E+04	**4.63E+06**	**7.19E+02**	4.45E+06	5.68E+03
	Std	1.33E+05	3.71E+04	3.27E+05	4.53E+02	2.96E+05	5.04E+03
MO-Ant-v2	Mean	5.53E+06	4.47E+04	**6.01E+06**	**8.19E+03**	5.76E+06	1.20E+04
	Std	4.91E+05	6.21E+04	3.27E+04	4.18E+03	9.17E+04	7.39E+03

---

[1] Codes can be found at https://github.com/amrzr/SA-MOEAMOPG.

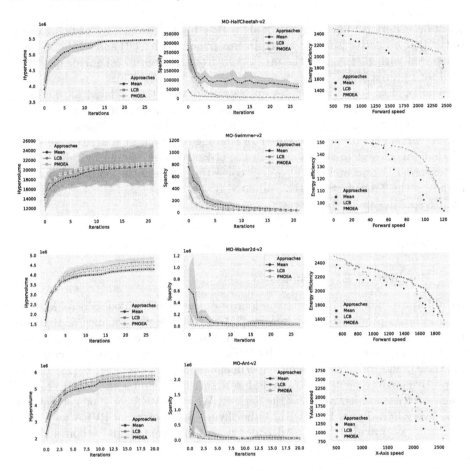

Fig. 2. Hypervolume (higher is better); on left, sparsity (lower is better); on middle, showing the variation over iterations and three different approaches for four environments. The final non-dominated solutions (all objectives maximized); on right, for the run with median hypervolume.

The proposed hybrid MOEA and MOPG with LCB and PMOEA was compared with a baseline hybrid approach that used the mean prediction (Mean) from the random forest surrogates as objectives that are optimized by the MOEA. We used hypervolume and sparsity indicators [17,27] to measure the performance of the approaches. Hypervolume measures both convergence and diversity of the solutions and a higher hypervolume value is considered better. The reference point chosen for hypervolume indicator was zero for all the objectives. The sparsity indicator represents the density of the solutions, and a lower value is considered better. Each approach was executed eleven times with random seed, and statistical tests were performed to determine the best performing approach. It should be noted that our tests did not consider the running times of the various approaches for comparison and only the final hypervolume and sparsity of the solutions.

The overall mean and standard deviation of hypervolume and sparsity of the solutions in EP_{MOEA} (after evaluating with the simulator) for three approaches across the tested benchmarks are shown in Table 1. The best performing approach for each indicator is shown in bold. It can be observed that LCB performs the best for both hypervolume and sparsity for almost all the benchmarks, with PMOEA coming second. The approach using mean prediction from surrogates was the worst for all benchmarks as the MOEA did not consider the uncertainty in the surrogates' prediction, which resulted in solutions with high approximation error. The progress of hypervolume and sparsity of EP_{MOEA} with iterations of PGMORL and the final nondominated solutions for the benchmark problems are in Fig. 2. It can be observed that both LCB and PMOEA produce a dense set of solutions with better objective values than the Mean approach. It can also be noticed that both hypervolume and sparsity improve much faster for LCB and PMOEA approaches compared to the Mean approach. Considering the MO-HalfCheetah environment, we can observe that both hypervolume and sparsity converge at around 10–15 iterations. Thus, the MOPG algorithm can be stopped when solutions in EP have converged with sufficient sparsity and overall sample efficiency can be further improved.

4.2   MORL for Autonomous Driving

Generally, an end-to-end autonomous driving reinforcement learning problem considers a single objective in the form of average speed or time consumed for travel to be optimized. Autonomous driving can be regarded as a MORL problem if fuel efficiency or CO_2 emissions is considered the second objective to be optimized in addition to the average speed. We used Gym HighwayEnv [16] and designed a two-lane one-way highway environment as shown in Fig. 3. The observations were the Cartesian coordinates and velocities of the four closest vehicles relative to the controlled vehicle, the presence that distinguished if less than four vehicles were observed and the vehicle's heading ($S \subseteq \mathbb{R}^{24}$). The actions were the steering angle and throttle ($A \subseteq \mathbb{R}^2$). The first formulated reward R_1 represents the average speed taken by the controlled vehicle:

$$R_1 = 5 \times speed \times \cos{(heading)} - 10 \times (off + col).$$

The second reward R_2 representing CO_2 emissions is:

$$R_2 = 5 \times (1 - CO_2) - 10 \times (off + col),$$

where *speed* is the scalarized speed of the vehicle, *heading* is the angle of the vehicle in radians (scaled from zero to one), *off* is the indicator for the vehicle going outside the road and is set to one is the vehicle goes offroad; otherwise it is set to zero. The variable *col* is the vehicle collision indicator and is set to one if the vehicle collides; otherwise, it is set to zero. The CO_2 emission is computed

Fig. 3. Two lane highway environment. MORL agent controls the yellow vehicle. (Color figure online)

using the NGM emissions model [7] and scaled from zero to one. The latter terms in both R_1 and R_2 are penalties when the vehicle collides or goes offroad.

We used the hybrid approaches demonstrated in the benchmark tests to solve the autonomous driving problem. For the PGMORL algorithm, the total number of environment steps was set to 5×10^6, with the number of warmup iterations set to 80 and evolutionary stage iterations set to 20. The settings for the MOEA were kept the same as in the benchmark tests. The hypervolume and sparsity of solutions in EP_{MOEA} and EP after the final iteration of all the hybrid approaches and PGMROL, respectively, are shown in Table 2. It can be observed that PMOEA produces solutions with the best hypervolume and sparsity. The final nondominated solutions are also shown in Fig. 4. The Pareto front is disconnected, and PGMORL does not produce a dense set of solutions like LCB and PMOEA. The controlled vehicle needs minimum threshold energy to achieve the running speed for overtaking other vehicles. The disconnect in the Pareto front is the region where the controlled vehicle tries overtaking other vehicles.

Testing MORL algorithms is challenging since there are a handful of simulation environments available. Most of these simulations are multiobjective variants of environments originally designed to be solved as single objective MORL problems. Benchmark problems with a scaleable and wide variety of Pareto front features [12] are required to test the performance of MORL algorithms thoroughly.

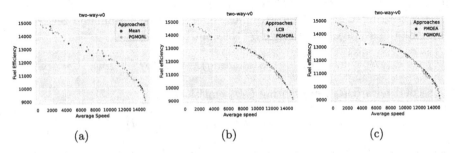

(a)   (b)   (c)

Fig. 4. The non-dominated policies obtained by the three different hybrid approaches compared with the non-dominated policies found by PGMORL.

Table 2. Comparison of hypervolume (higher is better) and sparsity (lower is better) for the three hybrid approaches and PGMORL while solving the two-way highway problem. Best performing approach has been shown in bold.

Indicators	PGMORL	Mean	LCB	PMOEA
Hypervolume	1.88E+08	1.86E+08	1.89E+08	**1.90E+08**
Sparsity	1.76E+05	4.85E+05	3.43E+04	**3.26E+04**

5   Conclusions

In this paper, we propose a hybrid approach that harnesses the capabilities of MOPG algorithms and surrogate assisted MOEAs to produce a dense set of Pareto optimal policies. We tested the proposed approach on MORL benchmark problems. The hybrid approach produced Pareto optimal policies with superior hypervolume and sparsity without reducing the sample efficiency. Finally, we applied our hybrid approach to solve a multiobjective autonomous driving problem where our approach performed better than the state-of-the-art PGMORL algorithm. Therefore, we conclude that using a hybrid surrogate assisted MOEA and MOPG approach with a robust selection mechanism such as LCB or PMOEA reduces the requirement of extensive search in the state space to find Pareto optimal policies for a diverse set of weight vectors.

Choosing an appropriate surrogate model to accurately approximate the returns of policies with a high dimensional parameter space is challenging. In addition, the model should be able to handle the discontinuities that exist in the Pareto optimal policies. A model management method can be integrated to select the best performing surrogates automatically. Tests should also be conducted to compare the proposed hybrid approaches with other state-of-the-art MORL algorithms and with a higher number of objectives. We used PGMORL, an MOPG algorithm that focuses on maximizing both the hypervolume and sparsity of the Pareto optimal policies. The MOPG can be tailored to focus on improving only the hypervolume since the surrogate assisted MOEA performs excellently in reducing the sparsity of the solutions. The proposed approach currently does not have a mechanism to transfer the optimal policies found by the MOEA to the MOPG. In the future, we plan to design a framework for seamlessly exchanging optimal policies found by MOPG and surrogate assisted MOEA with each other. Testing the proposed approach on real-world robotics problems will also be one of our future works.

Acknowledgements. The work is supported by Artificial Intelligence for Urban Low-Emission Autonomous Traffic (AIforLessAuto) funded under the Green and Digital transition call from the Academy of Finland. The research project has been granted funding from the European Union (NextGenerationEU) through the Academy of Finland under project number 347199.

References

1. Ao, Y., Li, H., Zhu, L., Ali, S., Yang, Z.: The linear random forest algorithm and its advantages in machine learning assisted logging regression modeling. J. Petrol. Sci. Eng. **174**, 776–789 (2019)

2. Arashi, M., Lukman, A.F., Algamal, Z.Y.: Liu regression after random forest for prediction and modeling in high dimension. J. Chemometr. **36**(4), e3393 (2022)

3. Bouhlel, M.A., Martins, J.R.R.A.: Gradient-enhanced kriging for high-dimensional problems. Eng. Comput. **35**(1), 157–173 (2018)

4. Chen, D., Wang, Y., Gao, W.: Combining a gradient-based method and an evolution strategy for multi-objective reinforcement learning. Appl. Intell. **50**(10), 3301–3317 (2020)

5. Cheng, R., Jin, Y., Olhofer, M., Sendhoff, B.: A reference vector guided evolutionary algorithm for many-objective optimization. IEEE Trans. Evol. Comput. **20**, 773–791 (2016)

6. Chugh, T., Sindhya, K., Hakanen, J., Miettinen, K.: A survey on handling computationally expensive multiobjective optimization problems with evolutionary algorithms. Soft. Comput. **23**, 3137–3166 (2019)

7. Conlon, J., Lin, J.: Greenhouse gas emission impact of autonomous vehicle introduction in an urban network. Transp. Res. Rec. **2673**(5), 142–152 (2019)

8. Deb, K., Jain, H.: An evolutionary many-objective optimization algorithm using reference-point-based nondominated sorting approach, part I: Solving problems with box constraints. IEEE Trans. Evol. Comput. **18**, 577–601 (2014)

9. Deb, K., Pratap, A., Agarwal, S., Meyarivan, T.: A fast and elitist multiobjective genetic algorithm: NSGA-II. IEEE Trans. Evol. Comput. **6**(2), 182–197 (2002)

10. Forrester, A., Sobester, A., Keane, A.: Engineering Design via Surrogate Modelling. John Wiley & Sons, Hoboken (2008)

11. Hayes, C.F., Reymond, M., Roijers, D.M., Howley, E., Mannion, P.: Risk aware and multi-objective decision making with distributional monte carlo tree search (2021). arXiv:2102.00966

12. Hayes, C.F., et al.: A practical guide to multi-objective reinforcement learning and planning. Auton. Agents Multi-Agent Syst. **36**(1), 26 (2022)

13. Jin, Y.: Surrogate-assisted evolutionary computation: recent advances and future challenges. Swarm Evol. Comput. **1**, 61–70 (2011)

14. Jin, Y., Wang, H., Chugh, T., Guo, D., Miettinen, K.: Data-driven evolutionary optimization: an overview and case studies. IEEE Trans. Evol. Comput. **23**, 442–458 (2019)

15. Knowles, J.D., Thiele, L., Zitzler, E.: A tutorial on the performance assessment of stochastic multiobjective optimizers (2006)

16. Leurent, E.: An environment for autonomous driving decision-making (2018). https://github.com/eleurent/highway-env

17. Li, M., Yao, X.: Quality evaluation of solution sets in multiobjective optimisation. ACM Comput. Surv. **52**(2), 1–38 (2019)

18. Mazumdar, A., Chugh, T., Hakanen, J., Miettinen, K.: Probabilistic selection approaches in decomposition-based evolutionary algorithms for offline data-driven multiobjective optimization. IEEE Trans. Evol. Comput. **26**, 1182–1191 (2022)

19. Parisi, S., Pirotta, M., Smacchia, N., Bascetta, L., Restelli, M.: Policy gradient approaches for multi-objective sequential decision making. In: 2014 International Joint Conference on Neural Networks (IJCNN), pp. 2323–2330 (2014)

20. Rodriguez-Galiano, V., Sanchez-Castillo, M., Chica-Olmo, M., Chica-Rivas, M.: Machine learning predictive models for mineral prospectivity: an evaluation of neural networks, random forest, regression trees and support vector machines. Ore Geol. Rev. **71**, 804–818 (2015)
21. Siddique, U., Weng, P., Zimmer, M.: Learning fair policies in multiobjective (deep) reinforcement learning with average and discounted rewards. In: Proceedings of the 37th International Conference on Machine Learning (2020)
22. Stork, J., et al.: Open issues in surrogate-assisted optimization. High-Performance Simulation-Based Optimization p. 225–244 (2019)
23. Xu, J., Tian, Y., Ma, P., Rus, D., Sueda, S., Matusik, W.: Prediction-guided multiobjective reinforcement learning for continuous robot control. In: Proceedings of the 37th International Conference on Machine Learning, pp. 10607–10616. PMLR (2020)
24. Yang, K., Emmerich, M., Deutz, A., Bäck, T.: Efficient computation of expected hypervolume improvement using box decomposition algorithms. J. Global Optim. **75**(1), 3–34 (2019)
25. Zapotecas Martínez, S., Coello Coello, C.A.: Moea/d assisted by RBF networks for expensive multi-objective optimization problems. In: Proceedings of the 15th Annual Conference on Genetic and Evolutionary Computation, pp. 1405–1412. Association for Computing Machinery (2013)
26. Zhang, Q., Li, H.: MOEA/D: a multiobjective evolutionary algorithm based on decomposition. IEEE Trans. Evol. Comput. **11**, 712–731 (2007)
27. Zitzler, E., Deb, K., Thiele, L.: Comparison of multiobjective evolutionary algorithms: empirical results. Evol. Comput. **8**, 173–195 (2000)

Towards Physical Plausibility
in Neuroevolution Systems

Gabriel Cortês$^{(\boxtimes)}$ ⓘ, Nuno Lourenço ⓘ, and Penousal Machado ⓘ

Department of Informatics Engineering, CISUC/LASI – Centre for Informatics
and Systems of the University of Coimbra, University of Coimbra, Coimbra, Portugal
{cortes,naml,machado}@dei.uc.pt

Abstract. The increasing usage of Artificial Intelligence (AI) models, especially Deep Neural Networks (DNNs), is increasing the power consumption during training and inference, posing environmental concerns and driving the need for more energy-efficient algorithms and hardware solutions. This work addresses the growing energy consumption problem in Machine Learning (ML), particularly during the inference phase. Even a slight reduction in power usage can lead to significant energy savings, benefiting users, companies, and the environment. Our approach focuses on maximizing the accuracy of Artificial Neural Network (ANN) models using a neuroevolutionary framework whilst minimizing their power consumption. To do so, power consumption is considered in the fitness function. We introduce a new mutation strategy that stochastically reintroduces modules of layers, with power-efficient modules having a higher chance of being chosen. We introduce a novel technique that allows training two separate models in a single training step whilst promoting one of them to be more power efficient than the other while maintaining similar accuracy. The results demonstrate a reduction in power consumption of ANN models by up to 29.2% without a significant decrease in predictive performance.

Keywords: Evolutionary Computation · Neuroevolution · Energy Efficiency

1 Introduction

As the demand for Machine Learning (ML) continues to grow, so does the electrical power required for training and assessment. According to Patterson et al., GPT-3, the model behind ChatGPT, consumes 1287 MWh, corresponding to approximately 552 tons of CO_2 equivalent emissions just for training during 15 days [16]. In addition to the environmental impacts of this power usage, it can also burden individual users and organizations, who may face high energy costs. Therefore, finding ways to reduce the power consumption of ML processes is becoming increasingly important.

Artificial Neural Networks (ANNs) are a type of ML model inspired by biological neural networks [19]. They consist of multiple layers of artificial neurons,

S. Smith et al. (Eds.): EvoApplications 2024, LNCS 14635, pp. 76–90, 2024.
https://doi.org/10.1007/978-3-031-56855-8_5

which are functions that take input data and produce an output based on it. The connections between neurons have an associated weight value modified in the training process to allow the network to "learn" how to solve a specific task. Deep Neural Networks (DNNs) are ANNs with a considerable number of hidden layers [9,10]. This allows them to avoid the feature engineering step, thus automatically discovering the representations needed for classification and achieving higher accuracy values. Training and executing ANNs is power-intensive due to the required computational resources.

Evolutionary Algorithms (EAs) are algorithms inspired by natural selection [6,17]. To evolve solutions over multiple generations, they utilize mechanisms, such as selection, crossover, and mutation. The process begins with a randomly initialized population whose evolution is steered by a fitness function that measures the quality of an individual. In conjunction with the mentioned evolutionary mechanisms, the process is predicted to culminate in near-optimal individuals.

Neuroevolution (NE) uses EAs to generate and optimize ANNs for a given task [7]. It can optimize the ANN's architecture and hyperparameters.

We hypothesise that we can address the energy inefficiency issue by using NE to search for well-suited models for a particular problem while being power-efficient. Fast Deep Evolutionary Network Structured Representation (Fast-DENSER) is a method that utilizes an Evolution Strategy (ES) to find optimal ANN models by using their accuracy as the fitness function, thus guiding the search towards accurate models [2].

In this work, we propose novel approaches integrated into Fast-DENSER to find power-efficient models. We have incorporated a new approach to measure the power consumption of a DNN model during the inference phase. This metric has been embedded into multi-objective fitness functions to steer the evolution towards more power-efficient DNN models. We also introduce a new mutation strategy that allows the reutilization of modules of layers with inverse probability to the power usage of a module, thus (re)introducing efficient sets of layers in a model. We propose the introduction of an additional output layer connected to an intermediate layer of a DNN model and posterior partitioning into two separate models to obtain smaller but similarly accurate models that utilize less power. To the best of our knowledge, no prior works employ a similar approach.

The experiments are analyzed through two metrics: accuracy and mean power usage during the validation step. The motive for using the power usage of the validation step instead of the training step is that the training is usually performed only once. Contrarily, the inference is executed multiple times. Moreover, inference does not necessarily occur on the machine where the training was conducted, which is vital since many devices are not optimized for these tasks.

The results of this work show that it is possible to have DNN models with substantially inferior power usage. The best model found regarding power consumes 29.18 W (29.2%) less whilst having a tiny decrease in performance (less than 1%).

This work is structured as follows: Sect. 2 provides background information on ANNs, and NE. Section 3 introduces our methodologies to enhance the power efficiency of ANN models. Section 4 outlines the experimental setup. Section 5 presents the experimental results. Finally, in Sect. 6, we provide our conclusions and prospects for future research.

2   Background

2.1   Artificial Neural Networks

Artificial Neural Networks are a type of supervised ML inspired by biologic neural networks [19]. An ANN consists of connected processing units known as neurons. The connections follow a specific topology to achieve the desired application. A neuron's input may be the output of other neurons, external sources, or itself. Every connection has an associated weight, allowing the system to simulate biological synapses. A weighted sum of the inputs is computed at a given instant, considering the connection weights. It is also possible to sum a bias value to this. An activation function is applied, and thus, the neuron's output is obtained.

DNNs are ANNs composed of many hidden layers. Due to this, DNNs can avoid the feature engineering step – which usually requires human expertise – by automatically discovering the representations needed for classification [9,10]. Thus, they can model more complex relationships and achieve higher accuracy on tasks requiring pattern recognition. The development and usage of DNNs have substantially increased due to the widespread deployment of more capable hardware, such as Graphics Processing Units (GPUs) [3].

2.2   Neuroevolution

NE is the application of evolutionary techniques to search for DNN models. It is used to optimize the structure and weights of DNNs to improve their performance on specific tasks, such as image classification and natural language processing. NE is a gradient-free method based on the concept of population [7]. It allows for the simultaneous exploration of multiple zones of the search space through parallelization techniques at the cost of taking a usually long time to execute since each individual of the population is a DNN that requires training and testing.

Deep Evolutionary Network Structured Evolution (DENSER) is a neuroevolutionary framework that allows the search of DNNs through a grammar-based neuroevolutionary approach that searches both network topology and hyperparameters [1].

The developed DNNs are structured according to a provided context-free grammar. DENSER uses Dynamic Structured Grammatical Evolution (DSGE) as the strategy that allows the modification of the network topology. DSGE is built upon Structured Grammatical Evolution (SGE), with the main differences of allowing the growth of the genotype and only storing encoded genes [11]. Allied with dynamic production rules, DSGE allows the creation of multiple-layer DNNs. SGE proves to perform better than Grammatical Evolution (GE), and DSGE proves to be superior to SGE [12]. The individuals of the evolutionary process are represented in two levels: the outer level encodes the topology of the ANN, and the inner one encodes its hyperparameters.

Fast-DENSER was developed to overcome some limitations verified on DENSER: evaluating the population consumes a considerable amount of time, and the developed DNNs are not fully trained [2]. Fast-DENSER is an extension of DENSER on which the evolutionary engine is replaced by a $(1 + \lambda)$-ES. This modification dramatically reduces the required number of evaluations per generation, enabling executions 20 times faster than the original version of DENSER.

Moreover, individuals are initialized with shallow topologies, and the stopping criterion is variable to allow an individual to be trained for a more extended time.

On the CIFAR-10 dataset [8], DENSER obtained models with an accuracy higher than most of the state-of-the-art results, and on the CIFAR-100 [8], it obtained the best accuracy reported by NE approaches. Fast-DENSER proves to be highly competitive relative to DENSER, achieving execution times far inferior to its predecessor. Additionally, Fast-DENSER can develop DNNs that do not require additional training after the evolutionary approach and are, therefore, ready to be deployed.

3 Approach

This section outlines the approaches developed to address the challenge of reducing power consumption in ANN models.

3.1 Power Measurement

Measuring the power a GPU consumes is fundamental when developing approaches that minimize a model's energetic footprint. The ecosystem of developing a DNN model mainly consists of three phases: design, training, and deployment.

The design phase uses some energy, be it with manual design techniques or automatic methods. DENSER is a NE framework and, as such, consumes energy in the search for optimal models, and such consumption might be on par with the energy used on manual, trial-and-error methods. Reducing the energy used in this phase is out of the scope of this work.

The training of a DNN model is an expensive process in which a model is trained on a large dataset to learn to predict unseen instances, taking a significant toll on technological companies' and individuals' power bills. While diminishing energy consumption during the training process remains a significant objective, it is worth noting that the inference phase in DNNs holds vital importance during software deployment, as the software obtains results through inference.

This becomes particularly relevant when considering the potential utilization of these models by millions of users. As such, tackling the minimization of energy consumption in this step is vital. For example, it is estimated that 80% to 90% of NVIDIA's ML computations are inference processing [13] and about 60% of Google's ML energy usage is for inference with the remaining portion being for training [16].

Considering this, our work focuses on the power consumption in the inference step to allow a large deployment, thus saving more computational resources and energy and, on another layer, reducing financial expenses and reducing environmental impact.

3.2   Model Partitioning

Training a DNN model requires a substantial amount of time and considerable energy. Creating a process on which a single model is trained but can be split posteriorly into two models would reduce the time spent on training two models by, at most, two times. Pushing one of those two models into being smaller than the other may produce a simpler, similarly accurate, yet more power-efficient model.

Following this line of reasoning, we propose a modification to Fast-DENSER on which an extra output layer is connected to an intermediate layer of the model. The two-output model (Fig. 1a) is trained to optimize for two outputs. At the validation step, it is split into left (Fig. 1b) and right (Fig. 1c) partitions. These partitions are disjoint and can be evaluated similarly to how the complete model is evaluated, and metrics such as accuracy and power consumption can be obtained.

The intermediate point is a marker for where the additional output is added at the model partitioning step. We can, for example, consider a model as an array of layers, and the mentioned marker is the index of the layer to which the additional output is connected. This point can be assigned to any intermediate layer of the model. The input and output layers are excluded to prevent useless and redundant partitions.

Since the maximum allowed value of the point is equal to the number of layers of the model minus one, the grammar initializer – which generates individuals according to the grammar – and the mutation mechanism for the macrostructure level of DENSER – which performs mutations on the hyperparameters of the individuals – were modified to consider the maximum number of layers of the model dynamically. To introduce the intermediate point in the evolutionary process, it was considered part of the macrostructure and, as such, as a rule of the grammar. The introduced rule is $<middle\_point> ::= [middle\_point, int, 1, 0, x]$, meaning that one integer value is obtained with the lower limit being zero. The upper limit is an arbitrary variable x that is replaced at any instance by the maximum number of layers of the model minus one.

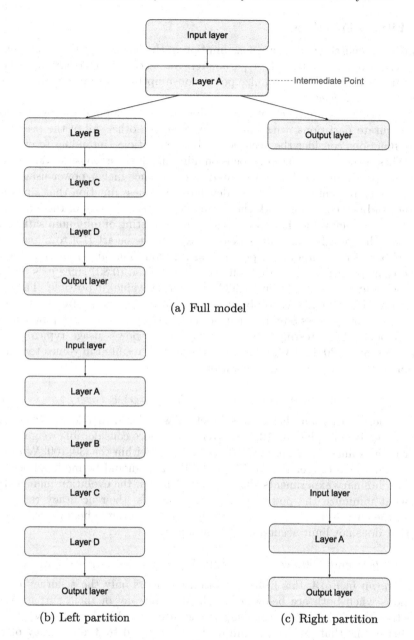

(a) Full model

(b) Left partition

(c) Right partition

Fig. 1. Example of a two-output model and its left and right partitions, with the layer marked by the intermediate point in red. (Color figure online)

3.3   Fitness Functions

To consider accuracy and power consumption in the fitness function, some functions were developed to take these parameters into account. Since our objective is to maximize accuracy but minimize power consumption, we consider the inverse of the latter, i.e., $power^{-1}$.

Considering our approach of the division of a DNN model into two comparably accurate partitions, with one smaller than the other, all of the presented fitness functions consider the accuracy of both partitions, intending to enhance both. These fitness functions only focus on minimizing power consumption within the larger partition, which is anticipated to experience higher power usage.

Firstly, as presented in Eq. 1, we developed a fitness function that sums the accuracy of both partitions with the inverse of the power usage of the left partition. The accuracy values have an upper limit, consisting of minimum satisfiable values for the models, i.e., values below the state-of-the-art [15] to allow some tradeoff between accuracy and power consumption. The upper limit is higher on the right partition (0.85) than on the left partition (0.80) since it is desired that the right partition obtains a higher accuracy value, if possible. The goal of this function design was to obtain satisfiable models and, after that, guide the evolutionary process only by their power usage to minimize the power usage of the models. After testing, we observed that the power usage typically falls within the range $[30, 100]$ W, which, when inverted, resulted in values too small to be able to properly steer the evolutionary process.

$$f_1 = \min(0.80, acc_{left}) + \min(0.85, acc_{right}) + power_{left}^{-1} \qquad (1)$$

Considering this, another fitness function was designed (Eq. 2), where the power usage is multiplied by 10, thus giving it a more considerable weight since power usage values for the used GPU typically fall within the $[30, 100]$ W range. This weight is closely related to the used GPU and should be modified accordingly. Preliminary experiments showed that although the evolution managed to somewhat minimize the power usage of the models, their accuracy remained around the chosen upper limits. Since this is not an optimal behaviour, a function that does not limit accuracy was developed.

$$f_2 = \min(0.80, acc_{left}) + \min(0.85, acc_{right}) + 10 * power_{left}^{-1} \qquad (2)$$

As shown in Eq. 3, this fitness function considers only the accuracy of the partitions when both are below a threshold. After any of them surpass their respective threshold, power consumption is also considered, with a weight of 10. This means that, at first, evolution is only steered by the accuracy of the models. When satisfiable models are obtained, power consumption starts being considered to evolve both accurate and energy-efficient models.

$$f_3 = \begin{cases} acc_{left} + acc_{right} & \text{if } acc_{left} \leq 0.80 \wedge acc_{right} \leq 0.85 \\ acc_{left} + acc_{right} + 10 * power_{left}^{-1}, & \text{otherwise} \end{cases}$$

$$(3)$$

3.4 Module Reutilization

Internally, Fast-DENSER considers modules of layers on each individual from which a DNN is then unravelled. One way to encourage the evolution of energy-efficient models is to provide an individual with a set of layers that are known to be efficient. As such, a scheme of module reutilization is proposed through the design of new mutation operators and the addition of an archive of modules and their respective power consumption.

Since this strategy only considers power consumption, it is expected that inaccurate models may sometimes be generated. Due to the nature of the evolutionary process and the used fitness function (Eq. 3), inaccurate models are intensely penalized and, as such, discarded in favour of better ones.

Whenever a module of layers is randomly generated or modified, its power consumption is measured. To do this, a temporary model is created, which consists of an input layer, the module's layers, and an output layer. Since the module's accuracy is irrelevant, this temporary network is neither trained nor fed with a proper dataset, i.e., it is given random values instead of a dataset.

An operator of mutation, *reuse module*, was introduced to take advantage of this information. It selects a module with a probability inversely proportional to its power consumption, i.e., modules with inferior power consumption have a superior probability of being chosen. As shown in Eq. 4, to obtain the probability of a module i being chosen, we divide the inverse of its power, $power_i$, by the sum of the inverse power of all modules, with n the number of saved modules. The selected module is introduced in a randomly chosen position. An operator that randomly removes a module from an individual is also introduced to counteract the described operator.

$$P(i) = \frac{\frac{1}{power_i}}{\sum_{j=0}^{n} \frac{1}{power_j}} \tag{4}$$

4 Experimental Setup

We performed two experiments: the baseline, which uses the plain version of Fast-DENSER with accuracy as the fitness function, and an experiment where our proposed approaches were applied, using the fitness function presented in Eq. 3. Table 1 presents the experimental parameters used across the experiments. Note that *DSGE-level rate* refers to the probability of a grammar mutation on the model's layers, the *Macro layer rate* pertains to the probability of a grammar mutation affecting the macrostructure, encompassing elements such as hyperparameters or intermediate point mutation, and the *Train longer rate* is the probability of allocating more time for an individual to be trained. The rates of reusing and removing modules do not apply to the baseline experiment. The experimental analyses consider the Mean Best Fitness (MBF) over 5 runs.

The experiments were performed on a server running Ubuntu 20.04.3 LTS with an Intel Core i7-5930K CPU with a clock frequency of 3.50 GHz, 32 GB of RAM, and an NVIDIA TITAN Xp with CUDA 11.2, CuDNN 8.1.0, Python 3.10.9, Tensorflow 2.9.1 and Keras 2.9.0 installed as well as the pyJoules 0.5.1 Python module with the NVIDIA specialization.

Table 1. Experimental parameters

Evolutionary Parameter	Value
Number of runs	5
Number of generations	150
Maximum number of epochs	10 000 000
Population size	5
Add layer rate	25%
Reuse layer rate	15%
Remove layer rate	25%
Reuse module rate	15%
Remove module rate	25%
Add connection	0%
Remove connection	0%
DSGE-level rate	15%
Macro layer rate	30%
Train longer rate	20%
Train Parameter	**Value**
Default train time	10 min
Loss	Categorical Cross-entropy

All experiments used the Fashion-MNIST dataset [18], which was developed as a more challenging replacement for the well-known MNIST dataset [5] by swapping handwritten digits with images of clothes such as shirts and coats, aiming at a more realistic and relevant benchmark. It is a balanced dataset consisting of a collection of 60 thousand examples for training and 10 thousand for testing, where each example is a 28×28 grey-scale image representing clothing items belonging to one of ten classes.

Since power usage is essential in making NE physically plausible, a function to measure power was developed using the *pyJoules* library. Its pseudocode can be analyzed in Algorithm 1, with *meter* being the library tool that facilitates the measurement of energy consumed, and *start* and *stop* the functions that allow controlling it. It wraps a function call (*func*, with corresponding arguments *args*) while measuring the GPU energetic consumption during its execution and the call's duration. This measurement is converted from milli-Joule to Watt and appended to the array of measures. These steps are performed $n\_measures$ times, and then the mean value is calculated. In our work, we considered $n\_measures = 30$. The described function was integrated with Fast-DENSER on the model's validation step to measure the power used in the inference phase.

Algorithm 1. Power Measure Algorithm

Require: func, args, $n\_measures$
   $measures \leftarrow \emptyset$
   $i \leftarrow 1$
   while $i \leq n\_measures$ **do**
      start(meter)
      $output \leftarrow func(args)$
      stop(meter)
      $(energy, duration) \leftarrow measure(meter)$
      $measure \leftarrow energy/1000/duration$ ▷ Convert mJ to W
      $measures \leftarrow measures \cup measure$
      $i \leftarrow i + 1$
   end while
   $mean\_power \leftarrow mean(measures)$
   return $(output, mean\_power)$

It should be noted that ambient conditions of the server's location, such as temperature and humidity, were not considered, as well as other external variables, and no other processes used the GPU during the execution of these experiments.

5 Results

This section compares the results from the baseline experiment and the experiment where our approaches were applied. The results show the mean accuracy and the mean power consumption, which are derived from the best individuals by fitness over 5 separate runs.

Since the results did not follow a normal distribution and the samples were independent, the Kruskal-Wallis non-parametric test was employed to determine if significant differences existed among the various approach groups. When significant differences were observed, the Mann-Whitney post-hoc test with Bonferroni correction was applied. We considered a significance level of $\alpha = 0.05$ in all statistical tests.

Figure 2 compares the accuracy obtained in the two experiments. The experiments present a similar accuracy until generation 70, where it becomes possible to observe a clear difference between them. The baseline experiment achieves a higher accuracy than the other experiment, and, relative to that experiment, it is visible that the smaller model obtains a marginally smaller accuracy than the larger one. Table 2 provides statistical analysis, and Table 3 showcases statistical values of the experiments. It is possible to see that, relative to the median values, the proposed method achieves inferior accuracy and that the smaller model obtains the worst accuracy.

Figure 3 presents a comparison of the power consumption measured in the two experiments. The baseline predominantly has an increasing behaviour, which

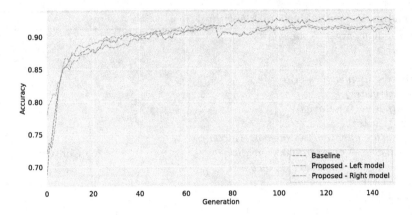

Fig. 2. Evolution of the accuracy over 150 generations.

Table 2. Pair-wise comparison of used groups on accuracy metric, using Mann-Whitney U post-hoc test with Bonferroni correction with bold values denoting statistically significant differences.

		Baseline	Proposed Method	
	Metric	$Accuracy$	$Accuracy_{left}$	$Accuracy_{right}$
Baseline	$Accuracy$			
Proposed Method	$Accuracy_{left}$	**1.09×10^{-4}**		
	$Accuracy_{right}$	**1.15×10^{-7}**	**1.16×10^{-4}**	

Table 3. Mean value, standard deviation, median and difference to baseline median of the accuracy of the experiments.

Experiment	Metric	Mean	SD	Median	Diff. to Baseline
Baseline	$Accuracy$	0.904	0.037	0.916	
Proposed Method	$Accuracy_{left}$	0.902	0.024	0.911	-0.005
	$Accuracy_{right}$	0.895	0.034	0.907	-0.009

can be explained by the fact that the evolution is only being guided by accuracy, i.e., there are no incentives to favour models that consume less power. Contrarily, the proposed method obtained relatively stable results over the evolutionary process, with the smaller model presenting marginally lower results than its counterpart. Table 4 provides statistical analysis, and Table 5 showcases statistical values of the experiments. We can conclude that relative to the median values, the proposed method achieves inferior power consumption and that the smaller model is the most power-efficient.

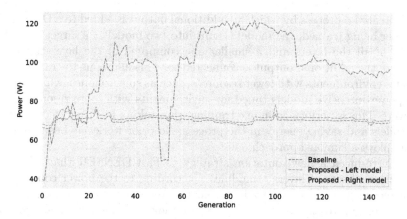

Fig. 3. Evolution of the power consumption over 150 generations.

Table 4. Pair-wise comparison of used groups on power metric, using Mann-Whitney U post-hoc test with Bonferroni correction with bold values denoting statistically significant differences.

		Baseline	Proposed Method	
	Metric	*Power*	$Power_{left}$	$Power_{right}$
Baseline	*Power*			
Proposed Method	$Power_{left}$	$\mathbf{2.72 \times 10^{-29}}$		
	$Power_{right}$	$\mathbf{8.84 \times 10^{-32}}$	$\mathbf{3.67 \times 10^{-19}}$	

Table 5. Mean value, standard deviation, median and difference to baseline median of the experiments power consumption.

Experiment	Metric	Mean	SD	Median	Diff. to Baseline
Baseline	*Power*	97.80 W	18.84 W	99.89 W	
Proposed Method	$Power_{left}$	71.92 W	1.60 W	72.20 W	−27.69 W
	$Power_{right}$	70.40 W	1.30 W	70.71 W	−29.18 W

6 Conclusion

In this work, we developed approaches integrated into Fast-DENSER, which empower it to generate DNN models with better power efficiency.

The most fundamental approach consists of measuring the power consumed by the GPU on the inference phase of the DNN. We use the measure provided by the GPU to do this. Using this metric, we developed multi-objective fitness functions that steer the evolutionary process in a path that minimizes power consumption.

We created a process by which an additional output is added to a DNN model and, after being trained, the model is split into two models – a larger one which consists of all the layers and a smaller one composed of the layers up to the one where the additional output is connected to. This allows us to create models tuned for environments with fewer resources, such as smartphones, while creating more power-intensive models tuned for environments with more resources, such as servers. This is performed in one training, thus taking less time to develop the two models and saving energy in the process. No prior work has been identified that employs a similar approach.

We introduced a new mutation strategy to Fast-DENSER that allows the reutilization of sets of layers – modules – according to the power consumption of the modules. We stochastically favour the reintroduction of modules in a model according to the inverse of the power they consume, thus incorporating power-efficient modules into a model.

The results obtained by our proposals show that we can reduce the power consumption of the ANNs without compromising their predictive performance, showing that it is possible to minimize power consumption while, at the same time, maximizing accuracy through the usage of NE frameworks such as Fast-DENSER. The best model found regarding power consumes 29.18 W (29.2%) less whilst having a tiny decrease in performance (less than 1%), proving that a small trade-off on accuracy can yield a considerable reduction in the power consumed by the model.

6.1   Future Work

We introduced novel approaches and performed a baseline experiment and an experiment where the mentioned strategies were applied. It could be valuable to explore other approaches and perform more experiments in the future.

To better understand the individual impact of each strategy on the efficiency of the models, it would be valuable to perform experiments with the employment of only one strategy at a time. It would also be interesting to vary the fitness functions (e.g., the weights used in them) and to vary evolutionary parameters such as the probabilities of the mutations.

One of the most important constraints of our work is GPU-time due to the amount of operations required to train every model of each generation. To minimize the required time, it would be noteworthy to research how to employ training-less strategies in Fast-DENSER, i.e., use strategies that estimate the accuracy of a model without training it [4,14]. Such strategies would allow us to perform more experiments in less time, saving energy in the design process.

Acknowledgments. This work was supported by the Portuguese Recovery and Resilience Plan (PRR) through project C645008882-00000055, Center for Responsible AI, by the FCT, I.P./MCTES through national funds (PIDDAC), by Project No. 7059 - Neuraspace - AI fights Space Debris, reference C644877546-00000020, supported by the RRP - Recovery and Resilience Plan and the European Next Generation EU

Funds, following Notice No. 02/C05-i01/2022, Component 5 - Capitalization and Business Innovation - Mobilizing Agendas for Business Innovation, and within the scope of CISUC R&D Unit - UIDB/00326/2020.

References

1. Assunção, F., Lourenço, N., Machado, P., Ribeiro, B.: DENSER: deep evolutionary network structured representation. Genet. Prog. Evolvable Mach. **20**(1), 5–35 (2019). https://doi.org/10.1007/S10710-018-9339-Y
2. Assunção, F., Lourenço, N., Ribeiro, B., Machado, P.: Fast-DENSER: fast deep evolutionary network structured representation. SoftwareX **14**, 100694 (2021). https://doi.org/10.1016/j.softx.2021.100694
3. Balas, V.E., Roy, S.S., Sharma, D., Samui, P. (eds.): Handbook of Deep Learning Applications. SIST, vol. 136. Springer, Cham (2019). https://doi.org/10.1007/978-3-030-11479-4
4. Chen, W., Gong, X., Wang, Z.: Neural architecture search on ImageNet in four GPU hours: a theoretically inspired perspective. In: 9th International Conference on Learning Representations, ICLR 2021, Virtual Event, Austria, 3–7 May 2021. OpenReview.net (2021). https://doi.org/10.48550/arXiv.2102.11535
5. Deng, L.: The MNIST database of handwritten digit images for machine learning research. IEEE Sig. Process. Mag. **29**(6), 141–142 (2012). https://doi.org/10.1109/MSP.2012.2211477
6. Eiben, A.E., Smith, J.E.: Introduction to Evolutionary Computing. NCS, Springer, Heidelberg (2015). https://doi.org/10.1007/978-3-662-44874-8
7. Galván, E., Mooney, P.: Neuroevolution in deep neural networks: current trends and future challenges. IEEE Trans. Artif. Intell. **2**(6), 476–493 (2021). https://doi.org/10.1109/TAI.2021.3067574
8. Krizhevsky, A., Hinton, G.: Learning multiple layers of features from tiny images. Technical report, University of Toronto (2009). https://www.cs.toronto.edu/kriz/learning-features-2009-TR.pdf
9. LeCun, Y., Bengio, Y., Hinton, G.E.: Deep learning. Nature **521**(7553), 436–444 (2015). https://doi.org/10.1038/nature14539
10. Liu, W., Wang, Z., Liu, X., Zeng, N., Liu, Y., Alsaadi, F.E.: A survey of deep neural network architectures and their applications. Neurocomputing **234**, 11–26 (2017). https://doi.org/10.1016/J.NEUCOM.2016.12.038
11. Lourenço, N., Assunção, F., Pereira, F.B., Costa, E., Machado, P.: Structured grammatical evolution: a dynamic approach. In: Ryan, C., O'Neill, M., Collins, J.J. (eds.) Handbook of Grammatical Evolution, pp. 137–161. Springer, Cham (2018). https://doi.org/10.1007/978-3-319-78717-6_6
12. Lourenço, N., Pereira, F.B., Costa, E.: SGE: a structured representation for grammatical evolution. In: Bonnevay, S., Legrand, P., Monmarché, N., Lutton, E., Schoenauer, M. (eds.) EA 2015. LNCS, vol. 9554, pp. 136–148. Springer, Cham (2016). https://doi.org/10.1007/978-3-319-31471-6_11
13. Luccioni, A.S., Viguier, S., Ligozat, A.L.: Estimating the carbon footprint of BLOOM, a 176B parameter language model (2022). https://doi.org/10.48550/ARXIV.2211.02001
14. Mellor, J., Turner, J., Storkey, A.J., Crowley, E.J.: Neural architecture search without training. In: Meila, M., Zhang, T. (eds.) Proceedings of the 38th International Conference on Machine Learning, ICML 2021, Virtual Event. Proceedings

of Machine Learning Research, 18–24 July 2021, vol. 139, pp. 7588–7598. PMLR (2021). https://doi.org/10.48550/arXiv.2006.04647

15. Meshkini, K., Platos, J., Ghassemain, H.: An analysis of convolutional neural network for fashion images classification (fashion-MNIST). In: Kovalev, S., Tarassov, V., Snasel, V., Sukhanov, A. (eds.) IITI 2019. AISC, vol. 1156, pp. 85–95. Springer, Cham (2020). https://doi.org/10.1007/978-3-030-50097-9_10

16. Patterson, D.A., et al.: The carbon footprint of machine learning training will plateau, then shrink. Computer 55(7), 18–28 (2022). https://doi.org/10.1109/MC.2022.3148714

17. Vikhar, P.A.: Evolutionary algorithms: a critical review and its future prospects. In: 2016 International Conference on Global Trends in Signal Processing, Information Computing and Communication (ICGTSPICC), pp. 261–265 (2016). https://doi.org/10.1109/ICGTSPICC.2016.7955308

18. Xiao, H., Rasul, K., Vollgraf, R.: Fashion-MNIST: a novel image dataset for benchmarking machine learning algorithms (2017). https://doi.org/10.48550/arXiv.1708.07747

19. Yegnanarayana, B.: Artificial Neural Networks. PHI Learning (2009)

Leveraging More of Biology
in Evolutionary Reinforcement Learning

Bruno Gašperov[✉], Marko Đurasević, and Domagoj Jakobovic

Faculty of Electrical Engineering and Computing, University of Zagreb, Zagreb,
Croatia
{bruno.gasperov,marko.durasevic,domagoj.jakobovic}@fer.hr

Abstract. In this paper, we survey the use of additional biologically
inspired mechanisms, principles, and concepts in the area of evolutionary reinforcement learning (ERL). While recent years have witnessed
the emergence of a swath of metaphor-laden approaches, many merely
echo old algorithms through novel metaphors. Simultaneously, numerous promising ideas from evolutionary biology and related areas, ripe
for exploitation within evolutionary machine learning, remain in relative
obscurity. To address this gap, we provide a comprehensive analysis of
innovative, often unorthodox approaches in ERL that leverage additional
bio-inspired elements. Furthermore, we pinpoint research directions in
the field with the largest potential to yield impactful outcomes and discuss classes of problems that could benefit the most from such research.

Keywords: evolutionary reinforcement learning · stochastic optimal
control · open-endedness · evolutionary algorithms · dynamic
environments · evolutionary machine learning · evolutionary biology

1   Introduction

Mimicking the process of biological evolution lies at the heart of the field of evolutionary computation (EC). However, caution must be exercised when trans-
ferring ideas from the biological into the algorithmic realm. Despite this, the
field remains inundated with various metaphor-based metaheuristics, inspired by
everything from flocks of birds to the COVID-19 pandemic. In reality, they often
amount to no more than repackaging existing algorithms under the guise of new
analogies, with effects that are cosmetic at best. This practice has been strongly
criticized [1,2], partly for being used to obfuscate the lack of genuine novelty.
Even worse, it risks prejudicing researchers against the very idea of transfer-
ring unexploited biological concepts into EC, which could lead to false negatives
- rejections of sound ideas whose adoption would genuinely advance the field.
At the same time, numerous promising biological principles and mechanisms,
as well as most of the modern evolutionary synthesis [3,4], remain unexploited
[5] within EC (albeit with some exceptions [6]). This primarily stems from the
fact that EC and evolutionary biology developed mostly independently of each
other [7]. Yet many findings seem (even uncannily) transferable across them. For

S. Smith et al. (Eds.): EvoApplications 2024, LNCS 14635, pp. 91–114, 2024.
https://doi.org/10.1007/978-3-031-56855-8_6

example, just like phenotypically different species in nature share a large part of the genome and thus occupy a small subset of the genotypic space [8], elite diverse solutions obtained via quality diversity (QD) algorithms tend to cluster in a small genotypic region, referred to as the elite hypervolume [9].

In this survey paper, we focus on the current and future use of additional bio-inspired mechanisms, principles, and concepts in the area of evolutionary reinforcement learning (ERL). ERL [10] is a fusion of EC and reinforcement learning (RL), in which evolutionary techniques (such as genetic algorithms or evolutionary strategies) are used to explore the search space of RL policies[1] and ultimately find the optimal ones, i.e., those that "solve" the underlying Markov decision process (MDP). The objectives of this paper are twofold. First, we offer a comprehensive analysis of innovative, often unorthodox approaches in ERL that utilize additional bio-inspired elements, which are shown in Fig. 1. This includes research perched at the intersection of ERL, artificial life, and research in evolutionary biology. Second, we identify research directions in the field with the potential to yield fruitful results, and we discuss two broad classes of problems and their application areas that could benefit the most from this research. We believe that the objectives are particularly timely and pertinent, given the scarcity of literature that amalgamates prior findings, and the possible impact of successfully incorporating additional bio-inspired elements into ERL. It is worth noting that recent surveys and perspectives by Yuen *et al.* [5], Vie *et al.* [11], Chelly Dagdia *et al.* [12], and Miikkulainen and Forrest [13] cover related topics. However, they do not emphasize ERL and, therefore, place less importance on concepts that are especially relevant in this context, like evolvability.

2   Evolutionary Reinforcement Learning

In recent years, deep RL (DRL) has been applied successfully to a wide range of complex decision-making tasks, including the game of Go [14], robot manipulation [15], and medical imaging [16]. While the majority of approaches in (D)RL are gradient-based, including renowned algorithms such as Proximal Policy Optimization (PPO) [17], there has been a growing interest in derivative-free and population-based methods [18]. This rise in popularity stems from earlier research indicating that both vanilla genetic algorithms [19] (GA) and evolutionary strategies [20] offer competitive alternatives to state-of-the-art gradient-based DRL algorithms across a plethora of tasks. ERL methods share the advantages of EC techniques, which particularly include their ability to a) tackle non-convex and non-differentiable objective functions due to their black-box function approximator nature, b) handle sparse RL rewards, c) jump over local optima by the use of variation operators, d) accommodate both deterministic and stochastic policies as they do not have to rely on stochasticity for exploration, e) leverage parallelization as they are population-based, f) avoid issues related to high variance and errors when estimating first or second-order derivatives of the reward function, and g) generate a range of diverse solutions through open-ended ERL

---

[1] It should be noted that ERL also includes approaches in which different state-action pairs are directly explored, as well as meta-RL methods.

variants. However, ERL methods also inherit certain disadvantages from EC, including a) sample inefficiency, b) the discarding of potentially useful gradient information, and c) the lack of utilization of the inner reward structure (since only the total return is considered, as in Monte Carlo RL methods).

2.1 Application Areas of Particular Importance

We anticipate certain classes of problems to especially benefit from the integration of additional bio-inspired principles, mechanisms, and concepts into ERL. One of these is **optimal stochastic control in dynamic (non-stationary) environments**, where swift reactions to environmental changes (i.e., changes in the underlying MDP) are essential. Such challenges frequently appear in real-life scenarios, many of which demand agents to dynamically adjust their behavior without being explicitly provided with the properties or timings of environmental changes. Examples abound in application areas such as finance, medicine, vehicle routing, and distributed communication systems [21]. Dynamic environments pose a significant challenge to DRL agents and are often tackled within the context of areas of meta-RL, online RL, and RL in the presence of partial observability. As a result, the most notable accomplishments achieved by DRL techniques have been confined to static , highly controlled, and even deterministic environments, such as mazes and video games [22]. Problems characterized by changing (dynamic) environments, i.e., dynamic fitness landscapes, have a long history of research in the EC community [23]. Traditionally, the main lines of such research endeavored to either a) track the moving optima by maintaining diversity through a large population of solutions, b) find solutions that remain robust in the face of new environmental conditions, or c) predict the environmental changes [24], i.e., the new locations of the moving optima.

Another class involves unlocking the creative and exploratory potential of evolution manifested by the **open-endedness** (OE) of biological evolution. While various definitions of OE exist, it generally refers to the ability of a process, system, or algorithm to continuously generate novel or increasingly diverse or complex solutions [25]. Given that biological evolution has yielded a colossal number of species (around 8.7 million eukaryotic species globally [26]), scattered around different ecological niches, it can be interpreted as an open-ended stochastic biological algorithm. Evolutionary processes in biology exhibit two main modes: optimization and expansion [27]. During optimization, the number of entities (variables) and their interactions are fixed, and the system as a whole is static. In expansion, novel entities and interactions emerge, which is crucial for OE. These two modes [28] can be seen as a type of exploitation-exploration trade-off [28]. Different ERL algorithms correspond to varying levels of expansion and optimization. For instance, novelty search (NS) [29] abandons classical objectives (optimization) and promotes expansion through its divergent search for behavioral diversity. It is particularly useful when dealing with sparse RL rewards or ill-defined problems (e.g., design). A more balanced approach is given by QD [30] which combines convergence (optimization) with divergence (exploration) by generating a wide range of mutually different yet high-performing solutions across different niches [31]. Its applications include generating levels in video games [32], evolutionary robotics [33], and many more.

It is also important to stress the interconnections between stochastic optimal control in dynamic environments, OE, and the leveraging of additional bio-inspired elements. Biological mechanisms guide the evolution of organisms (individuals) while they are situated in constantly changing, dynamic environments, with substantial noise (stochasticity), and relatively low levels of selective pressure [13], all while they are under the influence of feedback loops between themselves, their populations, and environments. This promotes the variance and diversity in organisms and ultimately leads to the emergence of OE [34].

3   Concepts, Principles, and Mechanisms

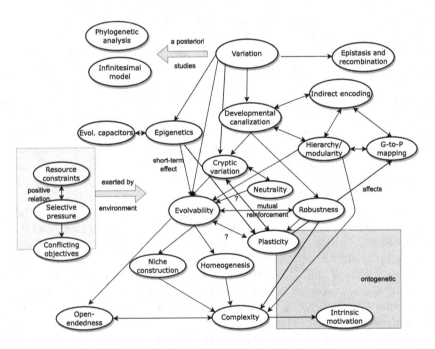

Fig. 1. A possible diagram of considered bio-inspired terms and their relationships. Two-headed (one-headed) arrows indicate bi-directional (one-directional) relationships, with question marks denoting poorly understood ones.

3.1   Evolvability and Robustness

Simply put, evolvability [35] refers to the ability to evolve further. It is a pivotal driver of the evolutionary process [36], and is said to be *"to evolution as generalisation is to learning"* [37]. Evolution itself is suspected to select for evolvability through mechanisms such as the niche founder effect [38]. A plethora of definitions of evolvability exist, mostly differing with respect to whether they only consider phenotypic variation, or also adaptive variation, as shown in Table 1.

Table 1. Characteristics of different evolvability definitions.

Type of variation	Underlying goal	Example	Evo. pressure	Problem
Phenotypic	divergence	novelty search	low or none	no adaptivity
Adaptive	convergence-divergence	multi-task opt	higher	fitness unknown

Focusing on phenotypic variation, Mengistu et al. [39] measure an individual's evolvability through the behavioral diversity of its immediate offspring. A direct search for the most evolvable individuals is then performed, resulting in adaptable RL agents capable of generalizing to unseen maze-based RL environments. Building upon this, Gajewski et al. [40] propose a new algorithm called evolvability-ES that optimizes an evolvability-based objective while significantly reducing the computational costs and enabling scalability to large deep neural networks (DNNs). The central idea is to estimate the gradients of evolvability. The experiments on two DRL locomotion benchmarks point to its competitiveness with Model-Agnostic Meta-Learning (MAML) [41], a staple meta-RL approach. Katona et al. [42] expand evolvability-ES by simultaneously optimizing for both fitness and evolvability, bringing the approach closer to QD optimization. The goal is to find a single individual whose distribution of offspring is both diverse and high-performing. The results, again on a set of RL locomotion tasks, show that the approach is capable of handling at least some level of RL reward deceptiveness. Gasperov and Djurasevic [43] demonstrate on a robotic arm task that divergent search improves evolvability even under low levels of selective pressure and propose a new phenotypic definition of evolvability that takes into account the evolvability of offspring as well. They also emphasize the concept of behavior landscapes, and describe its relation with evolvable solutions. Further research connecting evolvability and NS includes [44] and [45].

Some works employ the perspective of adaptive evolvability. Ferigo et al. use the mean difference in fitness between the parents and their offspring as a measure of evolvability. Their approach is tested on a locomotion task performed by Voxel-based Soft Robots, with the results indicating that evolvability depends on the type of evolutionary algorithm used. Outside ERL, other approaches adhering to the adaptive perspective study grammatical evolution [46] and multi-objective genetic programming for symbolic regression [47]. Finally, a number of ERL approaches indirectly rely on evolvability through parameter self-adaptation, exemplified by approaches like the covariance matrix adaptation evolution strategy (CMA-ES) and its expansions [48–51].

Evolvability is particularly important for optimal control in dynamic (non-stationary) RL environments. Highly evolvable solutions have genotypic neighbors with mutually diverse phenotypes, meaning that only a few mutations (or steps of gradient ascent) may be sufficient to find a solution that works well under the new conditions. Similar has also been studied in various applications of EC [52]. Evolvability also holds significant promise for creating agents capable of generalizing to a wider range of RL environments with only slight modifications. Exploring whether methods used to enhance generalization in RL, such as robust adversarial RL [53], also contribute to improving evolvability presents an interesting avenue for investigation. On the flip side, the key concerns across most of the listed approaches lie in the high computational costs of calculating evolvability. Novel methods for cheaply approximating evolvability are hence needed, perhaps also involving gradient-based techniques [40].

Robustness, the capacity of organisms to withstand genotypic or environmental changes, has been studied in conjunction with evolvability [54,55]. Research in biology has revealed that very limited amounts of information suffice to encode genetically strong organisms that can endure random genetic alterations. An open question is whether this property can be achieved within (E)RL as well, possibly through regulatory control mechanisms akin to those found in genetics.

3.2 Epistasis and Recombination

Recombination is known to facilitate both biological evolution [56] and EC methods [57] by enhancing exploration. Paixão and Barton [7,58], using ideas from quantitative and population genetics, decompose phenotypic variance V into multiple components: $V = V_A + V_I + V_D + V_E$. Here, V_A represents the additive variance corresponding to individual effects of the alleles (genes), V_I the variance due to epistatic [59] interaction effects between them , V_D the dominance component (zero in asexual reproduction), and V_E the variance that stems from environmental effects. The sum of additive genetic effects is also called the breeding value of an individual, measuring the value of its genes to progeny. Notably, when recombination is at play, interaction effects (V_I) are less inherited by offspring since recombination separates existing gene combinations. On this basis, the authors propose a novel GA variant in which the rate of recombination is reduced when epistatic effects prevail (and vice versa) and show its superiority over the vanilla GA variant on the royal road problem [60]. Polani and Miikkulainen [61,62] pioneer the use of epistasis in ERL through their Eugenic Algorithm (EuA) and showcase the algorithm's efficiency on the 2-pole-balancing benchmark task. They approximate epistasis as $E = 1 - D_{max}$, where D_{max} is the maximum difference in selection probabilities among alleles across all genes. When D_{max} is small (large), all alleles are (not) approximately equally beneficial, making epistasis more (less) important. Ventresca and Ombuki-Berman [63] delve into the theoretical aspects of epistatic interactions in evolutionary neuro-controllers, while using information theoretic measures.

Consideration of epistasis might lead to improved variants of ERL algorithms. For example, parent selection could be performed on the set of solutions with the largest pure breeding values (instead of fitnesses), as precisely these solutions are expected to yield the highest quality offspring. However, challenges lie in estimating the breeding values/epistasis levels, which is computationally costly, especially in DRL which employs highly complex DNNs. Also, ERL in uncertain domains [64] might benefit from the variance decomposition into V_E.

3.3 Developmental Canalization

The term developmental canalization describes the phenomenon in which certain dimensions of variation are explored whereas others are suppressed, resulting in mutations affecting only certain phenotypic properties. Canalization can enhance evolvability by restricting exploration to dimensions with a higher potential for novelty or complexity. Huizinga et al. [65] demonstrate the emergence of developmental canalization in Picbreeder, a system of interactive evolution of pictures, arguing that this is due to the system's open-ended and goal-free [34] nature. Katona et al. [66] discuss the links between canalization and indirect encoding, highlighting that the latter alters the type of phenotypic variations mutations can cause. They show that indirect encoding outperforms direct one when using MAML [41], in which the objective is the ability to adapt. Finally, Mengistu et al. [39] propose promoting canalization by directly maximizing behavioral distributions of all the offspring, i.e., the evolvability of the entire population.

Canalization might be a prerequisite for evolving agents that can tackle highly complex environments, as attempted by open-ended RL approaches, such as the Paired Open-Ended Trailblazer (POET) [67]. Generally, much remains ripe for discovery. Future endeavors might revolve around novel methods for parameter self-adaptation that encode the loci, type, and magnitude of mutations as part of the RL agent's genotype, thereby enabling their coevolution. Ideally, such methods would build upon the large body of research on self-adaptive parameter control in EC [68–71]. Moreover, the identification of promising dimensions of variation could inform the design of the behavior space in QD approaches, which is traditionally handcrafted [72]. Some open questions also remain about the choice and design of canalization metrics.

3.4 Epigenetics

In the biological realm, epigenetic layers above genes regulate gene expression based on the current environmental conditions. This regulation is done through methylation marks [73], which determine which genes are expressed (or muted) by modifying histone levels. Histones are proteins around which DNA is spooled [74,75]. Epigenetic mechanisms enable rapid phenotypic changes in response to non-stationary environmental dynamics, increasing the probability of survival. Information stored in the epigenetic layers, or simply epigenetic information, representing the memory of former environmental conditions, is inherited by the next generation [76]. Mukhlish et al. [77] put forth a novel epigenetic learning algorithm called EpiLearn, inspired by Lamarckian evolution, for use in swarm robotics. The algorithm is shown to be capable of handling deceptive problems. The same authors [78] then propose an ERL method RELEpi in which reward, temporal difference, methylation, and epigenetic inheritance are utilized to approximate optimal RL policies. In their approach, rewards provided by the

RL environment directly affect the histone values. Finally, the method is successfully demonstrated on a simulated search and rescue mission [79]. Sousa and Costa [80] introduce EpiAL, an artificial life epigenetic model. Their main finding is that the agents with epigenetic mechanisms find it much easier to flourish in dynamic environments, in which the attributes, such as temperature, light, and food, vary over time. Interestingly, this only applies to non-moving agents, as moving agents do not require epigenetic enhancements and can simply move to more favorable areas, thereby reflecting the biological findings in plants [81].

Similarly to evolvability, consideration of epigenetics is expected to be fruitful when addressing certain types of dynamic RL environments. For instance, in environments involving temporary shifts or changes, short-term epigenetic changes might be preferred over modifying the genotype itself, which could be optimal in the long term. In the framework of continual RL [82], epigenetic markers can be used to dynamically change or fine-tune an RL agent's behavior during its lifetime, and then inherited by its offspring, enabling their immediate adaptation to the ongoing environmental circumstances. More specifically, epigenetic effects could be implemented through binary masks overlaid on the underlying (D)RL controller's genotype (NN weights in DRL). This transforms (D)NN learning into a combinatorial optimization task. The existing research on DNN pruning through binary supermasks [83,84] and various forms of the lottery ticket hypothesis [84,85] might also provide insights here.

3.5   Neutrality

Neutrality, a term stemming from Kimura's neutral theory of molecular evolution [86], refers to the phenomenon where most evolutionary changes are a product of neutral mutations that have no effect on the phenotype or fitness. Galván [87] studies the role of neutrality when evolving DNNs, with possible applications in deep ERL. Neutrality is also discussed in [88], where experiments are performed on pole balancing RL benchmarks, but from the perspective of multi-task optimization. As there is currently no consensus [89] on whether neutrality hinders or expedites the evolutionary search process, this topic presents yet another possible research pathway, especially in relation to different encodings and variational operators used in ERL.

3.6   Niche Construction

Niche construction (NC) [90,91] is the process whereby organisms modify their environments through their own activities. Such activities are related to the concept of the extended phenotype [92], which defines phenotypes as encompassing not only biological mechanisms (e.g., protein synthesis) but also all the other effects that genes have on their environment. The process of NC, in turn, affects the selection pressures for organisms' (and other) species, i.e., alters the evolutionary fitness landscapes. Hence, NC is more than just a byproduct of evolution - it is also an evolutionary driver itself [93]. The investigation of NC is usually undertaken from a multi-agent RL perspective, with applications spanning diverse areas such as common-pool resource appropriation [94] and robotic learning of complex tool use skills [95]. Among pure ERL approaches, Hamon *et al.* [96] continuously evolve RL agents without resetting their environment, leading to complex eco-evolutionary feedback effects. Additionally, Chiba *et al.* [12] perform research at the intersection of evolutionary ecology and DRL, studying NC in a two-dimensional environment in which virtual organisms, represented by NNs, construct structures to avoid predation. It is concluded that such interactions between agents and their environment can substantially contribute to the emergence of open-ended evolutionary processes in real domains, such as the embodied evolution of robots. Lastly, Berseth *et al.* [97] introduce SMiRL - surprise minimizing RL, which shows that complex skills can be acquired by carving out niches that sustain a degree of predictability despite the surrounding entropy. This mirrors analogous findings from biology and cognitive sciences [98]. SMiRL represents a type of anti-NS; while in NS novelty is actively sought, here the open-world environments themselves are treated as sources of high novelty, and the challenge lies in maintaining a steady equilibrium amid environmental chaos. Concretely, the goal is to reduce the entropy of the states visited by the RL agent. In summary, alongside canalization, NC is an emergent phenomenon indicative of the increasing complexity and OE of an evolutionary system.

3.7   Hierarchy/Modularity

Hierarchy can be defined as *"the recursive composition of function and structure into increasingly larger and adapted units"*, called modules [99]. The modules themselves are highly connected clusters of entities, only sparsely connected to entities from other clusters [100]. It has been argued that the cost of neural connections promotes the emergence of both hierarchy [100] and modularity [101], which in turn drive the appearance of canalization. A related topic in (E)RL is that of hierarchical RL [102] where problems are broken down into sub-problems, which are solved separately and then combined by using a composition strategy. To this end, different forms of temporal and state abstractions are utilized. Abramowitz and Nitschke [103] merge Scalable Evolution Strategies (S-ES) with hierarchical RL and come up with a novel method - Scalable Hierarchical Evolution Strategies (SHES). It is used to train a two-level hierarchy of policies - a higher-level controller policy which sets goals, and a lower-level primitive policy which directs the agent to fulfill these goals. During training, the parameters associated with the controller and the primitive undergo coevolution. The experimental results on evolutionary robotics RL environments indicate competitive performance with state-of-the-art methods, with no hyperparameter tuning, and with high behavioral robustness, but at the price of sample (in)efficiency. A related topic in biology is given by genetic architecture and genotype-to-phenotype (G-P) mappings [104]. Wright and Laue [105] compare the properties of biological G-P mappings with those of the logic gate circuits, while focusing on robustness, evolvability, and redundancy. They finally suggest strategies for facilitating the evolution of complexity, concluding that performing evolutionary searches in the vicinity of complex phenotypes should result in complex genotypes as well.

3.8 Phylogenetic Analysis

Phylogenetic analysis [106,107] studies the evolutionary relationships between different species, individuals, or genes, aiming to understand their evolutionary past. In the field of EC, phylogenetics has been used to study speciation dynamics of algorithms [108], investigate links between phenotypic and phylogenetic diversity [109], and characterize evolutionary dynamics in artificial life systems [110]. Most of these analyses are *post-hoc* and used to delve deeper into the inner workings of existing algorithms. An exception to this is research by Lalejini *et al.* [111] in which runtime phylogeny tracking is leveraged to direct the evolutionary search process. More precisely, two methods of phylogeny-informed fitness estimation in lexicase selection are introduced and shown to improve diversity maintenance. The authors accentuate a possible use in ERL - in combination with QD algorithms, where behavioral diversity is required.

A substantial amount of related work also exists in ERL. In their Few-shot quality-diversity optimization (FAERY) approach, Salehi *et al.* [112] utilize paths taken in the parameter space, i.e., study evolution forests comprised of parent-offspring relationships between nodes to develop a few-shot QD algorithm. By doing this, the authors tacitly employ phylogenetic ideas. The experiments performed on navigation and robotic manipulation benchmarks point to the significantly improved sample efficiency of QD optimization when solving novel problems. Rainford and Porter [113] use phylogenetic analysis to study how different variational operators affect the fitness of a population. The adjustment of mutation types in accordance with the findings is then shown to improve the fitness of a genetic code improvement system by 20%. Knapp and Peterson [114] improve upon the seminal NeuroEvolution of Augmenting Topologies (NEAT) algorithm [115] by proposing a new speciation strategy based on cladistics, a phylogenetic analysis method. This is done by changing the way innovation numbers are allocated to new nodes and hence also the implementation of reproductive compatibility. The resulting natural evolution NEAT (NENEAT) algorithm is shown to demand fewer generations to solve the XOR problem and fewer evaluations to solve the double pole cart problem when compared to the vanilla NEAT algorithm. Lastly, in [116], phylogenetic trees are used primarily as a data visualization tool to show the evolution of behaviors during coordination learning in multi-agent systems.

3.9 Plasticity

In addition to phylogenesis, which involves developmental changes over evolutionary time, intelligent agents can be developed through lifetime learning or ontogenesis [117]. Ontogenetic approaches make up a significant portion of research in continual RL. A related term is phenotypic plasticity, the ability of an organism to adapt to its environment during its lifetime [118]. This can be done by altering the traits encoded by its fixed genotype in response to environmental cues. The relationship between plasticity and evolvability is fiercely debated, with several competing theories [35]. At the intersection of ontogenetic and phylogenetic approaches, Abrantes *et al.* [119] propose Evolution via Evolutionary Reward (EvER), which aligns the (ontogenetic) reward function with the (phylogenetic) fitness function and successfully demonstrate its performance on bio-inspired environments. Stanton and Clune [120] proposed curiosity search, a step towards plasticity in ERL. In curiosity search, intra-life behavioral novelty is promoted. Although the agents cannot learn within their lifetime and hence exhibit no genuine plasticity, they rely on an intra-life novelty "compass" that indicates the locations of new areas. The approach achieves excellent results on the notoriously difficult RL benchmark Montezuma's Revenge [121].

Schmidgall [122] demonstrates the use of evolved neuromodulated plasticity on "Crippled Ant", a high-dimensional continuous control RL task. Self-modifying plastic networks are shown to outperform a standard NN architecture. Yaman *et al.* [123] introduce an evolutionary approach to optimize and discover synaptic plasticity rules, enabling autonomous learning in dynamic conditions. The proposed algorithm was tested on agent-based foraging and prey-predator tasks, achieving results comparable to those obtained using hill climbing. Further research in this subfield could include reducing the cost of phenotypic plasticity for an individual [118], ideally through novel few-shot learning approaches, and developing learning systems that implicitly adjust the level of plasticity based on the current conditions [5].

3.10 Homeogenesis

Homeogenesis happens when organisms adapt to environmental changes by adding an "adapter" function to their existing functionality, rather than completely redeveloping the functionality [124], as the latter might be too expensive. The adapter, which "converts" current environmental conditions into former ones, represents a shortcut in the evolutionary fitness landscape. It is closely related to the chemistry conservation principle [125], which states that *"chemical traits or organisms are more conservative than the changing environment and retain information about ancient conditions"*. Along with "classical" adaptation and NC, it is one of the classes of adaptation mechanisms, and a driver of complexity through the accumulation of adapters. Homeogenesis is known to appear in domains such as metabolism and chemical reaction networks [124], but is also expected to appear generally in other evolutionary systems. In modular deep learning [126], various adapter layers (e.g., sequential and parallel bottleneck) are used to adapt pretrained models to novel tasks [127]. In ERL, homeogenesis may be combined with NEAT-like approaches [115,128], for example, by using adapters in the form of additional functions applied to the RL agent's actions.

3.11   Resource Constraints

Paradoxically at first glance, resource bottlenecks and constraints, such as energy or size constraints or various forms of parsimony pressure, are hypothesized to have contributed to the rise of complex and intelligent life. Tang *et al.* [129] limit the attention of RL agents by incorporating self-attention bottlenecks in their perception. This is argued to function similarly to indirect encoding, with results indicating competitive results in game domains with significantly fewer parameters. Similarly, constraints to RL agents have been introduced in the form of NN weight agnosticism [130] and observational dropout [131]. These and similar methods could serve as efficient regularizers guiding the exploration of RL policy space. Furthermore, exploring the relationship between homeogenesis and resource constraints might also be a worthwhile endeavor.

3.12   Other Mechanisms, Principles, and Concepts

A plethora of further biologically inspired mechanisms, principles, and concepts are available to ERL practitioners. Barton and Paixão [7] rely on the **infinitesimal model** [132] from population genetics and propose controlling population sizes and selection intensities accordingly, as well as the mutation rate. Smith *et al.* [133,134] employ **conflicting objectives** to promote phenotypic diversity, as an alternative to NS. This is analogous to the conflicting selective pressures faced by organisms. For example, there might be trade-offs between traits that boost survival chances and those conducive to reproductive success. When dealing with environmental stressors causing pressure in opposite directions, trade-offs are also inevitable [135]. On top of that, there is a certain tension between sensitivity to environmental response (plasticity) and robustness [136]. **Gender-specific** ideas, which have been proven to increase diversity in populations (reduce homogenization) in the context of GAs, might be used to create gendered RL agents, with gender definitions derived from either their behavioral or genotypic properties. This could be combined with the exploration of different mating systems (panmictic and other) and mate selection strategies. Genetic processes such as **gene duplication, horizontal gene transfer, or translocation** might be used to devise new variation operators in ERL. In a similar vein, Vassiliades and Mouret [9] studied the distribution of elite solutions in the genotypic space to design novel variational operators utilizing inter-elites correlations, hence touching upon **comparative genetics**. Utilization of **selective pressure** [137], which represents the accumulation of all the environmental forces that influence the survival and reproduction of individuals, is another possibility. Different selection operators in EC/ERL [138] (e.g. truncation strategies), move strategies in local search algorithms [139], or walks on the fitness landscape, can be constructed depending on the desired level of selective pressure. **Intrinsic motivations** (IMs) [140,141] holds promise for open-ended learning [142], and has been used to build autotelic [143] RL agents, i.e., agents that self-select their own goals and collect data by pursuing those goals. A variant called interactional motivation, espousing an RL paradigm in which the reward depends on the agent's own observation and action, rather than the environment [144], has also been proposed. **Genetic drift**, the change in allele frequencies due to chance events [119], and **evolutionary capacitors** [145, 146] have also been investigated in ERL and related fields. **Indirect encoding**, important for reducing the search space and ensuring scalability in DRL, has extensive applications in ERL, including HyperNEAT [147] and compositional pattern-producing networks, with links to artificial embryogeny [148]. Lastly, **complexity** and **entropy** [105, 149] and their interrelationship have also undergone analysis.

4   Discussion

The development of ERL methods entails numerous aspects, including the selection of fitness function or diversity metrics, and the construction of RL policy regularization techniques and exploration-exploitation strategies. While clearly valuable, many of the current approaches rely on *ad hoc* measures and artificial tricks, e.g., naive reward (fitness) function engineering, use of simple proxies, or different types of manual engineering. For example, in divergent search approaches in the spirit of NS, the k-NN distance is commonly used as the sparsity metric. But ideally, what we expect novel divergent search algorithms to do is bypass the need for the promotion of diversity through an explicit design of the fitness function. Diversity should arise as an emergent phenomenon, similar to how a simple form of developmental canalization appeared in [150] without directly optimizing for it. In the vernacular of evolutionary biology, we would like favorable properties to be indirectly rather than directly selected for, and discovered endogenously rather than chosen exogenously. Despite the striking diversity of species on Earth, natural evolution does not optimize for it directly; it is rather a reflection of the fact that invading empty phenotypic niches through evolutionary expansion [151] enables organisms to coexist with others with less or no local competition [152]. The result of all this is an abundance of promising but still limited approaches that fail to consider or properly integrate some of the core evolutionary principles, pointing to what seems to be a missing puzzle at the heart of current efforts. In the context of OE, the current methods harness its power only to a small degree, incomparable to that of natural evolution, even if simple organisms like *E. coli* are considered [153]. Using the classification system from [154], the majority of current methods primarily exhibit OE at the elementary level of variation, meaning that novelty is achieved only within the explicitly pre-defined state space. OE at higher levels of innovation, involving systems unexpectedly enlarging their state space or introducing new mechanisms during major transitions, has yet to be realized, perhaps by allowing the changes in the genomic and regulatory architecture, as it happens in the biological realm [35]. Importantly, biological evolution does not merely explore the predefined state space of all possible designs but rather expands this space on the go, opening floodgates for more complexity and even introducing novel replicators [155].

Among (E)RL methods designed to address dynamic RL environments [156], the impression is that they seldom leverage the power of OE, despite its potential. Moreover, these methods commonly rely on the meta-RL framework [41], which requires that the distribution of tasks is given together and in advance. In contrast to this, real-world environments involve unknown tasks that appear only sequentially, rendering the former assumption highly questionable. Therefore, combining ERL with the online learning (OL) framework [157], in which adaptations to the evolving environment are done continuously (i.e., without resetting the model and removing past information), may represent a promising way toward attaining true adaptivity. Findings on the learning processes of biological systems indicating they perform well when sequentially presented with novel environments (a key feature of intelligence), also resonate well with the OL framework. Various combinations of meta-RL and OL [158–160] also present topics worthy of investigation in the context of ERL. For instance, online meta-learning can overcome the weaknesses of both meta-RL (the need for the presentation of the set of tasks beforehand) and online RL algorithms (the inability to leverage past experiences due to catastrophic forgetting). Improving performance in non-stationary environments may involve ensuring the diversity of policies, which can be obtained in multiple ways, for example by QD-like approaches [161] combined with multi-armed bandit algorithms or by multi-agent ERL. Connections with related research areas such as artificial life [162], synthetic biology [163], and cultural evolution [164] should be leveraged as well.

The importance of developing realistic RL environments must also be emphasized. Instead of only having naively static environments, malleable environments that allow two-way interactions with RL agents should also be considered, leading to the dynamic fitness function landscapes. In current approaches, the interplay between organisms and their environments is neglected despite compelling evidence suggesting that effects such as NC can override external selection forces and steer the course of evolution in new directions [90]. Fortunately, recent years have seen the introduction of multiple RL environment frameworks, such as MiniHack [165] a sandbox for open-ended RL research consisting of game-like environments with varying levels of complexity, and escape room-based testbeds [166], which are aimed at hierarchical RL.

We, perhaps ambitiously, expect the next generation of ERL approaches to integrate biological mechanisms and principles more naturally, smoothly, and systematically into their structure and inner workings, such that the *desiderata* emerges through careful algorithm design, without resorting to *ad hoc* tricks. Integration and reimagining of ideas from related fields such as multi-armed bandits, complexity sciences, deep learning, and artificial life might also be fruitful in enriching ERL. Many of the results might be transferable to wider EC settings. With exceptions in the field of artificial life, the ultimate goal should not be simply to replicate natural evolution with more fidelity, but rather to naturally incorporate the principles that will lead to ERL approaches with improved performance, whether that means achieving OE, maximizing classical fitness functions, or achieving more sample-efficient or even few-shot learning [112].

Multiple underlying tensions and trade-offs (exploration vs exploitation, mutation vs stasis, risk aversion vs risk-seeking, etc.) must, however, be considered as part of this endeavor, as well as the thorny issue of sample inefficiency.

5 Conclusion

In this paper, we have explored and categorized a wide array of relevant methodologies and approaches that combine ERL with additional biologically-inspired mechanisms and principles. We have offered novel insights into research gaps, strengths, and weaknesses in current methods, as well as some avenues for further exploration. By harnessing the power of biological evolution to the fullest extent, and prioritizing first principles over cheap metaphors, we are convinced that multiple novel ERL approaches and frameworks rooted in the biological perspective await. Interdisciplinary efforts and collaborations covering related fields such as theoretical evolutionary biology [167] and evolutionary systems biology [168] and the cross-pollination of ideas between them and ERL are especially desirable. Hopefully, this will lead to significant progress in this challenging and exciting field, by narrowing the gap between what evolution entails and how it is modeled in the EC realm.

References

1. Aranha, C., et al.: Metaphor-based metaheuristics, a call for action: the elephant in the room. Swarm Intell. **16**(1), 1–6 (2022)
2. Sörensen, K.: Metaheuristics-the metaphor exposed. Int. Trans. Oper. Res. **22**(1), 3–18 (2015)
3. Kutschera, U., Niklas, K.J.: The modern theory of biological evolution: an expanded synthesis. Naturwissenschaften **91**, 255–276 (2004)
4. Barton, N.H.: The "new synthesis". Proc. Nat. Acad. Sci. **119**(30), e2122147119 (2022)
5. Yuen, S., Ezard, T.H.G., Sobey, A.J.: Epigenetic opportunities for evolutionary computation. R. Soc. Open Sci. **10**(5), 221256 (2023)
6. Grudniewski, P.A., Sobey, A.J.: cMLSGA: a co-evolutionary multi-level selection genetic algorithm for multi-objective optimization. arXiv preprint arXiv:2104.11072 (2021)
7. Barton, N., Paixão, T.: Can quantitative and population genetics help us understand evolutionary computation? In: Proceedings of the 15th Annual Conference on Genetic and Evolutionary Computation, pp. 1573–1580 (2013)
8. Pontius, J.U., et al.: Initial sequence and comparative analysis of the cat genome. Genome Res. **17**(11), 1675–1689 (2007)
9. Vassiliades, V., Mouret, J.-B.: Discovering the elite hypervolume by leveraging interspecies correlation. In: Proceedings of the Genetic and Evolutionary Computation Conference, pp. 149–156 (2018)
10. Khadka, S., Tumer, K.: Evolutionary reinforcement learning. arXiv preprint arXiv:1805.07917 (2018)
11. Vie, A., Kleinnijenhuis, A.M., Farmer, D.J.: Qualities, challenges and future of genetic algorithms: a literature review. arXiv preprint arXiv:2011.05277 (2020)

12. Dagdia, Z.C., Avdeyev, P., Bayzid, M.S.: Biological computation and computational biology: survey, challenges, and discussion. Artif. Intell. Rev. **54**, 4169–4235 (2021)
13. Miikkulainen, R., Forrest, S.: A biological perspective on evolutionary computation. Nat. Mach. Intell. **3**(1), 9–15 (2021)
14. Silver, D., et al.: Mastering the game of go without human knowledge. nature **550**(7676), 354–359 (2017)
15. Nguyen, H., La, H.: Review of deep reinforcement learning for robot manipulation. In: 2019 Third IEEE International Conference on Robotic Computing (IRC), pp. 590–595. IEEE (2019)
16. Zhou, S.K., Le, H.N., Luu, K., Nguyen, H.V., Ayache, N.: Deep reinforcement learning in medical imaging: a literature review. Med. Image Anal. **73**, 102193 (2021)
17. Schulman, J., Wolski, F., Dhariwal, P., Radford, A., Klimov, O.: Proximal policy optimization algorithms. arXiv preprint arXiv:1707.06347 (2017)
18. Qian, H., Yang, Yu.: Derivative-free reinforcement learning: a review. Front. Comp. Sci. **15**(6), 156336 (2021)
19. Such, F.P., Madhavan, V., Conti, E., Lehman, J., Stanley, K.O., Clune, J.: Deep neuroevolution: genetic algorithms are a competitive alternative for training deep neural networks for reinforcement learning. arXiv preprint arXiv:1712.06567 (2017)
20. Salimans, T., Ho, J., Chen, X., Sidor, S., Sutskever, I.: Evolution strategies as a scalable alternative to reinforcement learning. arXiv preprint arXiv:1703.03864 (2017)
21. Yang, S., Ong, Y.-S., Jin, Y.; Evolutionary Computation in Dynamic and Uncertain Environments, vol. 51. Springer, Heidelberg (2007). https://doi.org/10.1007/978-3-540-49774-5
22. Sun, H., Zhang, W., Runxiang, Yu., Zhang, Y.: Motion planning for mobile robots-focusing on deep reinforcement learning: a systematic review. IEEE Access **9**, 69061–69081 (2021)
23. Jin, Y., Branke, J.: Evolutionary optimization in uncertain environments - a survey. IEEE Trans. Evol. Comput. **9**(3), 303–317 (2005)
24. Jiang, M., Huang, Z., Qiu, L., Huang, W., Yen, G.G.: Transfer learning-based dynamic multiobjective optimization algorithms. IEEE Trans. Evol. Comput. **22**(4), 501–514 (2017)
25. Stanley, K.O., Lehman, J., Soros, L.: Open-endedness: the last grand challenge you've never heard of (2017)
26. Mora, C., Tittensor, D.P., Adl, S., Simpson, A.G.B., Worm, B.: How many species are there on earth and in the ocean? PLoS Biol. **9**(8), e1001127 (2011)
27. Rasmussen, S., Sibani, P.: Two modes of evolution: optimization and expansion. Artif. Life **25**(1), 9–21 (2019)
28. Packard, N., et al.: An overview of open-ended evolution: editorial introduction to the open-ended evolution ii special issue. Artif. Life **25**(2), 93–103 (2019)
29. Lehman, J., Stanley, K.O.: Novelty search and the problem with objectives. In: Riolo, R., Vladislavleva, E., Moore, J. (eds.) Genetic Programming Theory and Practice IX. Genetic and Evolutionary Computation. Springer, New York (2011). https://doi.org/10.1007/978-1-4614-1770-5_3
30. Pugh, J.K., Soros, L.B., Stanley, K.O.: Quality diversity: a new frontier for evolutionary computation. Front. Robot. AI **3**, 40 (2016)

31. Pugh, J.K., Soros, L.B., Szerlip, P.A., Stanley, K.O.: Confronting the challenge of quality diversity. In: Proceedings of the 2015 Annual Conference on Genetic and Evolutionary Computation, pp. 967–974 (2015)
32. Earle, S., Snider, J., Fontaine, M.C., Nikolaidis, S., Togelius, J.: Illuminating diverse neural cellular automata for level generation. In: Proceedings of the Genetic and Evolutionary Computation Conference, pp. 68–76 (2022)
33. Chand, S., Howard, D.: Path towards multilevel evolution of robots. In: Proceedings of the 2020 Genetic and Evolutionary Computation Conference Companion, pp. 1381–1382 (2020)
34. Stanley, K.O., Lehman, J.: Why Greatness Cannot Be Planned. The Myth of the Objective. Springer, Cham (2015). https://doi.org/10.1007/978-3-319-15524-1
35. Riederer, J.M., Tiso, S., van Eldijk, T.J.B., Weissing, F.J.: Capturing the facets of evolvability in a mechanistic framework. Trends Ecol. Evol. **37**(5), 430–439 (2022)
36. Dawkins, R.: The evolution of evolvability. In: Artificial Life, pp. 201–220. Routledge (2019)
37. Watson, R.A., Szathmáry, E.: How can evolution learn? Trends Ecol. Evol. **31**(2), 147–157 (2016)
38. Lehman, J., Stanley, K.O.: Evolvability is inevitable: increasing evolvability without the pressure to adapt. PLoS ONE **8**(4), e62186 (2013)
39. Mengistu, H., Lehman, J., Clune, J.: Evolvability search: directly selecting for evolvability in order to study and produce it. In: 2016 Proceedings of the Genetic and Evolutionary Computation Conference, pp. 141–148 (2016)
40. Gajewski, A., Clune, J., Stanley, K.O., Lehman, J.: Evolvability ES: scalable and direct optimization of evolvability. In: Proceedings of the Genetic and Evolutionary Computation Conference, pp. 107–115 (2019)
41. Finn, C., Abbeel, P., Levine, S.: Model-agnostic meta-learning for fast adaptation of deep networks. In: International Conference on Machine Learning, pp. 1126–1135. PMLR (2017)
42. Katona, A., Franks, D.W., Walker, J.A.: Quality evolvability ES: evolving individuals with a distribution of well performing and diverse offspring. In: The 2022 Conference on Artificial Life, ALIFE 2022. MIT Press (2021)
43. Gašperov, B., Đurasević, M.: On evolvability and behavior landscapes in neuroevolutionary divergent search. arXiv preprint arXiv:2306.09849 (2023)
44. Doncieux, S., Paolo, G., Laflaquière, A., Coninx, A.: Novelty search makes evolvability inevitable. In: Proceedings of the 2020 Genetic and Evolutionary Computation Conference, pp. 85–93 (2020)
45. Shorten, D., Nitschke, G.: How evolvable is novelty search? In: 2014 IEEE International Conference on Evolvable Systems, pp. 125–132. IEEE (2014)
46. Medvet, E., Daolio, F., Tagliapietra, D.: Evolvability in grammatical evolution. In: Proceedings of the Genetic and Evolutionary Computation Conference, pp. 977–984 (2017)
47. Liu, D., Virgolin, M., Alderliesten, T., Bosman, P.A.N.: Evolvability degeneration in multi-objective genetic programming for symbolic regression. In: Proceedings of the Genetic and Evolutionary Computation Conference, pp. 973–981 (2022)
48. Hansen, N., Müller, S.D., Koumoutsakos, P.: Reducing the time complexity of the derandomized evolution strategy with covariance matrix adaptation (CMA-ES). Evol. Computat. **11**(1), 1–18 (2003)
49. Shala, G., Biedenkapp, A., Awad, N., Adriaensen, S., Lindauer, M., Hutter, F.: Learning step-size adaptation in CMA-ES. In: Bäck, T., et al. (eds.) PPSN 2020,

Part I. LNCS, vol. 12269, pp. 691–706. Springer, Cham (2020). https://doi.org/10.1007/978-3-030-58112-1_48

50. Krause, O., Arbonès, D.R., Igel, C.: CMA-ES with optimal covariance update and storage complexity. In: Advances in Neural Information Processing Systems, vol. 29 (2016)

51. Heidrich-Meisner, V., Igel, C.: Uncertainty handling CMA-ES for reinforcement learning. In: Proceedings of the 11th Annual Conference on Genetic and Evolutionary Computation, pp. 1211–1218 (2009)

52. Branke, J., Mattfeld, D.C.: Anticipation and flexibility in dynamic scheduling. Int. J. Prod. Res. **43**(15), 3103–3129 (2005)

53. Pinto, L., Davidson, J., Sukthankar, R., Gupta, A.: Robust adversarial reinforcement learning. In: International Conference on Machine Learning, pp. 2817–2826. PMLR (2017)

54. Masel, J., Trotter, M.V.: Robustness and evolvability. Trends Genet. **26**(9), 406–414 (2010)

55. Wagner, A.: Robustness and evolvability: a paradox resolved. Proc. R. Soc. B Biol. Sci. **275**(1630), 91–100 (2008)

56. Spencer, C.C.A., et al.: The influence of recombination on human genetic diversity. PLoS Genet. **2**(9), e148 (2006)

57. Zainuddin, F.A., Samad, Md.F.A., Tunggal, D.: A review of crossover methods and problem representation of genetic algorithm in recent engineering applications. Int. J. Adv. Sci. Technol. **29**(6s), 759–769 (2020)

58. Paixão, T., Barton, N.: A variance decomposition approach to the analysis of genetic algorithms. In: Proceedings of the 15th Annual Conference on Genetic and Evolutionary Computation, pp. 845–852 (2013)

59. Rochet, S.: Epistasis in genetic algorithms revisited. Inf. Sci. **102**(1–4), 133–155 (1997)

60. Mitchell, M., Holland, J.H., Forrest, S.: The royal road for genetic algorithms: fitness landscapes and GA performance. Technical report, Los Alamos National Lab., NM (United States) (1991)

61. Polani, D., Miikkulainen, R.: Fast reinforcement learning through eugenic neuro-evolution, pp. 99–277. The University of Texas at Austin, AI (1999)

62. Polani, D., Miikkulainen, R.: Eugenic neuro-evolution for reinforcement learning. In: Proceedings of the 2nd Annual Conference on Genetic and Evolutionary Computation, pp. 1041–1046 (2000)

63. Ventresca, M., Ombuki-Berman, B.: Epistasis in multi-objective evolutionary recurrent neuro-controllers. In: 2007 IEEE Symposium on Artificial Life, pp. 77–84. IEEE (2007)

64. Flageat, M., Cully, A.: Uncertain quality-diversity: evaluation methodology and new methods for quality-diversity in uncertain domains. IEEE Trans. Evol. Comput. (2023). https://doi.org/10.1109/TEVC.2023.3273560

65. Huizinga, J., Stanley, K.O., Clune, J.: The emergence of canalization and evolvability in an open-ended, interactive evolutionary system. Artif. Life **24**(3), 157–181 (2018)

66. Katona, A., Lourenço, N., Machado, P., Franks, D.W., Walker, J.A.: Utilizing the untapped potential of indirect encoding for neural networks with meta learning. In: Castillo, P.A., Jiménez Laredo, J.L. (eds.) EvoApplications 2021. LNCS, vol. 12694, pp. 537–551. Springer, Cham (2021). https://doi.org/10.1007/978-3-030-72699-7_34

67. Wang, R., et al.: Enhanced poet: open-ended reinforcement learning through unbounded invention of learning challenges and their solutions. In: International Conference on Machine Learning, pp. 9940–9951. PMLR (2020)
68. Karafotias, G., Hoogendoorn, M., Eiben, Á.E.: Parameter control in evolutionary algorithms: trends and challenges. IEEE Trans. Evol. Comput. **19**(2), 167–187 (2014)
69. Rand, W.: Genetic Algorithms in Dynamic and Coevolving Environments. Ph.D. thesis. Citeseer
70. Bedau, M.A., Packard, N.H.: Evolution of evolvability via adaptation of mutation rates. Biosystems **69**(2–3), 143–162 (2003)
71. Aleti, A.: An adaptive approach to controlling parameters of evolutionary algorithms. Swinburne University of Technology (2012)
72. Xu, K., Ma, Y., Li, W.: Dynamics-aware novelty search with behavior repulsion. In: Proceedings of the Genetic and Evolutionary Computation Conference, pp. 1112–1120 (2022)
73. Weber, M., Schübeler, D.: Genomic patterns of DNA methylation: targets and function of an epigenetic mark. Curr. Opin. Cell Biol. **19**(3), 273–280 (2007)
74. Turner, B.M.: Histone acetylation and an epigenetic code. BioEssays **22**(9), 836–845 (2000)
75. Hu, T.: Evolvability and rate of evolution in evolutionary computation. Ph.D. thesis, Memorial University of Newfoundland (2010)
76. Wang, Y., Liu, H., Sun, Z.: Lamarck rises from his grave: parental environment-induced epigenetic inheritance in model organisms and humans. Biol. Rev. **92**(4), 2084–2111 (2017)
77. Mukhlish, F., Page, J., Bain, M.: Reward-based epigenetic learning algorithm for a decentralised multi-agent system. Int. J. Intell. Unmanned Syst. **8**(3), 201–224 (2020)
78. Mukhlish, F., Page, J., Bain, M.: From reward to histone: combining temporal-difference learning and epigenetic inheritance for swarm's coevolving decision making. In: 2020 Joint IEEE 10th International Conference on Development and Learning and Epigenetic Robotics (ICDL-EpiRob), pp. 1–6. IEEE (2020)
79. Page, J., Armstrong, R., Mukhlish, F.: Simulating search and rescue operations using swarm technology to determine how many searchers are needed to locate missing persons/objects in the shortest time. In: Naweed, A., Bowditch, L., Sprick, C. (eds.) ASC 2019. CCIS, vol. 1067, pp. 106–112. Springer, Singapore (2019). https://doi.org/10.1007/978-981-32-9582-7_8
80. Sousa, J.A.B., Costa, E.: Designing an epigenetic approach in artificial life: the EpiAL model. In: Filipe, J., Fred, A., Sharp, B. (eds.) ICAART 2010. CCIS, vol. 129, pp. 78–90. Springer, Heidelberg (2011). https://doi.org/10.1007/978-3-642-19890-8_6
81. Boyko, A., Kovalchuk, I.: Epigenetic control of plant stress response. Environ. Mol. Mutagen. **49**(1), 61–72 (2008)
82. Khetarpal, K., Riemer, M., Rish, I., Precup, D.: Towards continual reinforcement learning: a review and perspectives. J. Artif. Intell. Res. **75**, 1401–1476 (2022)
83. Zhou, H., Lan, J., Liu, R., Yosinski, J.: Deconstructing lottery tickets: zeros, signs, and the supermask. In: Advances in Neural Information Processing Systems, vol. 32 (2019)
84. Ramanujan, V., Wortsman, M., Kembhavi, A., Farhadi, A., Rastegari, M.: What's hidden in a randomly weighted neural network? In: Proceedings of the IEEE/CVF Conference on Computer Vision and Pattern Recognition, pp. 11893–11902 (2020)

85. Frankle, J., Carbin, M.: The lottery ticket hypothesis: finding sparse, trainable neural networks. arXiv preprint arXiv:1803.03635 (2018)
86. Kimura, M.: The Neutral Theory of Molecular Evolution. Cambridge University Press, Cambridge (1983)
87. Galván, E.: Neuroevolution in deep learning: the role of neutrality. arXiv preprint arXiv:2102.08475 (2021)
88. Dal Piccol Sotto, L.F., Mayer, S., Garcke, J.: The pole balancing problem from the viewpoint of system flexibility. In: Proceedings of the Genetic and Evolutionary Computation Conference Companion, pp. 427–430 (2022)
89. Galván-López, E., Poli, R., Kattan, A., O'Neill, M., Brabazon, A.: Neutrality in evolutionary algorithms... what do we know? Evol. Syst. **2**, 145–163 (2011)
90. Odling-Smee, F.J., Laland, K.N., Feldman, M.W.: Niche Construction: The Neglected Process in Evolution (MPB-37). Princeton University Press (2013)
91. Flynn, E.G., Laland, K.N., Kendal, R.L., Kendal, J.R.: Target article with commentaries: developmental niche construction. Dev. Sci. **16**(2), 296–313 (2013)
92. Dawkins, R.: The Extended Phenotype: The Long Reach of the Gene. Oxford University Press (2016)
93. Millhouse, T., Moses, M., Mitchell, M.: Frontiers in evolutionary computation: a workshop report. arXiv preprint arXiv:2110.10320 (2021)
94. Perolat, J., Leibo, J.Z., Zambaldi, V., Beattie, C., Tuyls, K., Graepel, T.: A multi-agent reinforcement learning model of common-pool resource appropriation. In: Advances in Neural Information Processing Systems, vol. 30 (2017)
95. Baker, B., et al.: Emergent tool use from multi-agent autocurricula. arXiv preprint arXiv:1909.07528 (2019)
96. Hamon, G., Nisioti, E., Moulin-Frier, C.: Eco-evolutionary dynamics of non-episodic neuroevolution in large multi-agent environments. In: Proceedings of the Companion Conference on Genetic and Evolutionary Computation, pp. 143–146 (2023)
97. Berseth, G., et al.: SMiRL: surprise minimizing reinforcement learning in unstable environments. arXiv preprint arXiv:1912.05510 (2019)
98. Friston, K.: The free-energy principle: a rough guide to the brain? Trends Cogn. Sci. **13**(7), 293–301 (2009)
99. Lipson, H., et al.: Principles of modularity, regularity, and hierarchy for scalable systems. J. Biol. Phys. Chem. **7**(4), 125 (2007)
100. Mengistu, H., Huizinga, J., Mouret, J.-B., Clune, J.: The evolutionary origins of hierarchy. PLoS Comput. Biol. **12**(6), e1004829 (2016)
101. Clune, J., Mouret, J.-B., Lipson, H.: The evolutionary origins of modularity. Proc. R. Soc. B Biol. Sci. **280**(1755), 20122863 (2013)
102. Hutsebaut-Buysse, M., Mets, K., Latré, S.: Hierarchical reinforcement learning: a survey and open research challenges. Mach. Learn. Knowl. Extr. **4**(1), 172–221 (2022)
103. Abramowitz, S., Nitschke, G.: Scalable evolutionary hierarchical reinforcement learning. In: Proceedings of the Genetic and Evolutionary Computation Conference Companion, pp. 272–275 (2022)
104. Hansen, T.F.: The evolution of genetic architecture. Annu. Rev. Ecol. Evol. Syst. **37**, 123–157 (2006)
105. Wright, A.H., Laue, C.L.: Evolving complexity is hard. In: Trujillo, L., Winkler, S.M., Silva, S., Banzhaf, W. (eds.) Genetic Programming Theory and Practice XIX. Genetic and Evolutionary Computation. Springer, Singapore (2023). https://doi.org/10.1007/978-981-19-8460-0_10

106. Smith, S.D., Pennell, M.W., Dunn, C.W., Edwards, S.V.: Phylogenetics is the new genetics (for most of biodiversity). Trends Ecol. Evol. **35**(5), 415–425 (2020)
107. Shonkwiler, R.W., Herod, J.: Phylogenetics. In: Mathematical Biology. UTM, pp. 497–537. Springer, New York (2009). https://doi.org/10.1007/978-0-387-70984-0_15
108. Cussat-Blanc, S., Harrington, K., Pollack, J.: Gene regulatory network evolution through augmenting topologies. IEEE Trans. Evol. Comput. **19**(6), 823–837 (2015)
109. Dolson, E., Ofria, C.: Ecological theory provides insights about evolutionary computation. In: Proceedings of the Genetic and Evolutionary Computation Conference Companion, pp. 105–106 (2018)
110. Moreno, M.A., Dolson, E., Rodriguez-Papa, S.: Toward phylogenetic inference of evolutionary dynamics at scale. In: Artificial Life Conference Proceedings 35, vol. 2023, p. 79 (2023)
111. Lalejini, A., Moreno, M.A., Hernandez, J.G., Dolson, E.: Phylogeny-informed fitness estimation. arXiv preprint arXiv:2306.03970 (2023)
112. Salehi, A., Coninx, A., Doncieux, S.: Few-shot quality-diversity optimization. IEEE Robot. Autom. Lett. **7**(2), 4424–4431 (2022)
113. Rainford, P.F., Porter, B.: Using phylogenetic analysis to enhance genetic improvement. In: Proceedings of the Genetic and Evolutionary Computation Conference, pp. 849–857 (2022)
114. Knapp, J.S., Peterson, G.L.: Natural evolution speciation for NEAT. In: 2019 IEEE Congress on Evolutionary Computation (CEC), pp. 1487–1493. IEEE (2019)
115. Stanley, K.O., Miikkulainen, R.: Evolving neural networks through augmenting topologies. Evol. Comput. **10**(2), 99–127 (2002)
116. Dixit, G.: Learning to coordinate in sparse asymmetric multiagent systems (2023)
117. Hannun, A.: The role of evolution in machine intelligence. arXiv preprint arXiv:2106.11151 (2021)
118. Turney, P., Whitley, D., Anderson, R.W.: Evolution, learning, and instinct: 100 years of the Baldwin effect. Evol. Comput. **4**(3), iv–viii (1996)
119. Abrantes, J.P., Abrantes, A.J., Oliehoek, F.A.: Mimicking evolution with reinforcement learning. arXiv preprint arXiv:2004.00048 (2020)
120. Stanton, C., Clune, J.: Curiosity search: producing generalists by encouraging individuals to continually explore and acquire skills throughout their lifetime. PLoS ONE **11**(9), e0162235 (2016)
121. Salimans, T., Chen, R.: Learning Montezuma's revenge from a single demonstration. arXiv preprint arXiv:1812.03381 (2018)
122. Schmidgall, S.: Adaptive reinforcement learning through evolving self-modifying neural networks. In: Proceedings of the 2020 Genetic and Evolutionary Computation Conference Companion, pp. 89–90 (2020)
123. Yaman, A., Iacca, G., Mocanu, D.C., Coler, M., Fletcher, G., Pechenizkiy, M.: Evolving plasticity for autonomous learning under changing environmental conditions. Evol. Comput. **29**(3), 391–414 (2021)
124. Davies, A.: On the interaction of function, constraint and complexity in evolutionary systems. Ph.D. thesis, University of Southampton (2014)
125. Macallum, A.B.: The paleochemistry of the body fluids and tissues. Physiol. Rev. **6**(2), 316–357 (1926)
126. Pfeiffer, J., Ruder, S., Vulić, I., Ponti, E.M.: Modular deep learning. arXiv preprint arXiv:2302.11529 (2023)

127. Stickland, A.C., Murray, I.: BERT and PALs: projected attention layers for efficient adaptation in multi-task learning. In: International Conference on Machine Learning, pp. 5986–5995. PMLR (2019)
128. Sunagawa, J., Yamaguchi, R., Nakaoka, S.: Evolving neural networks through bio-inspired parent selection in dynamic environments. Biosystems **218**, 104686 (2022)
129. Tang, Y., Nguyen, D., Ha, D.: Neuroevolution of self-interpretable agents. In: Proceedings of the 2020 Genetic and Evolutionary Computation Conference, pp. 414–424 (2020)
130. Gaier, A., Ha, D.: Weight agnostic neural networks. In: Advances in Neural Information Processing Systems, vol. 32 (2019)
131. Freeman, D., Ha, D., Metz, L.: Learning to predict without looking ahead: world models without forward prediction. In: Advances in Neural Information Processing Systems, vol. 32 (2019)
132. Fisher, R.A.: XV.-the correlation between relatives on the supposition of mendelian inheritance. Earth Environ. Sci. Trans. R. Soc. Edinburgh **52**(2), 399–433 (1919)
133. Smith, D., Tokarchuk, L., Wiggins, G.: Exploring conflicting objectives with MADNS: multiple assessment directed novelty search. In: Proceedings of the 2016 on Genetic and Evolutionary Computation Conference Companion, pp. 23–24 (2016)
134. Smith, D., Tokarchuk, L., Wiggins, G.: Harnessing phenotypic diversity towards multiple independent objectives. In: Proceedings of the 2016 on Genetic and Evolutionary Computation Conference Companion, pp. 961–968 (2016)
135. Uiterwaal, S.F., Lagerstrom, I.T., Luhring, T.M., Salsbery, M.E., DeLong, J.P.: Trade-offs between morphology and thermal niches mediate adaptation in response to competing selective pressures. Ecol. Evol. **10**(3), 1368–1377 (2020)
136. Walsh, B.: Crops can be strong and sensitive. Nat. Plants **3**(9), 694–695 (2017)
137. Ofria, C., Adami, C., Collier, T.C.: Selective pressures on genomes in molecular evolution. J. Theoret. Biol. **222**(4), 477–483 (2003)
138. Back, T.: Selective pressure in evolutionary algorithms: a characterization of selection mechanisms. In: Proceedings of the First IEEE Conference on Evolutionary Computation. IEEE World Congress on Computational Intelligence, pp. 57–62. IEEE (1994)
139. Tari, S., Basseur, M., Goëffon, A.: An extended neighborhood vision for hill-climbing move strategy design. In: Amodeo, L., Talbi, E.-G., Yalaoui, F. (eds.) Recent Developments in Metaheuristics. ORSIS, vol. 62, pp. 109–124. Springer, Cham (2018). https://doi.org/10.1007/978-3-319-58253-5_7
140. Gottlieb, J., Oudeyer, P.-Y.: Towards a neuroscience of active sampling and curiosity. Nat. Rev. Neurosci. **19**(12), 758–770 (2018)
141. Baldassarre, G.: Intrinsic motivations and open-ended learning. arXiv preprint arXiv:1912.13263 (2019)
142. Santucci, V.G., Oudeyer, P.-Y., Barto, A., Baldassarre, G.: Intrinsically motivated open-ended learning in autonomous robots. Front. Neurorobot. **3**, 115 (2020)
143. Colas, C., Karch, T., Sigaud, O., Oudeyer, P.-Y.: Autotelic agents with intrinsically motivated goal-conditioned reinforcement learning: a short survey. J. Artif. Intell. Res. **74**, 1159–1199 (2022)
144. Georgeon, O.L., Marshall, J.B., Gay, S.: Interactional motivation in artificial systems: between extrinsic and intrinsic motivation. In: 2012 IEEE International Conference on Development and Learning and Epigenetic Robotics (ICDL), pp. 1–2. IEEE (2012)

145. Reinitz, J., Vakulenko, S., Grigoriev, D., Weber, A.: Adaptation, fitness landscape learning and fast evolution. F1000Research **8**, 358 (2019)
146. Kouvaris, K.: How evolution learns to evolve: principles of induction in the evolution of adaptive potential. Ph.D. thesis, University of Southampton (2018)
147. Stanley, K.O., D'Ambrosio, D.B., Gauci, J.: A hypercube-based encoding for evolving large-scale neural networks. Artif. Life **15**(2), 185–212 (2009)
148. Bai, H., Cheng, R., Jin, Y.: Evolutionary reinforcement learning: a survey. Intell. Comput. **2**, 0025 (2023)
149. Gomez, F.J., Togelius, J., Schmidhuber, J.: Measuring and optimizing behavioral complexity for evolutionary reinforcement learning. In: Alippi, C., Polycarpou, M., Panayiotou, C., Ellinas, G. (eds.) ICANN 2009. LNCS, vol. 5769, pp. 765–774. Springer, Heidelberg (2009). https://doi.org/10.1007/978-3-642-04277-5_77
150. Draghi, J., Wagner, G.P.: Evolution of evolvability in a developmental model. Evolution **62**(2), 301–315 (2008)
151. Van Valen, L.: Two modes of evolution. Nature **252**(5481), 298–300 (1974)
152. Lehman, J., Stanley, K.O.: Evolving a diversity of virtual creatures through novelty search and local competition. In: Proceedings of the 13th Annual Conference on Genetic and Evolutionary Computation, pp. 211–218 (2011)
153. Lavin, A., et al.: Simulation intelligence: towards a new generation of scientific methods. arXiv preprint arXiv:2112.03235 (2021)
154. Banzhaf, W., et al.: Defining and simulating open-ended novelty: requirements, guidelines, and challenges. Theor. Biosci. **135**, 131–161 (2016)
155. Dawkins, R.: The Selfish Gene. Oxford University Press (2016)
156. Song, X., Gao, W., Yang, Y., Choromanski, K., Pacchiano, A., Tang, Y.: ES-MAML: simple hessian-free meta learning. arXiv preprint arXiv:1910.01215 (2019)
157. Cesa-Bianchi, N., Lugosi, G.: Prediction, Learning, and Games. Cambridge University Press (2006)
158. Finn, C., Rajeswaran, A., Kakade, S., Levine, S.: Online meta-learning. In: International Conference on Machine Learning, pp. 1920–1930. PMLR (2019)
159. Yao, H., Zhou, Y., Mahdavi, M., Li, Z.J., Socher, R., Xiong, C.: Online structured meta-learning. In: Advances in Neural Information Processing Systems, vol. 33, pp. 6779–6790 (2020)
160. Rajasegaran, J., Finn, C., Levine, S.: Fully online meta-learning without task boundaries. arXiv preprint arXiv:2202.00263 (2022)
161. Cully, A.: Multi-emitter map-elites: improving quality, diversity and data efficiency with heterogeneous sets of emitters. In: Proceedings of the Genetic and Evolutionary Computation Conference, pp. 84–92 (2021)
162. Mercado, R., Munoz-Jimenez, V., Ramos, M., Ramos, F.: Generation of virtual creatures under multidisciplinary biological premises. Artif. Life Robot. **27**(3), 495–505 (2022)
163. Stock, M., Gorochowski, T.: Open-endedness in synthetic biology: a route to continual innovation for biological design. Sci. Adv. **10**, eadi3621 (2023)
164. Borg, J.M., Buskell, A., Kapitany, R., Powers, S.T., Reindl, E., Tennie, C.: Evolved open-endedness in cultural evolution: a new dimension in open-ended evolution research. Arti. Life, 1–22 (2023)
165. Samvelyan, M., et al.: Minihack the planet: a sandbox for open-ended reinforcement learning research. arXiv preprint arXiv:2109.13202 (2021)
166. Menashe, J., Stone, P.: Escape room: a configurable testbed for hierarchical reinforcement learning. arXiv preprint arXiv:1812.09521 (2018)

167. Kaznatcheev, A.: Algorithmic biology of evolution and ecology. Ph.D. thesis, University of Oxford (2020)

168. Beslon, G., Liard, V., Parsons, D.P., Rouzaud-Cornabas, J.: Of evolution, systems and complexity. In: Crombach, A. (ed.) Evolutionary Systems Biology, pp. 1–18. Springer, Cham (2021). https://doi.org/10.1007/978-3-030-71737-7_1

A Hierarchical Dissimilarity Metric for Automated Machine Learning Pipelines, and Visualizing Search Behaviour

Angus Kenny[1]([✉])[iD], Tapabrata Ray[1][iD], Steffen Limmer[2][iD],
Hemant Kumar Singh[1][iD], Tobias Rodemann[2][iD], and Markus Olhofer[2][iD]

[1] University of New South Wales, Canberra, Australia
`angus.kenny@unsw.edu.au`
[2] Honda Research Institute Europe, Offenbach, Germany

Abstract. In this study, the challenge of developing a dissimilarity metric for machine learning pipeline optimization is addressed. Traditional approaches, limited by simplified operator sets and pipeline structures, fail to address the full complexity of this task. Two novel metrics are proposed for measuring structural, and hyperparameter, dissimilarity in the decision space. A hierarchical approach is employed to integrate these metrics, prioritizing structural over hyperparameter differences. The Tree-based Pipeline Optimization Tool (TPOT) is utilized as the primary automated machine learning framework, applied on the *abalone* dataset. Novel visual representations of TPOT's search dynamics are also proposed, providing some deeper insights into its behaviour and evolutionary trajectories, under different search conditions. The effects of altering the population selection mechanism and reducing population size are explored, highlighting the enhanced understanding these methods provide in automated machine learning pipeline optimization.

Keywords: AutoML · TPOT · Visualization · Search characteristics

1 Introduction

Automated machine learning (AutoML) is a rapidly growing field, focusing on automating the application of machine learning to classification and regression tasks [6]. In this domain, machine learning models can function independently or sequentially, with one model's output feeding into the next. Such configurations, called machine learning *pipelines* [9], harness the combined strengths of multiple models to enhance overall performance. Despite growing interest, fitness landscape analysis in AutoML, especially regarding pipeline optimization, remains an area with significant unresolved questions [18]. A primary issue is

---

Supplementary Information The online version contains supplementary material available at https://doi.org/10.1007/978-3-031-56855-8_7.

the development of an effective metric for measuring the dissimilarity between solutions in the decision space. This task is complex, as pipeline hyperparameters can take continuous, discrete or categorical values, are often conditional, and typically have hierarchical relationships.

State-of-the-art investigations in this area typically operate on a severely reduced operator set or rely on restricting pipeline complexity [3,11,13,15–17]. These approaches are very limiting, and do not generalize to arbitrary pipelines. So far, there are no attempts (to the authors' knowledge) to define a dissimilarity metric for arbitrarily complex machine learning pipelines, and addressing this gap in the literature would represent a significant contribution to both the AutoML and broader machine learning communities.

The tree-based pipeline optimization tool (TPOT) [10], a well known Python library for automating machine learning, exemplifies this. It employs genetic programming to optimize machine learning pipelines, searching for the best combination of data preprocessing and modeling steps. Section 2 explains how TPOT produces new pipelines through mutation and crossover, highlighting unique aspects of its methodology. Within TPOT, mutations can alter both pipeline operators and hyperparameters. Consequently, this paper proposes two metrics: one for measuring the dissimilarity between pipeline structures, and another for hyperparameter differences, as detailed in Sect. 3. These metrics are integrated using a hierarchical approach to emphasize structural over hyperparameter changes, reflecting their greater impact on pipeline behaviour. To this end, the concept of pipeline structure is formally defined.

In Sect. 4, novel methods of visually representing the behaviour and evolutionary trajectory of TPOT search are introduced. Experiments conducted on the well-studied *abalone* dataset [4] demonstrate the utility of these visualizations under various conditions, including the effects of changing the population selection mechanism to focus solely on cross-validation error and reducing population size. These experiments underscore the enhanced insights into TPOT's search dynamics offered by the proposed visual representations.

2   Background

2.1   TPOT Pipeline Representation

As its name suggests, the tree-based pipeline optimization tool (TPOT) uses a tree-based representation to manage its pipelines. The pipelines are composed of a combination of transformation operators—each with their own set of hyperparameters. These operators take the form:

```
OpA(input_matrix, OpA__paramA1=True, OpA__paramA2=2.7).
```

Here, the operator is called OpA and has an input and two hyperparameters. The input (input_matrix) is always first, followed by the associated hyperparameters, which use the naming convention operatorName__hyperparameterName.

The set of operators—along their associated hyperparameters and possible hyperparameter values—are maintained in an object called the pset. When the

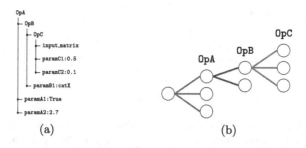

Fig. 1. Nested tree (a) and tree graph (b) representation of pipeline p.

pipeline is evaluated on some set of input data, the data is transformed by each operator in turn, passing the output of one operator to the input of the next, until the root of the tree is reached. During search, TPOT maintains the set of evaluated pipelines using a Python dictionary, which employs a nested bracket string representation of each pipeline as its keys, e.g.,

$$p = \texttt{OpA(OpB(OpC(input\_matrix, OpC\_\_paramC1=0.5, OpC\_\_paramC2=0.1),}$$
$$\texttt{OpB\_\_paramB1=catX), OpA\_\_paramA1=True, OpA\_\_paramA2=2.7).}$$

Figure 1 gives two visual representations of pipeline p: a nested tree representation showing the operators and their respective hyperparameter values; and an abstract tree graph representation, emphasizing the connections between operators. In the tree graph representation, the input nodes are always uppermost.

Although the pipeline representations in Fig. 1 have the appearance of being tree-like, the "branches" of these trees are simply the hyperparameters of each operator. Stripped of these hyperparameters, the pipeline structures shown are essentially linear. In order to allow for more complex structures, TPOT makes use of the CombineDFs operator. This operator puts the "tree" in TPOT, and has the form,

$$\texttt{CombineDFs(input\_matrix, input\_matrix).}$$

Each of the input_matrix nodes may contain an entire subtree, the output matrices of which are merged—by horizontal stacking—to produce a combined input for the next operator in the pipeline. Figure 2(a) illustrates this principle.

2.2 Producing New Pipelines

Throughout its execution, TPOT uses the genetic programming (GP) tools available in the DEAP Python library [2] to search the space of possible pipelines. With each generation, a population of N parent solutions is chosen using a non-dominated sort, with cross-validation error and number of operators as the selection criteria. A set of N offspring solutions is produced from this population, using standard one-point crossover, or one of three mutation operations.

The INSERT operation creates a new tree by slicing the main tree at an arbitrary input node and inserting a new operator between the two halves of the sliced tree.

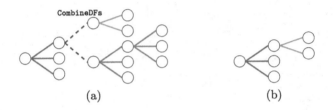

Fig. 2. The `CombineDFs` operator allows more complex pipeline structures to be constructed (a). However, when the SHRINK operation is applied to it (b), the entire subtree in its second input is removed as well.

The SHRINK operation reduces the size of the pipeline by arbitrarily removing one of its operators. It does this by slicing the pipeline tree at an arbitrary (non-leaf) input node, deleting the subtree of the operator that the slice occurred at, and repairing the tree.

Finally, the REPLACE operation can be used to substitute both operator and hyperparameter nodes in the pipeline tree. An arbitrary node in the pipeline is selected and, depending on whether it is an input or hyperparameter node, the `pset` is used to replace the existing operator subtree with a new operator, or the value of the hyperparameter node with another hyperparameter value.

The default probabilities of crossover and mutation are 0.1 and 0.9, respectively. In the case of crossover, two pipelines that share at least one operator are selected, but only the first offspring is retained.

A special case for `CombineDFs`**:** Typically, the mutation operations modify the pipeline by at most 1 operator. INSERT adds an operator, SHRINK removes an operator and REPLACE has no effect on the number of operators in the resulting pipeline. The exception to this is when the SHRINK, or REPLACE, mutation is applied to the `CombineDFs` operator. As the `CombineDFs` operator merges the output of two entire subtrees, there is no simple way to reconcile them, once `CombineDFs` is removed. TPOT addresses this issue by only treating the first input of `CombineDFs` as a "true" input, with the second input being considered a *de facto* hyperparameter. The consequence of this is that, when `CombineDFs` is removed from a pipeline, any subtree in its second input is also removed. Figure 2 illustrates the effect of removing `CombineDFs` on subtrees in both input positions.

Because the subtree in either input of `CombineDFs` can be arbitrarily large, arbitrarily large jumps in pipeline complexity can occur. A similar phenomenon is also observed for the REPLACE mutation operation.

2.3   The Tree Edit Distance Algorithm

In algorithmic graph theory, a common method of quantifying the dissimilarity between two arbitrary, labelled trees is through application of the tree edit distance algorithm [14]. Similar to the string edit distance algorithm, tree edit distance algorithms count the number of transformation operations needed to convert one tree to another. Typically, tree edit distance algorithms consider the following three operations:

- Insertion: Inserting a node into one of the trees.
- Deletion: Deleting a node from one of the trees.
- Substitution: Replacing a node in one tree with a differently labelled one.

Formally, let T_1 and T_2 be two labelled trees with m and n nodes, respectively. A tree edit script is a sequence of edit operations that transforms T_1 into T_2. Each edit operation has an associated cost. The tree edit distance between T_1 and T_2 is then defined as the minimum total cost of any tree edit script that transforms T_1 into T_2. Typically, the cost of inserting or deleting a node is 1 and the cost of substituting two nodes is 1 if their labels are different, and 0 otherwise.

3   A Metric for Pipeline Dissimilarity

3.1   Pipeline Structures

Although pipelines produced by crossover and mutation may be unique to each other, they are not necessarily *structurally* unique. During its operation, TPOT can be thought about as searching a hierarchy of two distinct classes of spaces. The first is the space of all possible combinations of operators, which it explores by using genetic programming to mutate and recombine tree representations of previously evaluated pipelines. The second is the subspace of all possible hyperparameter combinations for each unique combination of operators, which it explores using a grid-based search (having discretized any continuous parameter spaces). It is not possible to keep improving a pipeline by optimizing its hyperparameters alone; eventually a ceiling will be reached, and the only way to achieve further improvement is to change the combination of operators. This implies that search in the space of operator combinations is more influential than search in the space of hyperparameter combinations, so it is useful to have a method of grouping evaluated pipelines together by their so-called *pipeline structure*.

Let p, and q be two pipelines. Pipelines p and q are said to be unique to each other if they have at least one dissimilar hyperparameter value, or do not share the same configuration of operators. A pipeline structure is a subset in the set of evaluated pipelines, partitioned such that every pair of pipelines within a subset are unique and share the same set of operators, in the same configuration. As a given operator will have the same hyperparameters, regardless of its position in the pipeline, a pipeline structure can be represented by its configuration of operators and inputs alone. For example, let p and q be pipelines represented by the two TPOT nested bracket strings:

$p = $ OpA(OpB(input_matrix, OpB__paramB1=0.1, OpB__paramB2=0.5),

OpA__paramA1=True, OpA__paramA2=0.7);

$q = $ OpA(OpB(input_matrix, OpB__paramB1=0.6, OpB__paramB2=0.5),

OpA__paramA1=False, OpA__paramA2=0.2).

These pipelines are clearly unique to each other, as they do not share the same hyperparameter values. However, both have the same set of operators, organized in the same way, and therefore share the same structure. This is denoted with a bar over the pipeline symbol, and the tree-bracket string representation:

$$\bar{p} = \bar{q} = \{\text{OpA}\{\text{OpB}\{\text{input\_matrix}\}\}\}.$$

3.2 Quantifying Pipeline Dissimilarity

Comparing two machine learning pipelines in the decision space presents numerous challenges. When applied to a pipeline, the mutation operations provided by TPOT can modify either the entire structure or simply change the value of one or more hyperparameters. The effects of structural and hyperparameter mutations are typically asymmetrical, with structural changes having a much greater effect on the behaviour of the resulting pipeline than hyperparameter ones.

This section proposes metrics for quantifying both structural dissimilarity (denoted with $\bar{\delta}$) and hyperparameter dissimilarity (denoted with $\hat{\delta}$). These metrics are combined using a hierarchical approach to produce a general metric for tree-based machine learning pipeline dissimilarity (denoted with Δ).

Quantifying Structural Dissimilarity: With the exception of the CombineDFs operator, the INSERT operation typically adds one operator to a given pipeline, SHRINK removes one operator and REPLACE makes no change to its structure. In the tree edit distance algorithm, inserting a node into the tree increases its size by one, deleting a node from the tree reduces its size by one and substituting a node has no effect on the size of the tree. This indicates that there is a direct analogy between the INSERT, SHRINK and REPLACE mutation operations, and the insertion, deletion and substitution operations of tree edit distance algorithm—implying that tree edit distance is an appropriate metric by which to approximate the dissimilarity between pipeline structures, in the decision space.

Let p and q be the unique pipelines:

$p = $ OpA(OpB(input_matrix, OpB_paramB1=0.8, OpB_paramB2=catX),
          OpA_paramA1=True, OpA_paramA2=2.7);
$q = $ OpC(OpB(OpD(input_matrix, OpD_paramD1=6, OpD_paramD2=0.1),
          OpB_paramB1=0.8, OpB_paramB2=catX), OpC_paramC1=0.3).

The structural representations for p, q are:

$$\bar{p} = \{\text{OpA}\{\text{OpB}\{\text{input\_matrix}\}\}\};$$
$$\bar{q} = \{\text{OpC}\{\text{OpB}\{\text{OpD}\{\text{input\_matrix}\}\}\}\}.$$

The structural dissimilarity metric $\bar{\delta}$ uses the tree edit distance algorithm to count the minimum number of mutations required to transform pipeline structure $\bar{p}$ into $\bar{q}$. In this case $\bar{\delta} = 2$, as the transformation can be performed with a REPLACE mutation operation (OpA → OpC) and an INSERT operation (OpD).

Quantifying Hyperparameter Dissimilarity: The direct analogies that exist between its node operations and the TPOT mutation operations, suggest that tree edit distance is an appropriate metric to approximate the distance between pipeline structures in the decision space. However, this relationship does not extend to approximating the dissimilarity at the hyperparameter level. The reasons for this are illustrated through the following example. Let p and q be the pipelines:

$$p = \texttt{OpA(OpB(input\_matrix, OpB\_\_paramB1=0.8, OpB\_\_paramB2=catX),}$$
$$\texttt{OpA\_\_paramA1=True, OpA\_\_paramA2=2.7);}$$
$$q = \texttt{OpC(OpB(input\_matrix, OpB\_\_paramB1=0.8, OpB\_\_paramB2=catX),}$$
$$\texttt{OpC\_\_paramC1=0.3).}$$

Naively applying the tree edit distance metric in this situation suggests that two node substitutions (`OpA` $\rightarrow$ `OpC` and `OpA__paramA1=True` $\rightarrow$ `OpC__paramC1=0.3`) and one node deletion operation (`OpA__paramA2=2.7`) are required to transform p into q; when, in reality, this transformation could be achieved with a single REPLACE mutation operation (`OpA` $\rightarrow$ `OpC`). When `OpA` is replaced by `OpC`, all of the hyperparameters are automatically removed, and the hyperparameters for `OpC` are randomly assigned and inserted. The fact that changes at a structural level can have a large impact on the pipeline at a hyperparameter level, reinforces the notion that there is a hierarchical relationship between the two.

In the case where the structural dissimilarity between two unique pipelines is 0, then they must be distinguished by the differences in their hyperparameter values. Let r, s be the unique pipelines:

$$r = \texttt{OpC(OpD(input\_matrix, OpD\_\_paramD1=3, OpD\_\_paramD2=0.8),}$$
$$\texttt{OpC\_\_paramC1=1.4);}$$
$$s = \texttt{OpC(OpD(input\_matrix, OpD\_\_paramD1=1, OpD\_\_paramD2=0.3),}$$
$$\texttt{OpC\_\_paramC1=0.6).}$$

By inspection, both share the same structure: $\bar{r} = \bar{s} = \{\texttt{OpC}\{\texttt{OpD}\{\texttt{input\_matrix}\}\}\}$. Normalizing the hyperparameter values to the interval $[0, 1]$, these two pipelines can be represented as unique points contained within the unit hypercube, denoted with the notation, $\hat{r}, \hat{s}$. The hyperparameter dissimilarity between r and s is then the magnitude of the displacement vector between points $\hat{r}$ and $\hat{s}$:

$$\hat{\delta}(r, s) = ||\hat{r} - \hat{s}||,$$

where $||x||$ is the Euclidean norm of vector x. For individual hyperparameters r_i, s_i that take Boolean or categorical values, the component-wise difference, $\hat{r}_i - \hat{s}_i$, is set to 0 if the values are the same, and 1 otherwise.

A Hierarchical Dissimilarity Metric: While it is reasonable to directly compare two different pipeline structures, it does not make sense to compare the difference between the values of two different hyperparameters. Therefore, a hierarchical approach must be adopted when combining structural dissimilarity $\bar{\delta}$, and hyperparameter dissimilarity $\hat{\delta}$, to compute the overall dissimilarity between two pipelines, Δ. If any structural dissimilarity exists ($\bar{\delta} > 0$), then the overall dissimilarity is set to the structural dissimilarity ($\Delta = \bar{\delta}$). However, if there are no structural differences ($\bar{\delta} = 0$), then a scaled version of the hyperparameter dissimilarity is used.

Because the structural dissimilarity counts the minimum number of mutation operations required to transform one pipeline into another, $\bar{\delta}$ will always take a positive integer value. In general, the maximum hyperparameter dissimilarity between two pipelines occurs at the extremes of the hypercube, e.g., $\hat{t} = (0,0,\ldots,0)$, $\hat{u} = (1,1,\ldots,1)$. In this case, $\hat{\delta}(t,u) = \sqrt{n}$, where n is the number of hyperparameters under consideration, is the geometric length of the main diagonal of the hypercube, and is always greater than 1 for $n > 1$. This creates potential confusion as to whether a value of $\Delta > 1$ indicates that the change is hyperparameter-based, or structural. To address this ambiguity between hyperparameter and structural changes, a scaling factor can be introduced to adjust the hyperparameter dissimilarity relative to the number of hyperparameters under consideration:

$$\alpha = \left(1 - \frac{1}{n+1}\right).$$

Multiplying the hyperparameter dissimilarity by this term ensures the scaled hyperparameter dissimilarity will always satisfy $0 \leq \alpha \cdot \hat{\delta} < 1$. This means that the hierarchical dissimilarity metric explicitly indicates a structural transformation when $\Delta(p,q) \geq 1$, and explicitly indicates a hyperparameter transformation when $\Delta(p,q) < 1$.

Therefore, given two unique pipelines p,q, the formula for the hierarchical pipeline dissimilarity metric can be expressed as the piecewise function,

$$\Delta(p,q) = \begin{cases} \bar{\delta}(p,q) & \text{if } \bar{\delta}(p,q) > 0 \\ \alpha \cdot \hat{\delta}(p,q) & \text{if } \bar{\delta}(p,q) = 0. \end{cases}$$

4 Visual Representations of TPOT Search

Having established a metric for comparing pipelines in the decision space, it can be used to create a visual representation of the pipelines explored by TPOT throughout its search process, in the decision space. Experiments on the *abalone* [4] dataset are used to demonstrate these ideas.

4.1 Experimental Design

A modified version of TPOT v0.11.7 was employed for the experiments. The core algorithmic functionality of TPOT was retained; however, modifications crucial for enabling deeper insights into the evolutionary process were implemented. These adjustments allowed for the tracking of several key properties throughout the search process, such as the composition of the selected parent population for each generation, and the specific details of each mutation operation.

The default TPOT values for crossover and mutation rates (0.1 and 0.9, respectively) were used, along with the default population size and number of generations, 100. Additionally, parallel experiments with a reduced population size of 10 were conducted, with the aim of improving the readability, and interpretability, of the visual representations. The default time-out per evaluation of 5 min was used for all experiments, evaluated using mean-squared error and five-fold cross-validation error.

Moreover, the impact of altering the selection mechanism was explored. Typically, TPOT employs a multi-objective selection process, based on cross-validation (CV) error and pipeline complexity—i.e., number of operators in the pipeline. In these experiments, a variation was introduced where this was replaced by a single-objective criterion, based purely on CV performance. This modification allowed for the comparison of the evolutionary trajectories of TPOT search, under different selection pressures.

Finally, the search data was collected using the *abalone* dataset [4]. This choice was influenced by limitations in space and the intent to demonstrate techniques rather than perform an empirical evaluation. The *abalone* dataset, a well-studied regression benchmark dataset, comprises 4,177 data points, each with eight features including categorical, integer, and floating-point types. To ensure consistency, all examples in this section were initialized with the same random seed. Yet, the observations were consistent across multiple experiments with different initial seeds. Further examples, including those from experiments with randomly generated and synthetically produced data exhibiting a strong linear relationship, are provided in the supplementary materials, offering additional perspectives and validation of the techniques presented.

4.2 Results and Discussion

Figure 3 provides information about the pipeline structures explored during one such search execution. The top plot in this figure illustrates the distribution and quality of evaluated pipelines, by structure. The x axis gives the index for each unique structure, in the order it was explored during the search, with the number of unique pipelines for each structure on the y axis. Each point in the plot represents a unique pipeline, with colour indicating its evaluated CV error, sorted from worst at the bottom, to best at the top. To improve distinction in the CV ranges of interest, the colour map is cut off at the 75th percentile value. The bottom plot in this figure tracks the composition of the selected parent population, for each generation, over the course of the search. The x axis gives the structure indices, corresponding to the values in the top plot. The colour of each point in this plot represents the number of pipelines from each structure, selected in each generation, given on the y axis. Both plots have a red triangle indicating the structure that produced the pipeline with the lowest CV error.

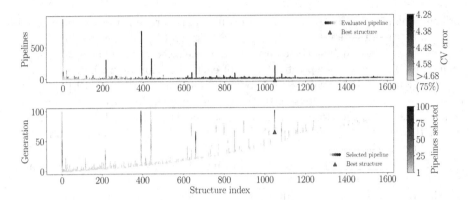

Fig. 3. Pipelines per structure (top) and selected pipeline structures per generation (bottom), for default population selection mechanism.

The frequency plot in Fig. 3 suggests that although TPOT is exploring over 1600 unique structures, its search is focused on 5 or 6 main ones. The most frequently evaluated structure contains 948 unique pipelines, with the average being 6.05 and the median being 1. This is supported by the population tracking plot underneath, where the spikes in the frequency plot correspond to structures which were selected from heavily at some stage in the search.

Performing a non-dominated sort on the evaluated pipelines enables TPOT to maintain control over the complexity of the pipelines. Minimizing both CV error, and number of operators, when selecting the parent population for each generation means that very complex pipelines are only selected when they are also high-performing. This is important, as "bloat" is a well-documented phenomena in many GP-based algorithms [12]; also, highly complex pipelines take longer to evaluate and are prone to over-fitting data [5,7]. However, one significant drawback of using this method to control pipeline complexity growth is exemplified by the largest structure in Fig. 3. Here, nearly 10% of the entire search budget was spent evaluating pipelines with this structure, but the best CV error that was achieved by any pipeline within it was around 4.56. The bracket representation for this pipeline structure is {RandomForestRegressor{input_matrix}}, which only contains a single operator and is therefore unlikely to ever be dominated, even when there are much better performing pipelines available. This can also be observed in the population tracking plot beneath, where this structure is selected from quite heavily for the first 50 or so generations, before fading out—but never totally disappearing—from the selected parent population.

Figure 4 provides the structure frequency and population tracking plots for a TPOT search where the selection pressure to minimize pipeline complexity has been removed, by choosing the parent population based on CV error alone. In these plots it can be seen that more than triple the number of structures were explored (4959 vs. 1630, in the previous example) when using this selection criteria. The budget is spread much more uniformly across the structures as

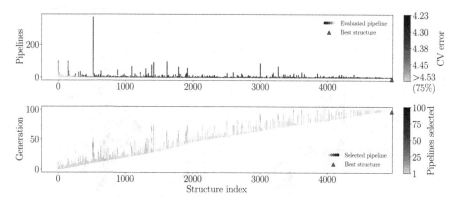

Fig. 4. Pipelines per structure (top) and selected pipeline structures per generation (bottom), for single objective population selection mechanism.

well, with the largest structure having 368 pipelines, and the average number of pipelines per structure being 2.0. The second example provided better quality pipelines over all as well, with the best CV error being 4.23—compared to 4.28 with multi-objective search—and the 75th percentile being 4.53—compared to 4.68. While employing a single objective approach to population selection does seem to yield pipelines with better CV errors, it also produced more complex pipelines, with the most complex pipeline comprising 20 operators—compared to a maximum complexity of 8 operators, in the previous example.

In order to gain further insight, the hierarchical pipeline dissimilarity metric as described in Sect. 3 can be used to create a visual representation of the evolutionary trajectory of the search, called a dissimilarity map. Figure 5 provides dissimilarity maps for both the default and single objective population selection mechanisms (larger versions are available as Figures S1 and S2 in the supplementary materials for this paper). Having partitioned the evaluated pipelines by structure, a pairwise structural dissimilarity ($\bar{\delta}$) matrix M is computed for each explored structure. The multi-dimensional scaling (MDS) algorithm [8] from the Scikit-learn Python package [1] is used to compute a 2D embedding of points, representing structures, such that the distances between the points preserve the values in M, as much as possible. The points are coloured based on the best CV error achieved for each structure, and their size is scaled relative to the number of pipelines each structure contains. The tracking data from the search is used to draw in directed connections between the structures, with the colour indicating how that structure was produced, and operations which transform a structure into an existing structure denoted with dashed lines. The result is a directed graph, which is acyclic when the dashed connections are removed. Inset into the corner is a subplot which provides a 2D embedding that preserves the pairwise hyperparameter dissimilarity ($\hat{\delta}$) for all the pipelines in the largest structure, coloured with respect to their CV errors. This serves as a reminder that each point in the dissimilarity map is representative of a set of pipelines.

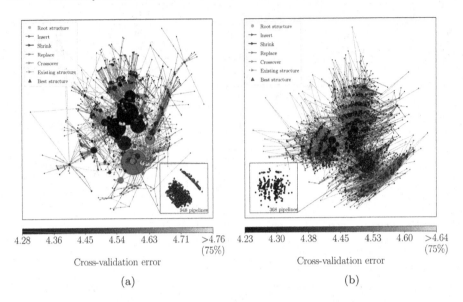

Fig. 5. Pipeline dissimilarity maps for default population selection mechanism (a) and single objective mechanism (b). Size indicates number of pipelines for each structure and colour indicates best CV. Inset illustrates hyperparameter dissimilarity and CV for all pipelines in largest structure.

The structures which comprise the initial population are highlighted with an orange outline, and a red triangle is used to indicate the structure containing the best pipeline over all.

The visual distinction between the pipeline dissimilarity maps in Fig. 5 supports observations made from Figs. 3 and 4. In Fig. 5(a), the default selection mechanism appears to focus on a few large structures in central locations. Conversely, Fig. 5(b) demonstrates that the single objective mechanism yields a greater number of structures, more evenly distributed in size. Notably, the individual transformation chains, defined as the longest, non-cyclic paths in the graph, vary in length, depending on the selection mechanism. Under the default mechanism, the longest transformation chain reaches 10 steps, with an average of 4.83 steps. In the single objective case, these numbers increase significantly, with the longest chain at 19 steps, and the average at 9.97.

These findings are consistent with the calculated pipeline dissimilarity metric. For the default mechanism, the maximum dissimilarity between any two pipelines is 11, while the single objective mechanism records a maximum dissimilarity of 22. However, it is important to recognize the limitations of the structural dissimilarity metric. This metric uses the tree edit distance to measure the shortest possible transformation chain between two pipelines, assuming a single transformation operation alters the pipeline complexity by at most one operator. Such an approach does not account for the pipeline complexity jumps achievable through crossover operations and the `CombineDFs` operator, as discussed in

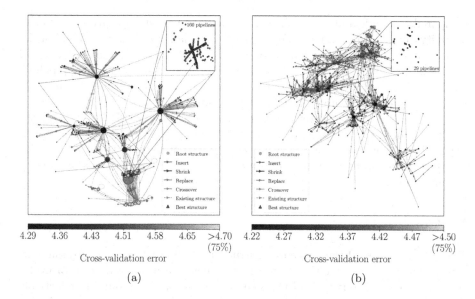

Fig. 6. Pipeline dissimilarity maps for default population selection mechanism (a) and single objective mechanism (b), with reduced population. Size indicates number of pipelines for each structure and colour indicates best CV. Inset illustrates hyperparameter dissimilarity and CV for all pipelines in largest structure.

Sect. 2. Consequently, the metric often overestimates the *actual* shortest transformation path. Since the metric must disregard the varied pipeline outcomes producible by crossover - contingent on the context of the other pipelines in the parent population - or by the `CombineDFs` operator, this overestimation becomes an intrinsic limitation of any metric that compares pipelines in isolation.

While the pipeline dissimilarity maps in Fig. 5 provide an intuitive global view of the evolutionary trajectory of TPOT search, there are so many elements, making it difficult to analyse the behaviour at a local level. To improve readability, Fig. 6 provides the pipeline dissimilarity maps for experiments conducted with a reduced population size of 10 (larger versions are available as Figures S7 and S8 in the supplementary materials). Similar patterns are observed in these experiments as for the full-size one. The default selection mechanism results in larger centrally-located structures, with shorter transformation chains. In the case of this reduced population experiment, the longest transformation chain produced by the default selection mechanism was 5 steps long, and the average was 3.3 steps, whereas the longest was still 19 for the single objective selection mechanism, with the average being 11.37. The dissimilarity metric correlated with this again, with the largest values for default and single objective selection mechanisms being 7 and 22, respectively.

Somewhat apparent in Fig. 5, but made much more clear in the reduced population size experiments, is the observation that a lot more unique structures were produced using crossover when the single objective selection mechanism is used, compared to the default. This is likely because the parent populations had more diversity across generations when using this mechanism, so more opportunities exist to make new combinations of operators; whereas selecting a similar set of (and fewer) structures each generation is more likely to produce new pipelines from existing structures. It can also be observed in both sets of dissimilarity maps that some structures produced by crossover appear to only have a single parent structure. When determining the crossover points, TPOT finds all the operators shared by each parent, and then randomly selects one. In the case where the parent structure contains two instances of the selected operator, this can sometimes produce two unique crossover points—resulting in the production of a new structure.

5   Conclusion

This paper combined the concepts of structural and hyperparameter dissimilarity to produce a hierarchical metric, providing an intuitive reflection of evolutionary changes throughout TPOT search. This metric was found to effectively distinguish between structural transformations and hyperparameter optimizations, providing clearer insights into the decision space navigated by TPOT. The importance of considering pipeline architectures in a holistic manner, as opposed to focusing solely on individual component adjustments, was underlined by these findings.

Through experiments on the *abalone* dataset, it was observed that TPOT's search behavior is predominantly influenced by the exploration of different operator combinations, rather than just hyperparameter tweaking. The utilization of the metric in visual representations for tracing and interpreting the evolution of pipeline configurations was also presented, providing deeper insights into TPOT's search process. These observations were consistent with those of the developed metric, suggesting it to be an appropriate approximation of pipeline dissimilarity—with some limitations, as discussed in Sect. 4.

Building on the findings and methodologies established in this paper, several key areas for future work have been identified. An extensive fitness landscape analysis using the developed hierarchical metric could provide deeper insights into the nature of evolutionary machine learning pipeline optimization. Additionally, the creation of an interactive tool for visualizing TPOT's search process would significantly enhance the usability and interpretability of the findings. Such a tool could allow users to dynamically explore the evolutionary trajectories of machine learning pipelines, offering a more intuitive understanding of the search process and its outcomes.

References

1. Buitinck, L., et al.: API design for machine learning software: experiences from the scikit-learn project. In: ECML PKDD Workshop: Languages for Data Mining and Machine Learning, pp. 108–122 (2013)

2. De Rainville, F.M., Fortin, F.A., Gardner, M.A., Parizeau, M., Gagné, C.: DEAP: a Python framework for evolutionary algorithms. In: Proceedings of the 14th Annual Conference Companion on Genetic and Evolutionary Computation, pp. 85–92 (2012)

3. Garciarena, U., Santana, R., Mendiburu, A.: Analysis of the complexity of the automatic pipeline generation problem. In: 2018 IEEE Congress on Evolutionary Computation (CEC), pp. 1–8. IEEE (2018)

4. Gijsbers, P., et al.: AMLB: an AutoML benchmark. arXiv preprint arXiv:2207.12560 (2022)

5. Hastie, T., Tibshirani, R., Friedman, J.H., Friedman, J.H.: The Elements of Statistical Learning. Data Mining, Inference, and Prediction, vol. 2. Springer, New York (2009). https://doi.org/10.1007/978-0-387-84858-7

6. Hutter, F., Kotthoff, L., Vanschoren, J.: Automated Machine Learning. Methods, Systems, Challenges. Springer, Cham (2019). https://doi.org/10.1007/978-3-030-05318-5

7. Kenny, A., Ray, T., Limmer, S., Singh, H.K., Rodemann, T., Olhofer, M.: Hybridizing TPOT with Bayesian optimization. In: Proceedings of the Genetic and Evolutionary Computation Conference, pp. 502–510 (2023)

8. Kruskal, J.B.: Multidimensional scaling by optimizing goodness of fit to a nonmetric hypothesis. Psychometrika **29**(1), 1–27 (1964)

9. Müller, A.C., Guido, S.: Introduction to Machine Learning with Python: A Guide for Data Scientists. O'Reilly Media, Inc. (2016)

10. Olson, R.S., Bartley, N., Urbanowicz, R.J., Moore, J.H.: Evaluation of a tree-based pipeline optimization tool for automating data science. In: 2016 Proceedings of the Genetic and Evolutionary Computation Conference, pp. 485–492 (2016)

11. Pimenta, C.G., de Sá, A.G.C., Ochoa, G., Pappa, G.L.: Fitness landscape analysis of automated machine learning search spaces. In: Paquete, L., Zarges, C. (eds.) EvoCOP 2020. LNCS, vol. 12102, pp. 114–130. Springer, Cham (2020). https://doi.org/10.1007/978-3-030-43680-3_8

12. Poli, R., Langdon, W.B., McPhee, N.F.: A Field Guide to Genetic Programming. Lulu Enterprises, UK Ltd. (2008)

13. Pushak, Y., Hoos, H.: AutoML loss landscapes. ACM Trans. Evol. Learn. **2**(3), 1–30 (2022)

14. Selkow, S.M.: The tree-to-tree editing problem. Inf. Process. Lett. **6**(6), 184–186 (1977)

15. Teixeira, M.C., Pappa, G.L.: Understanding AutoML search spaces with local optima networks. In: Proceedings of the Genetic and Evolutionary Computation Conference, pp. 449–457 (2022)

16. Teixeira, M.C., Pappa, G.L.: On the effect of solution representation and neighborhood definition in AutoML fitness landscapes. In: Pérez Cáceres, L., Stützle, T. (eds.) Evolutionary Computation in Combinatorial Optimization. EvoCOP 2023. LNCS, vol. 13987, pp. 227–243. Springer, Cham (2023). https://doi.org/10.1007/978-3-031-30035-6_15

17. Teixeira, M.C., Pappa, G.L.: Fitness landscape analysis of TPOT using local optima network. In: Naldi, M.C., Bianchi, R.A.C. (eds.) Intelligent Systems, BRACIS 2023. LNCS, vol. 14197, pp. 65–79. Springer, Cham (2023). https://doi.org/10.1007/978-3-031-45392-2_5

18. Tong, H., Minku, L.L., Menzel, S., Sendhoff, B., Yao, X.: What makes the dynamic capacitated arc routing problem hard to solve: insights from fitness landscape analysis. In: Proceedings of the Genetic and Evolutionary Computation Conference, pp. 305–313 (2022)

DeepEMO: A Multi-indicator Convolutional Neural Network-Based Evolutionary Multi-objective Algorithm

Emilio Bernal-Zubieta$^{(\boxtimes)}$ ⓘ, Jesús Guillermo Falcón-Cardona ⓘ, and Jorge M. Cruz-Duarte ⓘ

Tecnologico de Monterrey, School of Engineering and Sciences, Ave. Eugenio Garza Sada 2501, 64849 Monterrey, NL, México
{a01570751,jfalcon,jorge.cruz}@tec.mx

Abstract. Quality Indicators (QIs) have been used in numerous Evolutionary Multi-objective Optimization Algorithms (EMOAs) as selection mechanisms within the evolutionary process. Because each QI prefers specific point-distribution properties, an Indicator-based EMOA (IB-EMOA) that uses a single QI has an intrinsically limited scope of problems it can solve accurately. To overcome the issues that IB-EMOAs have, we present the first results of a new general multi-indicator-based multi-objective evolutionary algorithm, denoted as DeepEMO. It uses a Convolutional Neural Network (CNN) as a hyper-heuristic to choose, depending on the Pareto-front geometry, the appropriate indicator-based selection mechanism at each generation of the evolutionary process. We employ state-of-the-art benchmark problems with different Pareto front geometries to test our approach. Our experimental results show that DeepEMO obtains competitive performance across multiple QIs. This is because the CNN is employed to classify the geometry of the point cloud that approximates the Pareto front. Hence, DeepEMO compensates for the weaknesses of a single QI with the strengths of others, showing that its performance is invariant to the Pareto front geometry.

Keywords: Quality Indicators · Multi-Objective Optimization · Convolutional Neural Networks · Hyper-heuristics · Higher Education · Educational Innovation

1 Introduction

In many scientific, industrial, and engineering fields, some problems involve simultaneously optimizing m conflicting objective functions. These problems are known as Multi-objective Optimization Problems (MOPs) [8]. Unlike single-objective optimization problems, the solution to a MOP is a set denoted as the *Pareto set* of optimal solutions, and its corresponding image in objective space is the so-called *Pareto front* that shows the trade-off between the conflicting objectives. (In multi-objective optimization, it is expected to use the Pareto

S. Smith et al. (Eds.): EvoApplications 2024, LNCS 14635, pp. 130–146, 2024.
https://doi.org/10.1007/978-3-031-56855-8_8

dominance relation to induce a strict partial order and, thus, define an optimality criterion.) It is worth noting that the Pareto front is a manifold of dimension at most $m - 1$.

In the specialized literature, different techniques exist to solve MOPs, ranging from mathematical programming to bio-inspired metaheuristics [8]. Despite mathematical programming methods ensuring optimal solutions, they require the objectives to be differentiable once (or even twice), which is only possible if the objectives have an analytical definition. Another critical issue is that these techniques often generate a single solution per execution. In consequence, bio-inspired metaheuristics, such as Evolutionary Multi-objective Optimization Algorithms (EMOAs) [12,24,28,32,34], have emerged as promising methods to tackle MOPs. EMOAs are stochastic, population-based, and derivative-free methods that approximate the MOP's solution. Although they cannot ensure the optimality of solutions, EMOAs have been successfully applied to different complex real-world problems where mathematical programming techniques have difficulties.

In this regard, the output of an EMOA stands for a finite set of approximately optimal solutions whose image composes a Pareto front approximation. Such an approximation is a finite representation of the manifold associated with the Pareto front, i.e., an N-point cloud. Ideally, the Pareto front approximation should be as close to the true Pareto front as possible. Hence, these points should also cover the whole Pareto front, showing a good distribution regardless of the Pareto front shape [21]. Nevertheless, in recent years, Ishibuchi et al. emphasized that the performance of some EMOAs depends on the Pareto front shape [15]. Consequently, different approaches have been proposed to tackle this critical issue [12,24,28,32,34].

On the one hand, an effective approach to designing EMOAs with performance invariant to the Pareto front geometry is the use of multiple indicator-based selection mechanisms, giving rise to the Multi-Indicator-based EMOAs (MIB-EMOAs) [12,32]. Quality indicators (QIs) are the core of every MIB-EMOA [21]. A unary QI is a set function that evaluates a Pareto front approximation's quality (convergence, spread, or distribution) based on specific preferences. In other words, a QI assigns a real number to a Pareto front approximation. Hence, it is possible to search for the Pareto front approximation that optimizes a QI. That is, we can define an Indicator-based Subset Selection Problem (IBSSP) that, in terms of EMOAs, involves the selection of the fittest solutions according to the QI value. Thus, those objective vectors that approximate the solution of an IBSSP exhibit the preferences of the baseline QI; i.e., they approach the optimal μ-distribution of the QI. Considering the previous concepts, an MIB-EMOA exploits the strengths of a set of QIs to compensate for the weaknesses of a particular one. For instance, Wang et al. proposed the Two_Arch2 algorithm that uses two archives, each based on a specific QI, to improve the convergence and diversity properties of a Pareto front approximation [32]. Notwithstanding, another design strategy is conceptualized by the Island-based Multi-Indicator Algorithm (IMIA), where the cooperation between multiple Indicator-based EMOAs (IB-

EMOAs) is exploited [12]. In this strategy, each island of IMIA evolves a micropopulation using an IB-EMOA with a different QI. After some generations, some individuals migrate between islands, aiming to improve the diversity of the other islands.

On the other hand, data processing by learning models is at the heart of today's artificial intelligence revolution. Point clouds, like those produced by the EMOA approximation sets, are an essential data type that these models can process. Some applications of point clouds worth mentioning include robotics, indoor navigation, and self-driving vehicles. Plus, their analysis, namely point cloud classification and segmentation, has become relevant in recent years. Though traditional Deep Neural Networks (DNNs) require input data with a regular structure, point clouds have an irregular structure. Thus, it is clear that permutation invariance within the DNN is crucial due to point clouds' lack of topological information. Consequently, designing a DNN that can extract topological features from them is relevant. One can corroborate this claim from several point cloud classifiers proposed to tackle these issues. For instance, PointNet [6] uses the *max-pooling* symmetric function to deal with the unordered input set of points. Later, PointNet++ [26] builds upon PointNet's design and adds a local feature extractor by grouping points into neighborhoods similar to CNNs. Finally, Dynamic Graph CNN (DGCNN) [33] further exploits the CNNs implementation in point clouds by analyzing dynamically computed graphs in each network layer.

Despite EMOAs generating a point cloud in the objective space at each iteration, learning mechanisms do not exploit this information. In addition, we find no research done into using the type of geometry associated with the Pareto front approximation as a mechanism that selects from a pool the best-fitted indicator-based mechanism. So, exploiting geometric information from the point cloud can eliminate the need for sophisticated methods by leveraging the geometric biases inherent to QIs. In this regard, CNNs have yet to be used to classify Pareto front geometries and guide the selection process of an MIB-EMOA. Hence, our proposal is a pioneer work in this area. Geometric classification as a guide for MIB-EMOAs allows for exploiting the properties of individual indicator-based selection mechanisms as a hyper-heuristic. The main contributions of our work are the following.

- We propose the first CNN-based MIB-EMOA, called DeepEMO, that uses DGCNN to classify the geometry associated with the current Pareto front approximation at each generation. Then, DeepEMO chooses the best-fitted one from a pool of indicator-based selection mechanisms to guide the selection process. This is based on predefined rules that consider the effectiveness of indicator-based selection mechanisms on different geometries. For this proof-of-concept, we employed the Hypervolume Indicator (HV) [2], the discrete $R2$ indicator [4], and the Riesz s-energy (E_s) [3].
- We constructed a particular dataset to train DGCNN based on the Pareto fronts from several state-of-the-art benchmark problems. We also selected problems with different Pareto front geometries.

- We present a comprehensive study of the performance of DeepEMO, considering two- and three-objective problems with different Pareto front shapes. Moreover, we validate the performance of DeepEMO by comparing it to IB-EMOAs that use the baseline QIs, *i.e.*, HV, $R2$, and E_s. Based on different QIs, we realize that DeepEMO is a promising direction to combine EMOAs and Deep Learning.

The remainder of this paper is structured as follows. Section 2 provides the concepts that make this paper self-contained. Section 3 details DeepEMO, and Sect. 4 presents and analyzes the experimental results. Finally, Sect. 5 outlines the conclusions and possible improvements for future work.

2   Background

This section introduces some mathematical concepts that sustain our proposed approach. Thus, we start defining a MOP, then the notion of QI, HV, $R2$, and E_s, the generic IB-EMOA, and DGCNN.

2.1   Multi-objective Optimization Problem (MOP)

Throughout this paper, we focus on tackling, without loss of generality, unconstrained MOPs for minimization, which are defined as follows:

$$\min_{\vec{x} \in \Omega} \{f(x) := (f_1(\vec{x}), f_2(\vec{x}), \dots, f_m(\vec{x}))^{\mathsf{T}}\} \tag{1}$$

where $x = (x_1, \dots, x_n)^{\mathsf{T}}$ is an n-dimensional decision vector and $\Omega \subseteq \mathbb{R}^n$ is the decision space. $f : \Omega \mapsto \mathbb{R}^m$ is the objective vector of $m \geq 2$ conflicting objective functions $f_i : \Omega \mapsto \mathbb{R}, \forall i = 1, 2, \dots, m$.

The most common definition of optimality in multi-objective optimization is based on the Pareto dominance relation that induces a strict partial order among the decision vectors. Then, given two solutions $\vec{x}, \vec{y} \in \Omega$, $\vec{x}$ is said to Pareto dominate $\vec{y}$ (denoted as $\vec{x} \prec \vec{y}$) if $f_i(\vec{x}) \leq f_i(\vec{y}), \forall i = 1, 2, \dots, m$, and there exists at least an index $j \in \{1, 2, \dots, m\}$ such that $f_j(\vec{x}) < f_j(\vec{y})$. One can claim that $\vec{x}^* \in \Omega$ is a Pareto optimal solution if there is no other $\vec{x} \in \Omega$ such that $\vec{x} \prec \vec{x}^*$. Due to the conflict among the objectives, there is not a single Pareto optimal solution but a set of Pareto optimal solutions denoted as the Pareto set, whose image is the so-called Pareto front. Since the Pareto set cardinality could be infinite, some algorithms that tackle MOPs produce a finite approximation set $\mathcal{A} = \{\vec{a}_1, \vec{a}_2, \dots, \vec{a}_N\}$, where $\vec{a}_i \in \Omega$. Ideally, $\vec{a}_i \not\prec \vec{a}_j$ and $\vec{a}_j \not\prec \vec{a}_i$ for every $i \neq j$, *i.e.*, $\mathcal{A}$ has mutually non-dominated solutions. The Pareto front approximation is the image $f(\mathcal{A})$.

2.2   Quality Indicator (QI)

A QI ($\mathcal{I}$) is a set function that assigns a real value to a given number k of Pareto front approximations [21]. That is, a k-ary indicator is defined as $\mathcal{I} : \Psi^k \mapsto \mathbb{R}$, where Ψ is the set of all possible finite Pareto front approximations. When $k = 1$, the QI is known as a unary indicator. Currently, many QIs measure the three main properties of a Pareto front approximation, *i.e.*, convergence, uniformity, and spread [21]. In the following lines, we briefly describe three well-known indicators considered in this work.

The Hypervolume Indicator (HV) is the most popular QI due to its mathematical properties [2]. HV measures the region weakly dominated by $\mathcal{A}$ and bounded by an anti-optimal reference point $\vec{r}$. It simultaneously measures convergence and spread and is the only Pareto-compliant QI. Therefore, given an approximation set $\mathcal{A}$ and a reference point $\vec{r} \in \mathbb{R}^m$ dominated by all points in $\mathcal{A}$, HV is defined as:

$$ \mathrm{HV}(\mathcal{A}, \vec{r}) = \mathcal{L} \left(\bigcup_{\vec{a} \in \mathcal{A}} \left\{ \vec{b} \mid \vec{a} \prec \vec{b} \prec \vec{r} \right\} \right), \tag{2} $$

where $\mathcal{L}$ is the Lebesgue measure in $\mathbb{R}^m$. It is worth mentioning that we abuse notation since $\vec{r}$ is in the objective space. However, the Pareto dominance relation (defined above) induces a strict partial order in Ω by checking the objective vectors of the solutions. Thus, we can compare $f(\vec{a})$, $f(\vec{b})$, and $\vec{r}$.

Another well-known QI is the discrete $R2$ indicator [4]. $R2$ is a convergence-uniformity indicator that uses a set of weight vectors (W) in $\mathbb{R}^m$ to measure the average minimum utility value generated by a Pareto front approximation. Unlike HV, whose computational cost is high, the cost of $R2$ is $\mathcal{O}(m|\mathcal{A}||W|)$, but it is weakly Pareto-compliant. So, for a given set of m-dimensional weight vectors W and a utility function $u_{\vec{w}} : \mathbb{R}^m \mapsto \mathbb{R}$, the $R2$ indicator is defined as follows:

$$ R2(\mathcal{A}, W) = \frac{1}{|W|} \sum_{\vec{w} \in W} \min_{\vec{a} \in \mathcal{A}} u_{\vec{w}}(f(\vec{a})). \tag{3} $$

Lastly and more recently, the Riesz s-energy (E_s) has been employed in evolutionary multi-objective optimization to generate well-diversified solution sets [11]. E_s is a pair-potential energy function taken from physics that measures the interaction between pairs of particles in an N-point set. Despite E_s being used mainly for subset selection in EMO, it can also be used as a diversity indicator. Hence, given a Pareto front approximation $\mathcal{A}$ and $s > 0$, E_s is determined by:

$$ E_s(\mathcal{A}) = \sum_{i=1}^{N} \sum_{\substack{j=1 \\ j \neq i}}^{N} \frac{1}{\|f(\vec{a}_i) - f(\vec{a}_j)\|^s}. \tag{4} $$

2.3   Indicator-Based EMOA (IB-EMOA)

This section introduces a generic steady-state IB-EMOA, which is based on the framework of $\mathcal{S}$-Metric Selection EMOA (SMS-EMOA), that employs HV [2]. Regardless of the QI, the backbone of this generic IB-EMOA is the contribution

(C) of a single solution ($\vec{x} \in \mathcal{A}$) to the overall indicator value. This contribution value is calculated as:

$$C_{\mathcal{I}}(\vec{x}, \mathcal{A}) = |\mathcal{I}(\mathcal{A}) - \mathcal{I}(\mathcal{A} \setminus \{\vec{x}\})|. \tag{5}$$

Considering the contribution value, it is possible to define a heuristic method to approximate the solution of an indicator-based subset selection problem. In other words, given a Pareto front approximation of size $\mu + \lambda$, we aim to find $\mathcal{A}'$ such that $|\mathcal{A}'| = \mu$ and $\mathcal{I}(\mathcal{A}')$ is maximum. (Without loss of generality, we assume that maximizing $\mathcal{I}$ implies better quality.)

Algorithm 1 outlines the generic steady-state IB-EMOA whose main loop comprises lines 3 to 14. First, a new solution $\vec{y}$ is generated via variation operators and joined with the current population P_t to define a temporary population Q of size $N + 1$. Then, in line 6, Q is sorted using the non-dominated sorting algorithm [9] to define a set of layers $\{\mathcal{L}_1, \mathcal{L}_2, \ldots, \mathcal{L}_p\}$. It is worth noting that layer $\mathcal{L}_p$ contains a subset of solutions of Q, which are the worst regarding the Pareto dominance relation. If the cardinality of $\mathcal{L}_p$ is greater than 1, then we calculate which is the worst-contributing $\vec{x}_{\text{worst}}$ solution to $\mathcal{I}$ according to (5). Otherwise, $\vec{x}_{\text{worst}}$ is the sole solution in $\mathcal{L}_p$. In line 12, $\vec{x}_{\text{worst}}$ is deleted from Q to determine the population for the next iteration $t + 1$. The algorithm outputs the last population as the approximation set.

Algorithm 1. Generic Steady-State IB-EMOA

Input: Indicator $\mathcal{I}$
Output: Approximation set $\mathcal{A}$
1: Randomly initialize population P_0
2: Set $t \leftarrow 0$
3: **while** Stopping criterion is not fulfilled **do**
4:     $\vec{y} \leftarrow \text{VARIATION}(P_t)$
5:     $Q \leftarrow P_t \cup \{\vec{y}\}$
6:     $\{\mathcal{L}_1, \mathcal{L}_2, \ldots, \mathcal{L}_p\} \leftarrow \text{NONDOMINATEDSORTING}(Q)$
7:     **if** $|\mathcal{L}_p| > 1$ **then**
8:         $\vec{x}_{\text{worst}} \leftarrow \underset{\vec{x} \in \mathcal{L}_p}{\text{argmin}} \; C_{\mathcal{I}}(\vec{x}, \mathcal{L}_p)$
9:     **else**
10:         $\vec{x}_{\text{worst}} \leftarrow$ the sole individual in $\mathcal{L}_p$
11:     **end if**
12:     $P_{t+1} \leftarrow Q \setminus \{\vec{x}_{\text{worst}}\}$
13:     $t \leftarrow t + 1$
14: **end while**
15: **return** $\mathcal{A} \leftarrow P_t$

Algorithm 1 follows the framework of the SMS-EMOA, which is a steady-state IB-EMOA. To reproduce the SMS-EMOA behavior with Algorithm 1, we have to set $\mathcal{I} = \text{HV}$. So, HV is to be maximized; the worst-contributing solution to HV is the one with the minimum contribution value. Depending on the definition of $\vec{r}$, the preferences of SMS-EMOA may change. For instance, if $\vec{r}$ is approximately equal to the nadir point, SMS-EMOA generates uniform Pareto front approximations in linear triangular Pareto fronts, or it can produce solutions in the boundary and around the Pareto front's knee when the geometry is concave triangular. Since SMS-EMOA has to perform multiple calculations of HV (which increases super-polynomially with the number of objectives), it is computationally expensive. Other less computationally expensive but weaker QIs have been used to avoid this issue. For instance, Brockhoff *et al.* proposed $R2$-EMOA that uses the $\mathcal{I} = R2$ indicator [5]. Unlike SMS-EMOA, $R2$-EMOA generates uniform Pareto front approximations in both linear triangular and concave triangular Pareto fronts. However, it has issues when tackling disconnected or degenerate Pareto fronts. Finally, in case that $\mathcal{I} = E_s$, we can generate an IB-EMOA that will show the preferences of E_s, and we denote it as E_s-EMOA.

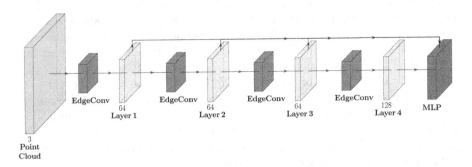

Fig. 1. Dynamic Graph Convolutional Neural Network (DGCNN) architecture.

2.4 Dynamic Graph Convolutional Neural Network (DGCNN)

DGCNN [33] is a point cloud classifier inspired by similar works like Point-Net [6]. Its main feature is its ability to capture local geometric structures while maintaining permutation invariance. This is achieved through an operation called *edge convolution* (EdgeConv). Given a point cloud, EdgeConv constructs a directed graph using the k-Nearest Neighbors (k-NN) algorithm, similar to graph CNNs. According to the authors, DGCNN outperforms other point cloud classifiers because the EdgeConv process is recomputed after each layer of the CNN. Hence, the graph is dynamically updated and not fixed like in traditional graph CNNs. [33]

Due to the DNN architecture employed, the hidden layers work in the feature space created by the previous layer. DGCNN features four hidden layers and the input and output layers, as shown in Fig. 1. The first three hidden layers are made up of 64 neurons, while the last hidden layer is made up of 128 neurons. The input layer of DGCNN consists of a set of N three-dimensional real-valued points. Hence, we could feed DGCNN with $f(\mathcal{A})$, where $\mathcal{A}$ is the approximation set generated by an EMOA for a three-objective MOP. At each layer of DGCNN, EdgeConv constructs a directed graph, extracting local geometric information by connecting neighboring points. The graph's edges are then used to compute *edge features* via a nonlinear function h_Θ with parameters Θ. The edge features are then fed into a *max-pooling* operation with a *ReLU* activation function that captures global shape structure and local neighborhood information. The features outputted by the last EdgeConv layer are then globally aggregated by another *max-pooling* operator, forming a 1D global descriptor used to generate the c classification label in the output layer.

3   Proposed Approach

Our proposal, called DeepEMO, is a steady-state MIB-EMOA that employs a heuristic selection mechanism (based on the classification label produced by DGCNN) to execute the best-fitted indicator-based selection mechanism according to specific rules. The following sections introduce DeepEMO's general framework and how we incorporate DGCNN into an EMOA.

Algorithm 2. DeepEMO General Framework

Input: Certainty threshold β
Output: Approximation set $\mathcal{A}$
 1: Randomly initialize population P_0
 2: Set $t \leftarrow 0$
 3: **while** Stopping criterion is not fulfilled **do**
 4:     $\vec{y} \leftarrow \text{VARIATION}(P_t)$
 5:     $Q \leftarrow P_t \cup \{q\}$
 6:     Normalize Q
 7:     $\{\mathcal{L}_1, \mathcal{L}_2, \ldots, \mathcal{L}_p\} \leftarrow \text{NONDOMINATEDSORTING}(Q)$
 8:     **if** $|\mathcal{L}_p| > 1$ **then**
 9:         geometry, certainty $\leftarrow$ DGCNN($f(Q)$)
10:         **if** geometry is *convex* and certainty $\geq \beta$ **then**
11:             $\vec{x}_{\text{worst}} \leftarrow \underset{\vec{x} \in \mathcal{L}_p}{\text{argmin}}\, C_{\text{HV}}(\mathcal{L}_p, \vec{z}_{\text{ref}})$
12:         **else if** geometry is *concave* and certainty $\geq \beta$ **then**
13:             $\vec{x}_{\text{worst}} \leftarrow \underset{\vec{x} \in \mathcal{L}_p}{\text{argmin}}\, C_{R2}(\mathcal{L}_p, W)$
14:         **else**
15:             $\vec{x}_{\text{worst}} \leftarrow \underset{\vec{x} \in \mathcal{L}_p}{\text{argmax}}\, C_{E_s}(\mathcal{L}_p)$
16:         **end if**
17:     **else**
18:         $\vec{x}_{\text{worst}} \leftarrow$ the sole individual in $\mathcal{L}_p$
19:     **end if**
20:     $P_{t+1} \leftarrow Q \setminus \{\vec{x}_{\text{worst}}\}$
21:     $t \leftarrow t + 1$
22: **end while**
23: **return** $\mathcal{A} = P_t$

3.1   General Framework

The general framework of DeepEMO is presented in Algorithm 2. It follows a similar structure to Algorithm 1. Lines 8 to 17 encompass the core idea of DeepEMO. Our proposed EMOA employs a hyper-heuristic that uses a set of predefined rules to select the best-fitted indicator-based density estimator. The selection rules are based on previous studies on the convergence and diversity properties of indicator-based density estimators [23]. We used HV, $R2$, and E_s for this proof-of-concept to define individual density estimators. According to the literature, we know that an HV-based density estimator has a good performance on MOPs whose Pareto front geometry is convex. This is because HV rewards solutions around the Pareto front's knee and on the boundaries. $R2$ is suitable for triangular concave Pareto front shapes because of the utilization of the simplex-like weight vectors. E_s is an appropriate strategy for other Pareto front geometries [11]. Hence, in line 9 of Algorithm 2, we feed a previously trained DGCNN (described in the next section) with the approximation set Q image. DGCNN returns the classification label and a `certainty` value. We use the degree of certainty in tandem with the geometric classification because the model might not be entirely sure of the Pareto front geometry. In such a case, applying a more general QI (*e.g.*, the Riesz s-energy) would be preferable to other more specialized indicators. If the `geometry` is convex and `certainty` is greater than or equal to a user-supplied threshold (β), then the HV-based density estimator is performed in line 11. In case the `geometry` is concave and `certainty`$\geq \beta$, the $R2$-based density estimator is executed in line 13. Otherwise, the E_s-based density estimator is performed by default in line 15. It is worth noting that we set $\beta = 10\%$ based on previous experiments. A limitation of DeepEMO is that it can only tackle two- and three-objective MOPs. This problem stems from using DGCNN, which can only classify two- and three-dimensional point clouds. This is unsurprising since point clouds usually represent real-world objects; therefore, DGCNN cannot classify point clouds of dimension four or more.

3.2   Using DGCNN in DeepEMO

To use DGCNN in DeepEMO, training the model with data related to Pareto front approximations is mandatory. Hence, we constructed a special dataset (using the format required by DGCNN) that contains m-dimensional points from normalized Pareto front approximations of size 50, varying the related geometries. We obtained the data from thirteen EMOAs, available in PlatEMO [29], with distinct preferences: NSGA-II [9], MOEA/D [36], MOEA/DD [18], MOMBI-II [13], AdaW [22], BiGE [20], SPEA2+SDE [19], RPEA [25], RVEA-iGNG [24], SRA [17], SPEA-R [16], t-DEA [35], and Two_Arch2 [32]. Aiming to maximize the range of geometries, we selected problems from the following test suites: Deb-Thiele-Laumanns-Zitzler (DTLZ) [10], Irregular MOPs (IMOPs) [30], Viennet test suite (VIE) [31], and the Walking-Fish-Group (WFG) [14]. Specifically, we chose the problems DTLZ1, DTLZ2, DTLZ5, DTlZ7, WFG1, WFG2, and WFG3 with two and three objectives, and IMOP1-IMOP8 and VIE1-VIE3 using the given fixed number of objectives. By default, DGCNN can only process three-dimensional point clouds; thus, we added a fictional variable with a zero value to two-objective Pareto front approximations to make them compatible with DGCNN. Finally, the dataset size was then augmented by rotating the Pareto fronts 360° in 10° intervals over the 45° azimuth. After data curation, we obtained a dataset of 75,600 Pareto front approximations. Then, we use a simple validation with 80% of the instances for the training set and the rest for the test set. The model we use in DeepEMO in line 9 of Algorithm 2 is produced using the training set.

4   Experimental Results

We compared DeepEMO with three IB-EMOAs resulting from setting $\mathcal{I} = \mathrm{HV}$, $R2$, or E_s in Algorithm 1. We denote these IB-EMOAs as SMS-EMOA, $R2$-EMOA, and E_s-EMOA. To determine if the DGCNN-based heuristic selection is better than a simple random selection, we conducted a comparative analysis of DeepEMO with a random version, which we denote as Rand-DeepEMO. Since the five algorithms are genetic steady-state EMOAs, we used the simulated binary crossover (SBX) and polynomial-based mutation (PBM). We set the crossover and mutation probabilities equal to 0.9 and $1/n$, where n is the number of decision variables, respectively. Both crossover and mutation distribution indexes are equal to 20. For a fair comparison, we employed a population size of 55 solutions and a stopping criterion of 50,000 function evaluations for all the algorithms. The population size equals the number of weight vectors $R2$-EMOA uses, employing the Simplex-Lattice-Design (SLD) method. To calculate $R2$, we implemented the Achievement Scalarizing Function (ASF). Plus, for E_s-EMOA, we set the parameter s to $m - 1$, and for DGCNN, we established a $g = 5$ parameter to construct the local graph via k-NN. For each algorithm in each instance, we performed 20 independent executions.

4.1   Test Problems

To test DeepEMO and the selected EMOAs, we used DTLZ1, DTLZ2, and DTLZ7 with three objective and their inverted variants, denoted as $\mathrm{DTLZ1}^{-1}$, $\mathrm{DTLZ2}^{-1}$, and $\mathrm{DTLZ7}^{-1}$ [15]. We used the inverted DTLZ problems because they were not employed when training the DGCNN model. We set $n = m + k - 1$ as the number of decision variables for these problems, where $k = 5$, 10, or 20 for DTLZ1, DTLZ2, and DTLZ7, and their corresponding inverted versions, respectively. The IMOP problems were also used in our comparative study because they test the ability of an EMOA to maintain diversified solutions. We employed ten decision variables for these problems, as suggested by the authors [30]. Finally, we also considered VIE1-VIE3 problems, with two-dimensional decision spaces. We must emphasize that all the selected problems have different Pareto front shapes. It is worth mentioning that DGCNN was trained using Pareto front approximations of the selected MOPs to classify the geometry of the point clouds. However, throughout the evolutionary process, DeepEMO feeds DGCNN with points not even close to the Pareto front. Hence, the training process of DGCNN does not provide DeepEMO and advantage over other EMOAs in terms of convergence behavior.

4.2   Performance Assessment

To measure the performance of the selected EMOAs, we used multiple QIs, *i.e.*, HV, $R2$, E_s, Inverted Generational Distance (IGD) [7], IGD^+, Averaged

Hausdorff Distance (Δ_p) [27], additive ϵ indicator (ϵ^+) [21], and the Solow-Polasky Diversity indicator (SPD) [1]. Table 1 specifies the reference point we used for HV. A set of 55 weight vectors produced by SLD was employed to define the same number of utility values based on the vector angle distance scaling function to calculate $R2$. Moreover, we considered $s = m - 1$ for E_s and $\theta = 10$ for SPD. Due to IGD, IGD$^+$, Δ_p, and ϵ^+ requiring a reference point set, we obtained the image of 500 Pareto optimal solutions for each problem from PlatEMO. Plus, we conducted a Wilcoxon rank-sum test with a significance level $\alpha = 0.05$ to get statistical confidence.

Table 2 shows the numerical comparison based on HV. Due to space limitations, Tables 2 to 9 from the Supplementary Material (freely available at https://github.com/eBernalZ/DeepEMO) show the numerical results of $R2$, E_s, IGD, IGD$^+$, Δ_p, ϵ^+, and SPD.

Table 1. Reference points employed for calculating HV per each MOP.

MOP	Reference point	MOP	Reference point
DTLZ1	$(1, 1, 1)$	DTLZ1^{-1}	$(0.1, 0.1, 0.1)$
DTLZ2	$(1.2, 1.2, 1.2)$	DTLZ2^{-1}	$(0.1, 0.1, 0.1)$
DTLZ7	$(1, 1, 21)$	DTLZ7^{-1}	$(0.1, 0.1, 0.1)$
IMOP1 IMOP2 IMOP3	$(2, 2)$	IMOP4 IMOP6 IMOP7	$(2, 2, 2)$
IMOP5	$(1.5, 1.5, 2)$	IMOP8	$(2, 2, 4)$
VIE1	$(5, 6, 5)$	VIE2	$(6, 6, 6)$
VIE3	$(19, 19, 19)$		

4.3 Discussion

An *a posteriori* EMOA should have a *robust* performance when tackling real-world problems. By robust performance, we mean that its performance should be good for different quality measures. This is why multiple QIs are used to evaluate the performance of DeepEMO. Moreover, the core idea of DeepEMO is to compensate for the weaknesses of a given QI with the strengths of others by using the DGCNN-based heuristic selector. Figure 2 depicts the number of times that each algorithm obtained either the first or second place in the comparison for all the selected QIs. This figure reveals that SMS-EMOA and E_s-EMOA often obtain the first position in the comparisons, followed by DeepEMO. Regarding the right-hand side of the figure, we can see that DeepEMO consistently obtains the second place for all QIs. From these observations, we can argue the following. First, the outstanding performance of SMS-EMOA comes with a high computational cost (as expected) and difficulty in setting the reference point to obtain uniform Pareto front approximations. Regarding E_s-EMOA, it produces Pareto front approximations with good diversity, but since E_s is a diversity indicator, E_s-EMOA would lose convergence pressure in MOPs with more than three objectives.

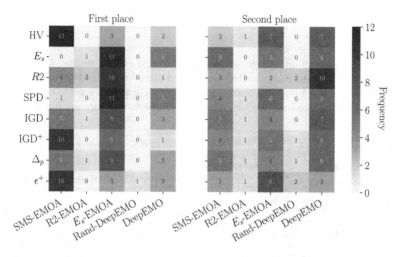

Fig. 2. Heatmap from the number of times an IB-EMOA was ranked first or second according to the HV, E_s, $R2$, SPD, IGD, IGD$^+$, Δ_p, and ϵ^+ indicators.

DeepEMO can be employed to compensate for the difficulties of always using a single QI in an IB-EMOA. By analyzing Table 2 related to the HV comparison, we can see that DeepEMO presents good convergence results. This is because DeepEMO crushes solutions towards the Pareto front by taking advantage of its baseline indicator-based mechanisms depending on the geometry classification of the current Pareto front approximation. Hence, in most cases, DeepEMO is less computationally expensive than SMS-EMOA because the probability of constantly applying the HV-based selection is close to zero. In this regard, due to the switching between selection mechanisms, DeepEMO generates more selection pressure, which makes it possible to scale its performance to MOPs with three or more objectives (once DGCNN scales too). By consistently obtaining the second place in the comparison as shown in Fig. 2, DeepEMO reveals that its Pareto front approximations are not biased to fulfill the preferences of a single QI (as in the case of SMS-EMOA or E_s-EMOA). This behavior is because Deep-EMO generates Pareto front approximations with good diversity as illustrated in Fig. 3 for the three-objective DTLZ1^{-1}. DeepEMO inherits this diversity property due to utilizing E_s, HV, and $R2$. Finally, by comparing DeepEMO and Rand-DeedEMO, we can conclude that using the rule-based heuristic selection in DeepEMO produces better results than randomly selecting indicator-based mechanisms.

Table 2. Mean and standard deviation (in parentheses) of HV results. A symbol # is placed when the outperforming EMOA performed significantly better than the other EMOAs based on a one-tailed Wilcoxon test using a significance level of $\alpha = 0.05$. The two best values are shown in grayscale, where the darkest tone corresponds to the best.

MOP	Dim.	SMS-EMOA	R2-EMOA	E_s-EMOA	Rand-DeepEMO	DeepEMO
DTLZ1	3	9.717413e−01[1] (6.986066e−05)	9.715930e−01[2]# (8.853934e−05)	9.715261e−01[4]# (1.570679e−04)	9.713076e−01[5]# (2.113887e−04)	9.715414e−01[3]# (1.349688e−04)
DTLZ1^{-1}	3	1.634825e+07[4]# (1.660599e+05)	1.398368e+07[5]# (1.232136e+06)	1.897918e+07[2]# (3.385136e+05)	1.786897e+07[3]# (5.839311e+05)	1.909096e+07[1] (2.877119e+05)
DTLZ2	3	7.408208e+00[1] (9.329663e−05)	7.394442e+00[4]# (3.066330e−04)	7.395424e+00[2]# (2.093433e−03)	7.395173e+00[3]# (2.893253e−03)	7.394185e+00[5]# (5.033179e−04)
DTLZ2^{-1}	3	5.750160e+01[1] (1.960825e−02)	4.866969e+01[5]# (9.853446e−01)	5.725490e+01[3]# (8.940140e−02)	5.483055e+01[4]# (7.272273e−01)	5.738631e+01[2]# (7.059462e−02)
DTLZ7	3	1.624788e+01[1] (1.716603e−01)	1.574660e+01[5]# (1.278959e−01)	1.619470e+01[3]# (1.926495e−01)	1.611579e+01[4]# (1.483026e−01)	1.619684e+01[2]# (1.378072e−01)
DTLZ7^{-1}	3	2.698252e+01[2] (1.027667 e−01)	2.679963e+01[5]# (1.101505e−01)	2.698817e+01[1] (3.117033e−03)	2.692513e+01[4]# (4.969799e−02)	2.695391e+01[3]# (6.691855e−02)
IMOP1	2	3.984844e+00[1] (8.313591e−06)	2.017798e+00[5]# (2.297379e−02)	3.983902e+00[3]# (7.384897e−05)	3.747995e+00[4]# (2.823840e−01)	3.984837e+00[2] (2.613447e−05)
IMOP2	2	3.065830e+00[1] (3.212894e−03)	2.000071e+00[4]# (4.775736e−05)	3.064269e+00[2]# (3.320674e−03)	2.094344e+00[3]# (2.732107e−01)	1.999785e+00[5]# (1.329149e−03)
IMOP3	2	3.639962e+00[1] (2.229995e−02)	2.379765e+00[5]# (2.478631e−02)	3.634140e+00[2]# (2.272153e−02)	2.636026e+00[4]# (8.415924e−02)	3.633433e+00[3] (1.747447e−02)
IMOP4	3	6.329276e+00[1] (3.019870e−02)	2.040948e+00[5]# (1.462829e−02)	6.327558e+00[2]# (2.733316e−02)	2.364078e+00[4]# (5.546803e−02)	6.324291e+00[3] (3.217523e−02)
IMOP5	3	6.860433e+00[1] (2.711298e−03)	6.192728e+00[5]# (1.151343e+00)	6.811442e+00[3]# (9.720637e−03)	6.681764e+00[4]# (2.125056e−02)	6.813510e+00[2]# (1.347183e−02)
IMOP6	3	6.796313e+00[3] (3.095751e−01)	3.906444e+00[5]# (4.870139e−01)	6.835016e+00[2]# (6.571779e−03)	3.981992e+00[4]# (6.573286e−01)	6.838866e+00[1] (3.501328e−03)
IMOP7	3	4.688555e+02[2]# (1.373135e+00)	4.001614e+00[5]# (1.939027e−03)	7.356681e+01[1] (4.169080e−03)	4.193473e+00[3]# (7.405203e−01)	4.187927e+00[4]# (7.393973e−01)
IMOP8	3	1.467495e+01[3] (1.703501e+00)	7.379639e+00[5]# (1.644353e+00)	1.490112e+01[1] (3.675314e−02)	1.211802e+01[4] (4.037208e+00)	1.485823e+01[2] (1.074160e−01)
VIE1	3	6.141673e+01[1] (9.511744e−03)	5.842468e+01[4]# (9.156291e−01)	6.072238e+01[2]# (7.808997e−02)	6.015306e+01[3]# (6.614788e−01)	6.141673e+01[1] (9.511744e−03)
VIE2	3	1.309430e+03[1] (3.082634e−03)	1.304450e+03[5]# (4.795875e+00)	1.309306e+03[2]# (2.438912e−02)	1.307448e+03[4]# (1.934688e+00)	1.309305e+03[3]# (2.176253e−02)
VIE3	3	1.420648e+03[1] (1.081860e−02)	1.192055e+03[5]# (1.719101e+02)	1.413235e+03[3]# (2.216500e+00)	1.410666e+03[4]# (3.997438e+00)	1.413311e+03[2]# (1.811991e+00)

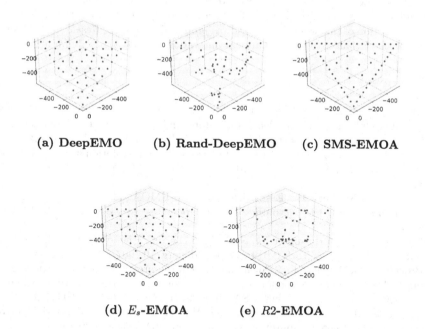

Fig. 3. Graphical performance comparison between **(a)** DeepEMO, **(b)** Rand-DeepEMO, **(c)** SMS-EMOA, **(d)** E_s-EMOA, and **(e)** $R2$-EMOA in DTLZ1^{-1}.

5   Conclusions

This paper proposed DeepEMO, the first Multi-Indicator-based EMOA that uses a CNN to detect the Pareto front geometry and choose the most appropriate indicator-based selection mechanism. Our proposal was compared with SMS-EMOA, $R2$-EMOA, E_s-EMOA, and a random version of DeepEMO. Our experimental results show that DeepEMO consistently obtains evenly distributed approximation sets, regardless of the Pareto front shape, with good convergence regarding multiple state-of-the-art QIs. These results prove that Deep-EMO can compensate for the weaknesses of a single indicator-based selection method with the strengths of others. In other words, DeepEMO can tackle different MOPs without sacrificing convergence and diversity performance across different QIs. A current drawback of DeepEMO is that its CNN can only classify three-dimensional point clouds, making it unable to scale in objective space naturally. For future work, we plan to refine the rule-based hyper-heuristic method of DeepEMO to improve its performance in more MOPs. Furthermore, because of our current limitation to two- and three-objective MOPs, we are interested in expanding the capabilities of DeepEMO to four or more dimensional MOPs, *i.e.*, the so-called Many-objective Optimization Problems (MaOPs). We believe this will allow DeepEMO to outperform the E_s-EMOA, as the Riesz s-energy function loses selection pressure when tackling MaOPs.

Acknowledgments. The authors wish to acknowledge the financial support of the Writing Lab, Institute for the Future of Education, Tecnológico de Monterrey, Mexico, in the production of this work. This work was produced during the Research Internship of Tec Semester thanks to the educational innovation of Tecnológico de Monterrey.

References

1. Basto-Fernandes, V., Yevseyeva, I., Deutz, A., Emmerich, M.: A Survey of Diversity Oriented Optimization: Problems, Indicators, and Algorithms. In: Emmerich, M., Deutz, A., Schütze, O., Legrand, P., Tantar, E., Tantar, A.-A. (eds.) EVOLVE – A Bridge between Probability, Set Oriented Numerics and Evolutionary Computation VII, pp. 3–23. Springer, Cham (2017). https://doi.org/10.1007/978-3-319-49325-1_1

2. Beume, N., Naujoks, B., Emmerich, M.: SMS-EMOA: Multiobjective selection based on dominated hypervolume. Eur. J. Oper. Res. 181(3), 1653–1669 (16 Sept 2007)

3. Borodachov, S.V., Hardin, D.P., Saff, E.B.: Discrete Energy on Rectifiable Sets. SMM, Springer, New York (2019). https://doi.org/10.1007/978-0-387-84808-2

4. Brockhoff, D., Wagner, T., Trautmann, H.: On the properties of the $R2$ indicator. in: 2012 genetic and evolutionary computation conference (GECCO'2012). pp. 465–472. ACM Press, Philadelphia, USA (July 2012), iSBN: 978-1-4503-1177-9

5. Brockhoff, D., Wagner, T., Trautmann, H.: R2 Indicator-based multiobjective search. evolutionary computation Vol. 23(3), pp. 369–395 (Fall 2015)

6. Charles, R., Su, H., Kaichun, M., Guibas, L.J.: Pointnet: deep learning on point sets for 3d classification and segmentation. In: 2017 IEEE Conference on Computer Vision and Pattern Recognition (CVPR), pp. 77–85. IEEE Computer Society, Los Alamitos, CA, USA (July 2017). https://doi.org/10.1109/CVPR.2017.16, https://doi.ieeecomputersociety.org/10.1109/CVPR.2017.16

7. Coello Coello, C.A., Cruz Cortés, N.: Solving multiobjective optimization problems using an artificial immune system. Genet. Program Evolvable Mach. 6(2), 163–190 (2005)

8. Coello Coello, C.A., Lamont, G.B., Van Veldhuizen, D.A.: Evolutionary Algorithms for Solving Multi-Objective Problems. Springer, New York, second edn. (September 2007), iSBN 978-0-387-33254-3

9. Deb, K., Pratap, A., Agarwal, S., Meyarivan, T.: A fast and elitist multiobjective genetic algorithm: NSGA-II. IEEE Trans. Evol. Comput. 6(2), 182–197 (2002). https://doi.org/10.1109/4235.996017

10. Deb, K., Thiele, L., Laumanns, M., Zitzler, E.: Scalable Test Problems for Evolutionary Multiobjective Optimization. In: Abraham, A., Jain, L., Goldberg, R. (eds.) Evolutionary Multiobjective Optimization. Theoretical Advances and Applications, pp. 105–145. Springer, USA (2005)

11. Falcón-Cardona, J.G., Covantes Osuna, E., Coello Coello, C.A., Ishibuchi, H.: On the utilization of pair-potential energy functions in multi-objective optimization. Swarm and Evolutionary Computation 79, 101308 (2023). https://doi.org/10.1016/j.swevo.2023.101308, https://www.sciencedirect.com/science/article/pii/S2210650223000810

12. Falcón-Cardona, J.G., Ishibuchi, H., Coello Coello, C.A., Emmerich, M.: On the Effect of the Cooperation of Indicator-Based Multiobjective Evolutionary Algorithms. IEEE Trans. Evol. Comput. 25(4), 681–695 (2021). https://doi.org/10.1109/TEVC.2021.3061545

13. Hernández Gómez, R., Coello Coello, C.A.: Improved Metaheuristic Based on the $R2$ Indicator for Many-Objective Optimization. In: 2015 Genetic and Evolutionary Computation Conference (GECCO 2015). pp. 679–686. ACM Press, Madrid, Spain (July 11–15 2015), iSBN 978-1-4503-3472-3

14. Huband, S., Hingston, P., Barone, L., While, L.: A Review of Multiobjective Test Problems and a Scalable Test Problem Toolkit. IEEE Trans. Evol. Comput. **10**(5), 477–506 (2006)

15. Ishibuchi, H., Setoguchi, Y., Masuda, H., Nojima, Y.: Performance of Decomposition-Based Many-Objective Algorithms Strongly Depends on Pareto Front Shapes. IEEE Trans. Evol. Comput. **21**(2), 169–190 (2017)

16. Jiang, S., Yang, S.: A strength pareto evolutionary algorithm based on reference direction for multiobjective and many-objective optimization. IEEE Trans. Evol. Comput. **21**(3), 329–346 (2017). https://doi.org/10.1109/TEVC.2016.2592479

17. Li, B., Tang, K., Li, J., Yao, X.: Stochastic ranking algorithm for many-objective optimization based on multiple indicators. IEEE Trans. Evol. Comput. **20**(6), 924–938 (2016). https://doi.org/10.1109/TEVC.2016.2549267

18. Li, K., Deb, K., Zhang, Q., Kwong, S.: An evolutionary many-objective optimization algorithm based on dominance and decomposition. IEEE Trans. Evol. Comput. **19**(5), 694–716 (2015). https://doi.org/10.1109/TEVC.2014.2373386

19. Li, M., Yang, S., Liu, X.: Shift-based density estimation for pareto-based algorithms in many-objective optimization. IEEE Trans. Evol. Comput. **18**(3), 348–365 (2014). https://doi.org/10.1109/TEVC.2013.2262178

20. Li, M., Yang, S., Liu, X.: Bi-goal evolution for many-objective optimization problems. Artificial Intelligence 228, 45–65 (2015). https://doi.org/10.1016/j.artint.2015.06.007, https://www.sciencedirect.com/science/article/pii/S0004370215000995

21. Li, M., Yao, X.: Quality evaluation of solution sets in multiobjective optimisation: A survey. ACM Computing Surveys 52(2), 26:1–26:38 (Mar 2019)

22. Li, M., Yao, X.: What weights work for you? adapting weights for any pareto front shape in decomposition-based evolutionary multiobjective optimisation. Evolutionary Computation **28**(2), 227–253 (Jun 2020). https://doi.org/10.1162/evco_a_00269

23. Liefooghe, A., Derbel, B.: A Correlation Analysis of Set Quality Indicator Values in Multiobjective Optimization. In: 2016 Genetic and Evolutionary Computation Conference (GECCO'2016). pp. 581–588. ACM Press, Denver, Colorado, USA (20–24 July 2016), iSBN 978-1-4503-4206-3

24. Liu, Q., Jin, Y., Heiderich, M., Rodemann, T., Yu, G.: An Adaptive Reference Vector-Guided Evolutionary Algorithm Using Growing Neural Gas for Many-Objective Optimization of Irregular Problems. IEEE Transactions on Cybernetics **52**(5), 2698–2711 (2022). https://doi.org/10.1109/TCYB.2020.3020630

25. Liu, Y., Gong, D., Sun, X., Zhang, Y.: Many-objective evolutionary optimization based on reference points. Applied Soft Computing 50, 344–355 (2017). https://doi.org/10.1016/j.asoc.2016.11.009, https://www.sciencedirect.com/science/article/pii/S1568494616305786

26. Qi, C.R., Yi, L., Su, H., Guibas, L.J.: Pointnet++: Deep hierarchical feature learning on point sets in a metric space. In: Proceedings of the 31st International Conference on Neural Information Processing Systems. p. 5105–5114. NIPS'17, Curran Associates Inc., Red Hook, NY, USA (2017)

27. Schütze, O., Esquivel, X., Lara, A., Coello Coello, C.A.: Using the Averaged Hausdorff Distance as a Performance Measure in Evolutionary Multiobjective Optimization. IEEE Trans. Evol. Comput. **16**(4), 504–522 (2012)

28. Tian, Y., Cheng, R., Zhang, X., Cheng, F., Jin, Y.: An Indicator-Based Multiobjective Evolutionary Algorithm With Reference Point Adaptation for Better Versatility. IEEE Trans. Evol. Comput. **22**(4), 609–622 (2018). https://doi.org/10.1109/TEVC.2017.2749619

29. Tian, Y., Cheng, R., Zhang, X., Jin, Y.: PlatEMO: A MATLAB Platform for Evolutionary Multi-Objective Optimization. IEEE Comput. Intell. Mag. **12**(4), 73–87 (2017)

30. Tian, Y., Cheng, R., Zhang, X., Li, M., Jin, Y.: Diversity Assessment of Multi-Objective Evolutionary Algorithms: Performance Metric and Benchmark Problems. IEEE Comput. Intell. Mag. **14**(3), 61–74 (2019)

31. Veldhuizen, D.A.V.: Multiobjective Evolutionary Algorithms: Classifications, Analyses, and New Innovations. Ph.D. thesis, Department of Electrical and Computer Engineering. Graduate School of Engineering. Air Force Institute of Technology, Wright-Patterson AFB, Ohio, USA (May 1999)

32. Wang, H., Jiao, L., Yao, X.: Two_Arch2: An Improved Two-Archive Algorithm for Many-Objective Optimization. IEEE Trans. Evol. Comput. **19**(4), 524–541 (2015). https://doi.org/10.1109/TEVC.2014.2350987

33. Wang, Y., Sun, Y., Liu, Z., Sarma, S.E., Bronstein, M.M., Solomon, J.M.: Dynamic graph CNN for learning on point clouds. ACM Transactions on Graphics **38**(5), 1–12 (Oct 2019). https://doi.org/10.1145/3326362, https://doi.org/10.1145/3326362

34. Yuan, J., Liu, H.L., Gu, F., Zhang, Q., He, Z.: Investigating the Properties of Indicators and an Evolutionary Many-Objective Algorithm Using Promising Regions. IEEE Trans. Evol. Comput. **25**(1), 75–86 (2021). https://doi.org/10.1109/TEVC.2020.2999100

35. Yuan, Y., Xu, H., Wang, B., Yao, X.: A new dominance relation-based evolutionary algorithm for many-objective optimization. IEEE Trans. Evol. Comput. **20**(1), 16–37 (2016). https://doi.org/10.1109/TEVC.2015.2420112

36. Zhang, Q., Li, H.: MOEA/D: A Multiobjective Evolutionary Algorithm Based on Decomposition. IEEE Trans. Evol. Comput. **11**(6), 712–731 (2007)

A Comparative Analysis of Evolutionary Adversarial One-Pixel Attacks

Luana Clare[(⊠)][iD], Alexandra Marques[iD], and João Correia[iD]

University of Coimbra, CISUC/LASI – Centre for Informatics and Systems of the University of Coimbra, Department of Informatics Engineering, Coimbra, Portugal
{luanasantos,icarregado}@student.dei.uc.pt, jncor@dei.uc.pt

Abstract. Adversarial attacks pose significant challenges to the robustness of machine learning models. This paper explores the one-pixel attacks in image classification, a black-box adversarial attack that introduces changes to the pixels of the input images to make the classifier predict erroneously. We use a pragmatic approach by employing different evolutionary algorithms - Differential Evolution, Genetic Algorithms, and Covariance Matrix Adaptation Evolution Strategy - to find and optimise these one-pixel attacks. We focus on understanding how these algorithms generate effective one-pixel attacks. The experimentation was carried out on the CIFAR-10 dataset, a widespread benchmark in image classification. The experimental results cover an analysis of the following aspects: fitness optimisation, number of evaluations to generate an adversarial attack, success rate, number of adversarial attacks found per image, solution space coverage and level of distortion done to the original image to generate the attack. Overall, the experimentation provided insights into the nuances of the one-pixel attack and compared three standard evolutionary algorithms, showcasing each algorithm's potential and evolutionary computation's ability to find solutions in this strict case of the adversarial attack.

Keywords: Adversarial Attacks · Genetic Algorithms · Covariance Matrix Adaptation Evolution Strategy · Differential Evolution

1 Introduction

The susceptibility of Deep Neural Networks, including Convolutional Neural Networks (CNNs), to adversarial attacks has received significant attention in recent years, leading researchers to investigate several approaches for evaluating and improving their robustness [15]. One particular type of attack is the one-pixel attack [14], a simple yet highly effective technique that manipulates a minimal set of pixels to mislead the neural network's predictions. The attack aims to manipulate a single pixel in the input image, imperceptible to the human eye, in order to mislead the neural network's classification output. The one-pixel attack presents a significant challenge to the resilience of CNNs since it highlights the models' susceptibility to minor alterations in the input of the network. The

S. Smith et al. (Eds.): EvoApplications 2024, LNCS 14635, pp. 147–162, 2024.
https://doi.org/10.1007/978-3-031-56855-8_9

success of such attacks raises concerns about the generalisation and reliability of deep learning models in real-world scenarios, driving researchers to explore and develop countermeasures to enhance the security of these models.

In the work of Su et al. [14], Differential Evolution (DE) was used to perform the one-pixel attack as the optimisation and search algorithm. Unlike traditional gradient-based approaches, evolutionary algorithms like DE excel in the context of the one-pixel attack due to their ability to navigate large and non-linear search spaces effectively. In the context of one-pixel attack perturbations, where the goal is to subtly modify the input image to deceive neural networks, the inherent exploration-exploitation balance of evolutionary algorithms becomes advantageous. E.g. the DE's population-based strategy enables it to explore diverse regions of the solution space simultaneously, facilitating the discovery of imperceptible yet effective perturbations. This characteristic, coupled with the algorithm's ability to escape local optima, makes DE well-suited for the intricate optimisation demands posed by the one-pixel attack, outperforming gradient-based methods in scenarios where the objective function is non-differentiable or that exhibits challenging characteristics for optimisation [2].

Despite DE being a state of the art approach to the problem in terms of efficiency in finding a one-pixel attack, other standard evolutionary algorithms have properties worth analysing in how they behave in such a problem. In this paper, we compare three single objective simple off-the-shelf optimisation algorithms employed to generate one-pixel untargeted attacks: DE, Covariance Matrix Adaptation Evolution Strategy (CMA-ES), and Genetic Algorithms (GA). These optimisation algorithms represent diverse strategies in solving optimisation problems, and understanding their relative performance can provide valuable insights into their suitability as an adversarial attack. DE, CMA-ES, and GAs each bring unique strengths and weaknesses to the table, and a comprehensive comparison can help identify which algorithm is more suitable for navigating the search space of the one-pixel attack. Secondly, this study contributes to the broader field of optimisation under specific conditions, providing a deeper understanding of optimisation techniques' applicability and limitations to this problem.

The contributions are as follows: (i) experimentation with three different evolutionary algorithms; (ii) extensive testing with networks from the state of the art of adversarial learning, normal and distilled networks (neural networks with defensive mechanisms) (iii) comparative analysis between the different algorithms discussing the overall performance of the different approaches according to different adversarial learning and optimisation criteria. The remainder of this paper is divided as follows. In Sect. 2, we cover approaches to the one-pixel adversarial attack. In Sect. 3, we describe the optimisation problem. We present our experimental setup in Sect. 4. The analysis of the results is presented in Sect. 5, and our final conclusions and future work are in Sect. 6.

2   Related Work

An adversarial attack typically consists in the introduction of subtle changes to original inputs, generating an adversarial example that leads a given model to misclassify the altered version. The adversarial attack can be untargeted or targeted. In an untargeted attack, the primary objective is to induce a misclassification, regardless of the specific output class. A targeted attack requires the misclassification to be directed toward a predefined label.

Methods of adversarial examples generation have been widely researched and are typically separated into two scenarios: white-box and black-box attacks. In the white-box scenario, the method has full access to the parameters of the target model and relies on gradient information to generate a perturbation to the original input. Notable state of the art white-box attacks include the Fast Gradient Sign Method [7], GreedyFool [6], and the Carlini and Wagner L2 attack [4]. In contrast, in the black-box attack, the method has limited information on the target model. The method of attack can only query the model and obtain its output, but it has no access to the parameters.

Recent research has shown the efficacy of evolutionary algorithms in navigating black-box scenarios, in both targeted and untargeted attacks. GAs have played a pivotal role in generating adversarial examples. Chen et al. [5] implemented a GA to optimise the perturbation to be added to the original image to generate high-quality adversarial examples with a fitness function based on attack success, perturbation size and a novel perturbation metric described in the paper. Alzantot et al. [1] generated visually imperceptible adversarial examples for targeted attacks, with a GA and a fitness function that aims to increase the confidence in the target class and decrease the confidence in the other classes. Bradley and Blossom [3] also used a GA to generate adversarial examples, focusing on the visual similarity, integration with neural networks, and the optimisation of algorithm parameters, and elaborated a survey for humans to confirm the similarity between the original and perturbed images. Wu et al. [17] explored GAs with multiple fitness functions, changing them in different evolutionary stages to avoid falling into local optima. DE is also an evolutionary algorithm capable of generating adversarial examples. Su et al. [14] proposed a method for generating adversarial examples by perturbing one pixel of the original image, with a search for the pixel being done by a DE. Jere et al. [10] found perturbations in the shape of starches to generate adversarial examples and showed the superiority of DE in relation to CMA-ES. Lin et al. [12] approximated gradients with DE and used it to construct adversarial samples. Additionally, new evolutionary algorithms have been implemented, such as Query-Efficient Evolutionary Attack - QuEry Attack [11], EVOBA [9], and Art-Attack [16].

Unlike the previous approaches, this paper aims to compare the performance of DE, CMA-ES, and GA in the context of generating adversarial examples for an untargeted black-box scenario. Inspired by Su et al. [14], the perturbation is executed by modifying only one pixel in contrast with most state of the art approaches described that generate perturbations that can affect multiple pixels.

3   One-Pixel Attack

The one-pixel attack can be formulated as an optimisation problem with a defined objective function. Consider that an original image is denoted as I with dimensions $W \times H \times C$, where W is the width, H is the height, and C is the number of channels. The perturbation P of dimensions $C \times W \times H$ represents changes to each pixel in the image. The perturbed image I' is obtained by adding the perturbation to the original image:

$$I' = I + P \tag{1}$$

Let f be the targeted model, i.e. the classifier being attacked, and I an original image correctly classified by f. The neural network outputs a confidence vector, which is a probability distribution over the possible classes c. Therefore, true label y is given by:

$$y = \operatorname*{argmax}_{c} f(I)_c \tag{2}$$

A misclassification of the perturbed image happens if:

$$\operatorname*{argmax}_{c} f(I)_c \neq \operatorname*{argmax}_{c} f(I + P)_c \tag{3}$$

The objective function is typically a measure of the misclassification error of a neural network f on the perturbed image I'. Considering $f(I + P)_y$ the confidence of the network in assigning $I + P$ the true label, the optimisation problem can be written as:

$$\min_{P} \ f(I + P)_y \tag{4}$$

subject to constraints on the perturbation vector to ensure its imperceptibility, such as bounding each element of P between the domain of available pixel values. This formulation captures the essence of the one-pixel attack, where the objective is to find one minimal perturbation that misleads the neural network while remaining visually imperceptible.

3.1   Evolutionary One-Pixel Attack

Due to the optimisation characteristics of the problem, in this paper, we assess the contribution of different evolutionary approaches for the one-pixel attack: DE, GA and CMA-ES. The evolutionary algorithm's goal is to find a one-pixel attack for an image. Based on Su et al. [14] the representation is an array containing the coordinates of the pixel and the RGB values: [x, y, r, g, b]. Let I_{ind}, representing an individual, be a new image equal to I, but with RGB values r, g, b in the pixel at x, y instead of the original value in I. With n representing the channels in the image and (x, y) the coordinates of the modified pixel, the perturbation caused by an individual is given by:

$$P_{ind} = \sum_{i=1}^{n} |I_i(x,y) - I_{ind_i}(x,y)| \tag{5}$$

The individual's success in triggering an attack happens when the target model does not classify the image as belonging to the true label. A boolean value to represent success is given by:

$$s(ind) = \underset{c}{\text{argmax}} f(I)_c \neq \underset{c}{\text{argmax}} f(I_{ind})_c \tag{6}$$

The fitness function evaluates the solution that satisfies the above conditions. The new pixel must be a minimal perturbation and simultaneously be a misclassification by the target model. In addition, we want to minimise the perturbation and the model's confidence in the true label y, while also boosting successful individuals. Therefore, we maximise our fitness function F given by:

$$F(ind) = \frac{1}{P_{ind} + 1} + s(ind) + \frac{1}{f(I_{ind})_y + 1} \tag{7}$$

The fitness function is composed of three components. As the perturbation is a non-negative value, (ind) is a boolean and $f(I_{ind})_y$ represents a probability, the ranges of the three components are $(0, 1]$, $[0, 1]$, $[0.5, 1]$. Next, we go into the details of the instantiation of the different evolutionary algorithms.

Genetic Algorithms. We used a standard GA for this problem, where the individuals are integer coordinates values in the interval $[0, 31]$ and integer RGB values in $[0, 255]$. After evaluation, we use one-point crossover and Gaussian mutation as variation operators to generate the offspring. The parent selection is done by a tournament and we have elitism. We acknowledge that other evolutionary methods could be more appropriate, such as arithmetic crossover, but chose to implement an off-the-shelf GA.

Differential Evolution. DE is also an evolutionary algorithm used for complex optimisation problems [13]. It is similar to the GA in its concept of generating offspring from the parents at each generation with variation operators. But differ in the variation operators and in the selection of individuals to make the next population.

The DE is based on the one presented by Su et al. [14]. We initialise our population of individuals with the same bounds as in the GA, but the x,y coordinates are initialised using a uniform distribution $U(0, 31)$ and the RGB values by a normal gaussian distribution $N(\mu = 128, \sigma = 127)$. Then, at every generation, a mutant population of the same size is created. For the generation of each mutant, different individuals from the population are selected - x_{r1}, x_{r1} and x_{r3}, different from each other. The generation is done by the DE formula, with F being a scale:

$$x_{mutant} = x_{r1} + F \cdot (x_{r2} - x_{r3}), r1 \neq r2 \neq r3 \qquad (8)$$

Then, each mutant competes with its corresponding original individual, where the candidate solution with the best fitness survives for the next generation.

Covariance Matrix Adaptation Evolution Strategy. CMA-ES is an evolutionary strategy often used for optimisation tasks, particularly in continuous domains [8]. Similarly to GA and DE, it starts with a population of candidate solutions (individuals) and evolves over generations. It evaluates the individuals with a fitness function. At each generation, a new population is generated from a multivariate normal distribution, with a mean (centroid), a deviation (σ), and a covariance matrix. The centroid and covariance matrix are updated after each iteration. In the update, the μ best individuals (parents) have a higher contribution.

Since CMA-ES works better with continuous domains, we had to make adaptations of the individuals considering that for GA and DE we used integers in intervals [0, 31] for x,y and [0, 255] for RGB values. Therefore, we mapped the integer values to float numbers in the interval [0, 1] for each gene and kept a separate phenotype associated with the individual. The phenotype is equal to the previously described individuals. Another adaptation was done for the initialisation of the first population. For each dimension i, we divided the interval [0, 1] with steps of $\frac{1}{upperbound[i]+1}$ size. Then, the individuals were randomly generated by choosing values from the list of each dimension. The first centroid was set as the mean of the first population.

4   Experimental Setup

Our goal is to generate adversarial attacks by modifying only one pixel of the original images. A successful attack consists of introducing the change to one pixel that leads the model under attack to make a wrong prediction of the original label corresponding to the original image. The evolutionary algorithms are set to maximise and optimise the fitness function in search of the best candidate solution - a pixel with coordinates and RGB values [x,y, red, green, blue] - that can turn an image correctly classified by a target model into an adversarial image for said model. The fitness function also takes into account the magnitude of the perturbation - the difference between the original and the candidate solution. We perform the attack in a set of images. The algorithm under evaluation optimises the pixel for each individual image.

The evolutionary attacks were performed on CIFAR-10 images. The CIFAR-10 dataset is composed of 32 × 32 RGB images of 10 different classes - airplane, automobile, bird, cat, deer, dog, frog, horse, ship and truck. The test set contains 10,000 images, with each class equally represented. There are 16,979,328,000 possible combinations for one-pixel modifications with no guaranteed solution, i.e.

Table 1. Evolutionary algorithms parameter values.

Parameter	DE	GA	CMA-ES
population size	400	400	400 (λ)
generations	100	100	100
mutation scale (F)	0.5	–	–
elite size	–	1	–
mutation per gene rate	–	0.25	–
gaussian mutation standard deviation	–	3	–
crossover rate	–	0.9	–
tournament size	–	2	–
σ	–	–	0.05
μ	–	–	$\frac{\lambda}{2}$

no one-pixel solution exists that can turn the original image into an adversarial one.

The target models defined are the regular and the distilled networks presented in the work of Carlini and Wagner, extensively utilised for adversarial learning benchmark [4]. The regular model is traditionally trained and the distilled model is trained with model distillation defence mechanism [4]. The implementation of the models was taken from the Carlini and Wagner Github[1], as well as the CIFAR-10 images. We obtained an accuracy of 79% and 76% on the test set, respectively. The models were trained using images with the shape $32 \times 32 \times 3$, and the pixel values are in between the interval $[-0.5, 0.5]$. It was necessary to convert the images to the RGB bounds of our individuals $[0, 255]$. For width and height, the bounds are $[0, 31]$. However, to obtain the original performance of the models during evolution, whenever a classification is needed the RGB values are converted back to their original bounds.

As attacking the entire CIFAR-10 test set is computationally expensive, for each target model we randomly selected 500 correctly classified images from the test set, as was previously done in Su et al. [14]. We also guaranteed that the 10 classes were equally represented in the selected subset, i.e. the subset is composed of 50 samples from each class.

The parameter values for DE were taken from the setup of Su et al. [14]. To compare the algorithms, the same budget of evaluations was given to GA and CMA-ES. Table 1 outlines the parameter configurations employed for each algorithm. Specifically, the μ parameter aligns with the value employed by Hansen [8], while the remaining parameters have been set through empirical considerations. In addition to the evolutionary algorithms, we also use a random one for benchmarking. It generates 40,000 solutions that are then evaluated by the fitness function.

---

[1] https://github.com/carlini/nn_robust_attacks.

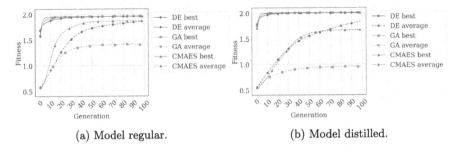

(a) Model regular.   (b) Model distilled.

Fig. 1. Mean best and average fitness for the four algorithms. The results presented are the averages of 10 runs. The images with no adversarial examples were note considered for the statistics presented in these graphics.

Our implementations of the evolutionary algorithms and the setup for the attacks are available in the paper repository[2]. We performed 10 independent runs so that each algorithm attacks an image 10 times with different random generation initialisations. The experiments were done using a PC with a AMD Ryzen 5 5600X 6-Core Processor and Nvidia RTX 3080ti GPU.

5   Experimental Results

In this section, we compare the performance of each evolutionary algorithm. We also added a random search algorithm to serve as a baseline for the problem. It uses the same number of evaluations keeping the best solution encountered during that search. The analysis performed comprises different aspects: (i) **Fitness optimisation**: an analysis of how the fitness function is optimised along the generations; (ii) **Number of Evaluations**: Represents the number of candidate solutions created before discovering an adversarial attack; (iii) **Success Rate**: Denotes the percentage of the 500-image set that were successfully attacked; (iv) **Number of Adversarial attacks found per Image**: Indicates the number of adversarial examples generated for each image; (v) **Covered Pixels**: Represents the number of distinct pixels covered by the algorithm in the search space; (vi) **Distortion**: Measures the dissimilarity between an adversarial image and its original counterpart where, since we are changing one pixel only, it can be computed as $dist = (|I_R - P_R| + |I_G - P_G| + |I_B - P_B|)$.

We conducted tests to ascertain statistical differences between the metric values (except fitness) obtained by the four algorithms across a sample size of 10 runs. For each metric, we evaluated the values by pairs of algorithms. Lilliefors test was used to assess the normality of the distributions. For the cases where the normality assumption holds, we performed paired dependent t-tests, and Wilcoxon tests were employed as an alternative. The significance level was set as $\alpha = 0.01$. The values for each run and other extensive results are presented in the supplementary work.

---

[2] https://github.com/luanaclare/evo_one_pixel_attack.

The fitness progression over generations is presented in Fig. 1 for both the target models, regular and distilled. On average, the CMA-ES algorithm presents itself as the fastest to optimise the problem and converge, reaching its optimal solutions early in the generational process and showing minimal improvement in the second half. In comparison to the other algorithms, its population is less diverse at the fitness level, as shown by the average fitness. On the other side, DE is the slowest to optimise the problem and the average fitness does not reach a plateau. The GA is competitive with the others in terms of the best fitness, but the average fitness is lower. Nevertheless, despite the speed, all algorithms demonstrate the capability to achieve competitive fitness values. As anticipated, the distilled model proves to be more challenging to attack, evident in its lower average fitness values, due to its defensive training.

Table 2. Mean number of evaluations required to find an adversarial example for each of the four algorithms, excluding the images with no success. The results presented are the averages of 10 runs.

(a) Model regular. All means are statistically different.

Algorithm	Mean	STD
DE	662.73	136.15
GA	374.02	99.70
CMA-ES	190.60	33.36
Random	2086.81	218.30

(b) Model distilled. Means are statistically different, with the exception of the pair: GA and CMA-ES.

Algorithm	Mean	STD
DE	444.72	130.14
GA	248.76	82.18
CMA-ES	188.52	27.79
Random	2193.39	375.35

Table 3. Mean number of evaluations required to find an adversarial example for each of the four algorithms, counting images with no success as 40,000 evaluations. The results presented are the averages of 10 runs.

(a) Model regular. All means are statistically different.

Algorithm	Mean	STD
DE	23643.28	386.61
GA	25853.92	258.88
CMA-ES	26321.44	219.53
Random	24630.73	167.75

(b) Model distilled. Means are statistically different, with the exception of the pairs: GA and CMA-ES, DE and Random.

Algorithm	Mean	STD
DE	28623.76	168.01
GA	31055.84	214.35
CMA-ES	30859.28	170.35
Random	28598.44	100.86

In our setup, each algorithm attacking an image has a budget of 40,000 evaluations. This metric aims to compare how fast the algorithm can find an

Table 4. Mean success rate in 500 images for the four algorithms. The results presented are the averages of 10 runs.

(a) Model regular. All means are statistically different.

Algorithm	Mean	STD
DE	42.58%	0.0094
GA	35.70%	0.0070
CMA-ES	34.36%	0.0054
Random	40.54%	0.0054

(b) Model distilled. Means are statistically different, with the exception of the pair: GA and CMA-ES.

Algorithm	Mean	STD
DE	28.76%	0.039
GA	22.50%	0.0052
CMA-ES	22.96%	0.0043
Random	30.16%	0.0036

adversarial example. An adversarial example occurs when the new pixel - coded into the individual - can turn the original image into an adversarial, regardless of the fitness. The first individual capable of successfully causing an attack can be a poor adversarial, with a high perturbation size for example, but it is successful nonetheless. As we have 500 images, we summarise the number of evaluations of a run as the mean number of evaluations between the images. The algorithms were not capable of generating adversarial examples for all images, so we make a distinction between two types of number of evaluations: skipping images with no adversarial examples and counting images with no adversarial examples as 40,000 evaluations. The mean between runs and the standard deviation (std) are presented in Table 2 for the former version, and in Table 3 for the latter.

In the case of excluding images for which no adversarial examples were found, for model regular, The CMA-ES algorithm requires fewer evaluations to find an adversarial example than its evolutionary counterparts. Since it optimises the fitness faster, as shown in Figs. 1a, this behaviour was anticipated. Despite taking more evaluations than CMA-ES, the other evolutionary algorithms are faster at finding adversarial examples than the random baseline. When it comes to the target model distilled, GA is, statistically, equally fast. Again, all evolutionary algorithms are better than the baseline.

The scenario changes when counting images with no success as 40,000 evaluations. Now, CMA-ES is the algorithm which takes the most evaluations to find an adversarial example for target model regular, and DE is the only algorithm to outperform random. For model distilled, DE and Random perform equally.

As previously mentioned, the algorithms were not able to find adversarial examples for all the selected 500 images through the one-pixel attack. With success rate, we compute the percentage that were successfully attacked in a run. In Table 4 we show the mean success rate between runs for each model. Again, as expected, the distilled model is more robust to adversarial attacks. In this metric, the baseline random algorithm is competitive in comparison to the evolutionary algorithms and the best for model distilled.

In the quantity of adversarial per image metric we measure the amount of adversarial examples that an algorithm can find. In a run of an algorithm, we compute how many adversarial examples were found for each image and compute the mean. The means between runs are presented in Table 5 for each target model. Although random is competitive in success rate, it is not when it comes to the quantity of adversarial examples. This indicates that it can find solutions, but the evolutionary algorithms are better at exploring and exploiting them. Comparing the evolutionary algorithms, it is possible to see that despite GA's weakness in attacking the selected set of images, when it is successful it finds more adversarial examples than the other algorithms.

Table 5. Mean quantity of adversarial examples per image for the four algorithms. The results presented are the averages of 10 runs.

(a) Model regular. All means are statistically different.

Algorithm	Mean	STD
DE	885.61	13.04
GA	4944.83	104.59
CMA-ES	977.17	20.80
Random	508.23	1.96

(b) Model distilled. All means are statistically different.

Algorithm	Mean	STD
DE	818.10	23.55
GA	1621.12	56.17
CMA-ES	515.10	20.4110
Random	277.05	0.99

Figure 2 shows the number of new adversarial samples per generation. Images with no adversarial examples were removed to highlight the algorithms' performance in terms of quantity when they are successful in attacking an image. Overall, GA discovers more adversarial examples than its counterparts, and continues to find new ones throughout the entire evolution despite a decline in the second half. In contrast, CMA-ES finds most of its adversarial examples early on, subsequently stabilising as it ceases to discover additional ones.

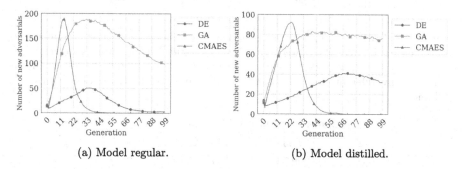

(a) Model regular.   (b) Model distilled.

Fig. 2. Mean quantity of new adversarial examples per generation per run for the three algorithms. The results presented are the averages of 10 runs. The images with no adversarial examples were not considered from the statistics presented.

Although the evolutionary algorithms are guided by the same fitness function, they differ in the way they operate in the search space. The covered pixels metric aims to quantify how many unique individuals, i.e. pixels, were evaluated for an image. We calculated the mean number of covered pixels in the 500 attacks. The mean values between runs are presented in Table 6 for each target model.

We compiled a set of adversarial examples from successful attacks by every algorithm for the model regular (Fig. 3) and distilled (Fig. 4). The figures contain the original image with the information about the true label and confidence, the adversarial image, the confidence on the new class and its distortion. As stated,

Table 6. Mean quantity of covered pixels images for the four algorithms. The results presented are the averages of 10 runs.

(a) Model regular. All means are statistically different.

Algorithm	Mean	STD
DE	25947.46	10994.72
GA	29399.90	3446.14
CMA-ES	8204.64	147.31
Random	40000	0

(b) Model distilled. All means are statistically different.

Algorithm	Mean	STD
DE	30437.02	10865.40
GA	29559.90	3867.28
CMA-ES	8605.44	164.92
Random	40000	0

Table 7. Mean distortion of images for the four algorithms. The results presented are the averages of 10 runs. The target model is the regular model. Means (means) are statistically different, with the exception of the pair: CMA-ES and GA. Means (min) are statistically different, with the exception of the pair: CMA-ES and Random.

Algorithm	Mean (mean)	STD	Mean (min)	STD
DE	494.28	3.50	242.54	7.19
GA	470.59	4.72	192.99	3.51
CMA-ES	468.73	4.48	213.24	5.53
Random	383.44	2.46	218.28	3.69

Table 8. Mean distortion of images for the four algorithms. The results presented are the averages of 10 runs. The target model is the distilled model. Means (means) are statistically different, with the exception of the pair: CMA-ES and DE. Means (min) are statistically different, with the exception of the pairs: CMA-ES and DE, CMA-ES and Random.

Algorithm	Mean (mean)	STD	Mean (min)	STD
DE	323.38	6.74	236.26	6.49
GA	269.33	4.17	184.81	2.28
CMA-ES	319.76	6.02	231.049	7.30
Random	393.19	1.90	240.65	3.60

the distortion is the difference between the original image and the perturbed image. We can observe in both figures that in most of the cases the pixel attacks are not perceptible to the human eye (e.g. last line of Fig. 3). Overall, for the regular and distilled models, we can observe that in the GA and CMA-ES in most of the cases, the value of distortion is lower than the other approaches as Tables 7 and 8 also suggest. The pixels on the distilled model are a bit more noticeable than the regular and the values for distortion are in general higher. Although the distilled model is more resilient to an attack, it seems that in the event of a successful attack, the pixel takes an atypical colour.

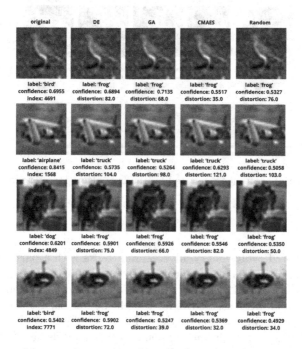

Fig. 3. Adversarial examples generated by the four algorithms for model regular.

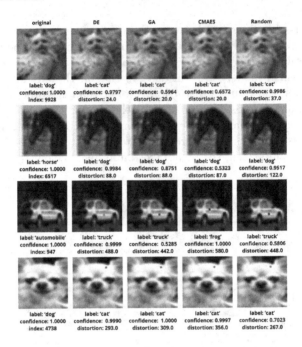

original	DE	GA	CMAES	Random
label: 'dog' confidence: 1.0000 index: 9928	label: 'cat' confidence: 0.9797 distortion: 24.0	label: 'cat' confidence: 0.5964 distortion: 20.0	label: 'cat' confidence: 0.6572 distortion: 20.0	label: 'cat' confidence: 0.9986 distortion: 37.0
label: 'horse' confidence: 1.0000 index: 6517	label: 'dog' confidence: 0.9984 distortion: 88.0	label: 'dog' confidence: 0.8751 distortion: 88.0	label: 'dog' confidence: 0.5323 distortion: 87.0	label: 'dog' confidence: 0.9517 distortion: 122.0
label: 'automobile' confidence: 1.0000 index: 947	label: 'truck' confidence: 0.9999 distortion: 488.0	label: 'truck' confidence: 0.5285 distortion: 442.0	label: 'frog' confidence: 1.0000 distortion: 580.0	label: 'truck' confidence: 0.5806 distortion: 448.0
label: 'dog' confidence: 1.0000 index: 4738	label: 'cat' confidence: 0.9990 distortion: 293.0	label: 'cat' confidence: 1.0000 distortion: 309.0	label: 'cat' confidence: 0.9997 distortion: 356.0	label: 'cat' confidence: 0.7023 distortion: 267.0

Fig. 4. Adversarial examples generated by the four algorithms for model distilled.

6   Conclusion

This paper explored evolutionary adversarial attacks with a particular focus on the one-pixel attack in image classification. By employing Differential Evolution, Genetic Algorithms, and Covariance Matrix Adaptation Evolution Strategy, we have explored these attacks, analysing the efficacy of each evolutionary algorithm. Through experimentation on the CIFAR-10 dataset, a widely recognized benchmark in image classification, our comparative analysis has shown the effectiveness of the chosen evolutionary algorithms in generating one-pixel attacks.

The experimentation consisted of testing the algorithms in their ability to generate adversarial examples by changing one pixel in each image. We attacked 500 correctly classified test images - 50 from each class. We selected two target models, regular and distilled from Carlini and Wagner [4]. The regular model was traditionally trained whereas the distilled one was trained with defensive strategies. Aside from the evolutionary algorithms, random search was used to serve as a baseline of comparison. The algorithms had a budget of 40,000 evaluations for each image and target model pair. The experimental results were judged in fitness optimisation, number of evaluations to generate an adversarial attack, success rate, number of adversarial per image, solution space coverage and level of distortion to the original image.

Based on the results, not all images were successfully attacked by the strict one-pixel attack, with DE (42%) being the algorithm with the most successful attacks to the 500 images. The fitness progress shows that the CMA-ES is the fastest to converge, followed by GA and DE. According to the number of adversarial encountered, we concluded that the GA was the approach that encountered more adversarial attacks. In terms of coverage, GA also comes ahead for model regular, and in close second to DE for model distilled. In terms of minimal distortion to the original image, GA has lower results than the other algorithms, in comparison to CMA-ES and DE which have similar results to the random search baseline. The competitive performance by random in some metrics might me explained by the lack of tuning for the algorithms.

For future work, we plan to extend the analysis to attacks with multiple pixels, to develop an attack capable of creating a perturbation that can affect multiple images at once, and to perform a hyper-parameter search to optimise the algorithms.

Acknowledgements. This work has been partially supported by Project "NEXUS Pacto de Inovação - Transição Verde e Digital para Transportes, Logística e Mobilidade". ref. No. 7113, supported by the Recovery and Resilience Plan (PRR) and by the European Funds Next Generation EU, following Notice No. 02/C05-i01/2022.PC645112083-00000059 (project 53), Component 5 - Capitalization and Business Innovation - Mobilizing Agendas for Business Innovation; by Project No. 7059 - Neuraspace - AI fights Space Debris, reference C644877546-00000020, supported by the RRP - Recovery and Resilience Plan and the European Next Generation EU Funds, following Notice No. 02/C05-i01/2022, Component 5 - Capitalization and Business Innovation - Mobilizing Agendas for Business Innovation and; by the FCT - Foundation for Science and Technology, I.P./MCTES through national funds (PIDDAC), within the scope of CISUC R&D Unit - UIDB/00326/2020 or project code UIDP/00326/2020.

References

1. Alzantot, M., Sharma, Y., Chakraborty, S., Zhang, H., Hsieh, C.J., Srivastava, M.B.: GenAttack: practical black-box attacks with gradient-free optimization. In: Proceedings of the Genetic and Evolutionary Computation Conference, GECCO 2019, pp. 1111–1119. Association for Computing Machinery, New York, NY, USA (2019). https://doi.org/10.1145/3321707.3321749
2. Banzhaf, W., Machado, P., Zhang, M.: Handbook of Evolutionary Machine Learning. Genetic and Evolutionary Computation. Springer, Cham (2023). https://doi.org/10.1007/978-981-99-3814-8, https://books.google.pt/books?id=fGLuzwEACAAJ
3. Bradley, J.R., Blossom, A.P.: The generation of visually credible adversarial examples with genetic algorithms. ACM Trans. Evol. Learn. Optim. **3**(1), 1–44 (2023). https://doi.org/10.1145/3582276
4. Carlini, N., Wagner, D.A.: Towards evaluating the robustness of neural networks. CoRR abs/1608.04644 (2016). http://arxiv.org/abs/1608.04644
5. Chen, J., Su, M., Shen, S., Xiong, H., Zheng, H.: POBA-GA: perturbation optimized black-box adversarial attacks via genetic algorithm. Comput. Secur. **85**, 89–106 (2019). https://doi.org/10.1016/j.cose.2019.04.014, https://www.sciencedirect.com/science/article/pii/S0167404818314378

6. Dong, X., et al.: GreedyFool: distortion-aware sparse adversarial attack (2020). https://doi.org/10.48550/ARXIV.2010.13773, https://arxiv.org/abs/2010.13773

7. Goodfellow, I.J., Shlens, J., Szegedy, C.: Explaining and harnessing adversarial examples (2014). https://doi.org/10.48550/ARXIV.1412.6572, https://arxiv.org/abs/1412.6572

8. Hansen, N.: The CMA evolution strategy: a tutorial. CoRR abs/1604.00772 (2016). http://arxiv.org/abs/1604.00772

9. Ilie, A., Popescu, M., Stefanescu, A.: EvoBA: an evolution strategy as a strong baseline for black-box adversarial attacks. In: Mantoro, T., Lee, M., Ayu, M.A., Wong, K.W., Hidayanto, A.N. (eds.) Neural Information Processing, pp. 188–200. Springer, Cham (2021). https://doi.org/10.1007/978-3-030-92238-2_16

10. Jere, M., Rossi, L., Hitaj, B., Ciocarlie, G., Boracchi, G., Koushanfar, F.: Scratch that! An evolution-based adversarial attack against neural networks. arXiv preprint arXiv:1912.02316 (2019)

11. Lapid, R., Haramaty, Z., Sipper, M.: An evolutionary, gradient-free, query-efficient, black-box algorithm for generating adversarial instances in deep convolutional neural networks. Algorithms **15**(11), 407 (2022). https://doi.org/10.3390/a15110407, https://www.mdpi.com/1999-4893/15/11/407

12. Lin, J., Xu, L., Liu, Y., Zhang, X.: Black-box adversarial sample generation based on differential evolution. J. Syst. Softw. **170**, 110767 (2020). https://doi.org/10.1016/j.jss.2020.110767, https://www.sciencedirect.com/science/article/pii/S0164121220301850

13. Storn, R., Price, K.: Differential evolution - a simple and efficient heuristic for global optimization over continuous spaces. J. Global Optim. **11**(4), 341–359 (1997). https://doi.org/10.1023/A:1008202821328

14. Su, J., Vargas, D.V., Sakurai, K.: One pixel attack for fooling deep neural networks. IEEE Trans. Evol. Comput. **23**(5), 828–841 (2019). https://doi.org/10.1109/TEVC.2019.2890858

15. Szegedy, C., et al.: Intriguing properties of neural networks (2013). https://doi.org/10.48550/ARXIV.1312.6199, https://arxiv.org/abs/1312.6199

16. Williams, P., Li, K.: Art-attack: black-box adversarial attack via evolutionary art (2022)

17. Wu, C., Luo, W., Zhou, N., Xu, P., Zhu, T.: Genetic algorithm with multiple fitness functions for generating adversarial examples. In: 2021 IEEE Congress on Evolutionary Computation (CEC), pp. 1792–1799 (2021). https://doi.org/10.1109/CEC45853.2021.9504790

Robust Neural Architecture Search Using Differential Evolution for Medical Images

Muhammad Junaid Ali[✉], Laurent Moalic, Mokhtar Essaid,
and Lhassane Idoumghar

Université de Haute-Alsace, IRIMAS UR 7499, 68093 Mulhouse, France
{muhammad-junaid.ali,laurent.moalic,mokhtar.essaid,
lhassane.idoumghar}@uha.fr

Abstract. Recent studies have demonstrated that Convolutional Neural Network (CNN) architectures are sensitive to adversarial attacks with imperceptible permutations. Adversarial attacks on medical images may cause manipulated decisions and decrease the performance of the diagnosis system. The robustness of medical systems is crucial, as it assures an improved healthcare system and assists medical professionals in making decisions. Various studies have been proposed to secure medical systems against adversarial attacks, but they have used handcrafted architectures. This study proposes an evolutionary Neural Architecture Search (NAS) approach for searching robust architectures for medical image classification. The Differential Evolution (DE) algorithm is used as a search algorithm. Furthermore, we utilize an attention-based search space consisting of five different attention layers and sixteen convolution and pooling operations. Experiments on multiple MedMNIST datasets show that the proposed approach has achieved better results than deep learning architectures and a robust NAS approach.

Keywords: Differential Evolution · Evolutionary AutoML ·
evolutionary NAS · Neural Architecture Search

1 Introduction

Deep Neural Networks (DNNs) have been widely used to solve medical imaging tasks, such as classification, localization, registration, and segmentation [4,26]. The robustness of these systems is crucial to secure them from vulnerability issues [2]. The latter can be caused by adversarial attacks, which are deliberate attempts to deceive or manipulate a machine-learning model by adding perturbations to the input data [9]. The real-life threats of these attacks are considered essential; for instance, in medical systems, they can exhibit misleading diagnostic outcomes.

Different types of adversarial attacks include black-box, white-box, and targeted attacks. In a black-box attack, the attacker cannot access the model parameters as it generates adversarial images without knowledge of the model.

S. Smith et al. (Eds.): EvoApplications 2024, LNCS 14635, pp. 163–179, 2024.
https://doi.org/10.1007/978-3-031-56855-8_10

However, in white-box attacks, the attacker can access the model parameters. Furthermore, targeted attacks aim to cause the model to predict incorrect class labels [5]. Some of the famous adversarial attack techniques are the Fast Gradient Sign Method (FGSM) [6], Projected Gradient Descent (PGD) [7] and their different variants. PGD is a powerful adversarial attack that uses randomized initialization and multi-step attacks to perturb input data iteratively along the gradient's direction, aiming to find effective adversarial examples. Following the same context, FGSM is a one-step adversarial attack that perturbs the input data by adding little noise in the direction.

For effective defense mechanisms against adversarial attacks, numerous approaches have been proposed in the literature, such as adversarial training [11,13], Denoising Convolutional Neural Networks (DCNN)'s [12], model ensemble techniques [10]. These approaches aim to enhance the robustness of deep learning models against adversarial perturbations. While adversarial robustness methods enhance a model's robustness against adversarial attacks, they may reduce performance on clean(original and unaltered) data samples due to overfitting towards adversarial examples.

Moreover, these techniques use hand-crafted architectures. Recent studies have shown that the choice of deep learning architecture directly impacts the performance of adversarial training [15]. Manual designing of deep learning architectures is a complex and time-consuming task. Neural Architecture Search (NAS) overcomes this problem by automatically searching for an efficient architecture for a specific task. Evolutionary NAS approaches are effective in the automated search of robust architectures for image classification [3]. These approaches efficiently search for robust and lightweight architectures. In these studies, each sampled architecture is trained with adversarial training and evaluated on accuracy scores obtained after applying different adversarial attacks on the validation set [3,20,24,25].

Studies have shown that DNNs trained on medical images can be more vulnerable to adversarial attacks than natural images. One of the main reasons is due to the complex biological textures in medical images, which may lead to more vulnerable regions, and deep neural networks designed for large-scale natural images can be overparametrized for medical image tasks, which leads to high vulnerability to adversarial attacks [1,19]. Therefore, there is a need for a separate adversarial NAS approach for medical images that focus only on medical images. A specific DNN learned on medical image data may be more robust against perturbations than existing deep learning architectures.

Motivated by the above studies, we have proposed an evolutionary approach for searching robust architectures against adversarial attacks on medical images. This study is about the architectural perspective of adversarial robustness for medical images. Besides, it shows how evolutionary approaches can be helpful to search for an architectural topology that is both lightweight and robust against adversarial attacks. We used Differential Evolution (DE) as a search algorithm for our proposed approach. The motivation behind adopting DE as a search algorithm is its simplicity and effectiveness in population explorations. It has been widely adopted by different studies and achieved state-of-the-art performance on various tasks [32–36]. In summary, the main contributions of this paper are as follows:

- We have proposed an approach for automatically searching robust architectures against adversarial attacks for medical images using an evolutionary algorithm.
- A fitness function is designed to aggregate the accuracy scores from multiple adversarial attacks.
- We designed an attention-based search space consisting of multiple convolution and pooling layers and different attention layers.

The rest of the article is structured as follows. Section 2 discusses the related work. We present the proposed methodology in Sect. 3. Section 4 discusses the results and experimental settings. Finally, we conclude this study in Sect. 5.

2   Related Work

Numerous studies have been proposed to defend deep learning models against adversarial attacks on medical images, specifically for medical image classification tasks [37–41]. These studies have compared the performance of different Convolutional Neural Network (CNN) architectures under adversarial attacks. A study by Hirano et al. proposed to evaluate the vulnerability of seven CNN models on medical image classification task. They have performed experiments on three medical image datasets (skin cancer, pneumonia classification, and diabetic retinopathy) against Universal Adversarial Attack (UAP) [40].

Moreover, Ma et al. compared the responses of natural and medical images on adversarial perturbations, finding that medical images are more vulnerable due to their complex biological structures and unique properties [19]. Similarly, Xu et al. conducted experiments on three models in multi-label, multi-class, and binary classification tasks, finding that models are unreliable when subjected to adversarial examples [41]. They also developed two types of defense techniques for dealing with adversarial instances. Denoising operations are also found to be quite effective against adversarial attacks. Xue et al. proposed enhancing the denoising ability of CNN classifiers for medical image classification using an auto-encoder to make the model robust against adversarial attacks [12].

Furthermore, various countermeasures have also been proposed to enhance the resilience of deep learning architectures against adversarial attacks. The most effective and popular defense technique is adversarial training. It improves the model's robustness by adding adversarial examples during the training [6,7]. In adversarial training, adversarial examples are fed to the network during training to increase its robustness against these attacks. Multiple studies have been proposed to improve the adversarial training process for increasing the robustness accuracy [16–18]. Adaptive adversarial training is also proposed, which uses a loss-defined margin to overcome performance degradation caused by high noise levels [11]. During adversarial training, the most widely used method is to replace the input data or combine it with adversarial examples generated by PGD to make the architecture more resilient to other attacks [7]. It was noticed that Defensive distillation is only effective when dealing with some gradient-based attack methods such as FGSM and PGD [31].

Similarly, obfuscated gradient only resists specific gradient-based attacks, as it does not enhance the robustness of architecture [28]. Adversarial training is the most effective approach for defense against adversarial attacks [42]. However, the performance of adversarial training is also dependent on the network architecture [27]. As manually designing a neural network architecture is difficult, NAS overcomes the issue by automatically searching an architecture topology. Therefore, adopting NAS for building robust architectures would be helpful. Accordingly, a variety of robust NAS approaches have been proposed to automatically search for adversarial robust architectures for image classification [14,24,25]. These approaches achieve robustness from an architectural design perspective.

Studies have shown that medical images exhibit certain characteristics different from natural images and are more sensitive against adversarial attacks [8]; therefore, there is a need for an adversarial NAS approach specifically designed for medical images. To the best of our knowledge, there are no previous studies for searching robust architectures against adversarial attacks in medical images. This study aims to automatically search for a robust, resilient architecture against multiple adversarial attacks by leveraging an evolutionary search approach and adversarial training.

3   Proposed Methodology

This section presents the overall methodology of the proposed approach to search for robust architectures against adversarial attacks on medical images. The overview of the proposed framework is shown in Fig. 1. The proposed NAS-based framework consists of three main steps: (i) population initialization, (ii) crossover and mutation operations, and (iii) fitness evaluation.

The individual network is represented using a vector consisting of continuous and discrete values of fixed length as a genotype. The proposed approach directly applies the reproduction operators on the genotype while decoding the genotype only for evaluation purposes. The cost of each architecture is computed using the proposed fitness function, and the search algorithm maximizes the fitness to search for new architecture with better robustness. Instead of simple training, the individual neural networks are trained with adversarial training.

3.1   Search Space Design

In this study, we have used cell-based search space for our NAS approach, in which the neural network architecture is constructed using a set of repeating modules or "cells" [21,23]. Each cell consists of several layers and their connections. This search space is frequently adopted by different studies thanks to its

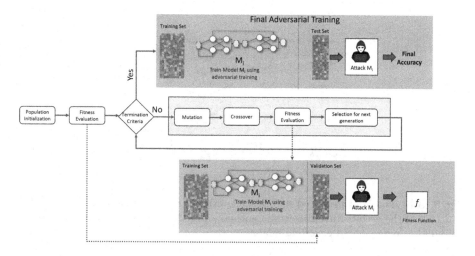

Fig. 1. System diagram of the proposed methodology

simplicity and flexible nature, which makes it easy to modify. This search space is frequently adopted by different studies thanks to its simplicity and flexibility, making it easy to modify. This search space categorizes the cells into two different types: normal and reduction cells. The normal cells process the input and keep the same resolution of the feature map.

In contrast, the reduction cell reduces the feature map's resolution using down-sampling operations such as pooling and stride-2 convolution. The network

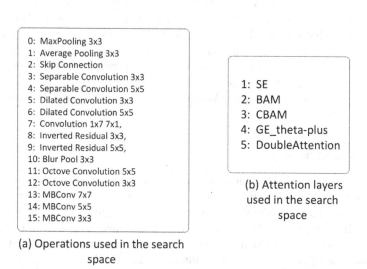

0: MaxPooling 3x3
1: Average Pooling 3x3
2: Skip Connection
3: Separable Convolution 3x3
4: Separable Convolution 5x5
5: Dilated Convolution 3x3
6: Dilated Convolution 5x5
7: Convolution 1x7 7x1,
8: Inverted Residual 3x3,
9: Inverted Residual 5x5,
10: Blur Pool 3x3
11: Octove Convolution 5x5
12: Octove Convolution 3x3
13: MBConv 7x7
14: MBConv 5x5
15: MBConv 3x3

(a) Operations used in the search space

1: SE
2: BAM
3: CBAM
4: GE_theta-plus
5: DoubleAttention

(b) Attention layers used in the search space

Fig. 2. Operations and attention layers used in the search space to design CNN architectures

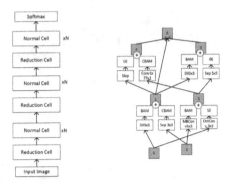

Fig. 3. An example of proposed search space utilized in this approach. Left: The full structure of the architecture consists of normal and reduction cells. Right: Example of cell build using proposed encoding scheme

is formed by stacking these cells one after another. Each cell consists of several predefined operations. Figure 3 shows a visual example of cell-based search space. This study uses 16 different operations and 5 different attention layers, as shown in Fig. 2. We have used these attention layers from the Attention-DARTS [29] study. This latter proposed to search attention layers along convolution and pooling operations. The objective is to use each attention layer after the candidate operation to focus more on salient regions. Various efficient and lightweight CNN components from different architectures, such as MobileNet, OctoveNet, and InceptionNets, are selected in our search space. These operations consist of dilated convolution, mobile convolution block, octave convolutions, separable convolutions, and inverted residual blocks with different kernel sizes. Moreover, five different kinds of attention layers were used. They are applied after each operation. Studies have shown that attention-based models are more robust against adversarial attacks [30]. These layers include Squeeze and Excitation (SE), Gather-Excite (GE) Attention, Bottleneck Attention Module (BAM), Convolution Bottleneck Attention Module (CBAM), and Double Attention (DA) blocks.

3.2 Encoding Scheme

In the search space, each candidate operation is represented by a real value between 0 and 1, while a value between 1 and 5 represents the associated attention layer, mapping to different attention types. Figure 4 shows an encoding scheme's visual representation. Each value of the encoding vector corresponds to a tuple consisting of a real value representing some network operation and an attention layer represented by an integer value. Similarly, each value of the decoding vector represents a tuple consisting of an operation and an attention layer defined in the search space (convolution or pooling layers), as illustrated above.

| 0.15394,3 | 0.73577,3 | 0.01724,1 | 0.07689,5 | 0.89058,0 | 0.37734,2 | 0.00395,3 | 0.70345,4 |
| 0.69953,4 | 0.13817,0 | 0.56487,4 | 0.98290,3 | 0.59894,4 | 0.80453,1 | 0.39271,2 | 0.09398,2 |

(a) Genotype Representation

| Skip Connection,CBAM | Octave Conv 5x5,CBAM | MaxPool 3x3,SE | AvgPool 3x3,Double Attention | MBConv 5x5 t1, Identity | DilConv5x5,BAM | MaxPool3x3,CBAM | Octove Conv 5x5,GE_theta-plus |
| Octove Conv 5x5,GE_theta-plus | Skip Connection,Identity | Inverted Residual 5x5, GE_theta-plus | MBConv 3x3 t1,CBAM | Inverted Residual 5x5,GE_theta-plus | Octove Conv 3x3,SE | DilatedConv 5x5, BAM | Avg Pool 3x3, BAM |

(b) Phenotype Representation

Fig. 4. Genotype and phenotype representation of individuals consisting of 8 genes

The mapping of an encoding vector $\bar{E}$ to a decoding vector $\bar{D}$ in this approach is given as:

$$\bar{E} = [(e_1, a_1), (e_2, a_2), (e_3, a_3), (e_4, a_4), \ldots, (e_n, a_n)] \tag{1}$$

where the operation gene e_i and attention gene a_i are defined as:

$$e_i = P_{ri} \qquad \text{where } 0 \leq P_{ri} \leq 1 \tag{2}$$
$$a_i = R_{ri} \qquad \text{where } 1 \leq R_{ri} \leq 5 \tag{3}$$

P_r: Random probability value for each candidate convolution or pooling operation

R_r: Random value for each attention operation

A_r: Total number of available attention operations

O_N: Total number of available convolution or pooling operations

The decoding vector is given as:

$$\bar{D} = [d_1, d_2, d_3, d_4, \ldots, d_n] \tag{4}$$

Each gene of a decoded individual represents:

$$d_i = (\text{Operations}[\lfloor O_N \times e_i \rfloor], \text{Attentions}[a_i])$$

3.3   Fitness Function

This study proposes a fitness function to evaluate an individual's robustness against multiple adversarial attacks. The fitness function consists of the harmonic mean of the accuracy scores of three different adversarial attacks (FGSM, Basic Iterative Method (BIM), and PGD) as a measure to evaluate the performance of an individual. The model's predictions are computed using perturbed validation data to compute these accuracy scores. The intuition behind using harmonic mean is to keep the fitness function having non-zero values. The fitness function is given as:

$$FitnessFunction = \frac{3}{\frac{1}{A_{FGSM}} + \frac{1}{A_{PGD}} + \frac{1}{A_{BIM}}} \tag{5}$$

where A_{FGSM}, A_{PGD}, and A_{BM} are the accuracy scores on the searched model after applying FGSM, PGD, and BIM attacks on the validation set. The pseudo-code of performance evaluation of an individual neural network is given in Algorithm 1

3.4   Adversarial Training

Adversarial training feeds adversarial samples during training to increase the model's robustness against perturbations. In this study, PGD-based adversarial training is used. In the case of multi-class classification, a multi-class input X and output Y are considered. The output Y is one-hot encoded, representing a corresponding label for each class in X. The goal is to train a classifier f_θ with some parameters θ that accurately predicts the class label from unseen data. For adversarial training in the case of multi-class, the adversarial perturbations are applied to each class separately. The original classification loss is given by:

$$\mathcal{L}(\theta) = \frac{1}{S} \sum_{i=1}^{S} \text{loss}(f_\theta(X_i), Y_i) \tag{6}$$

Where S represents the number of samples in the dataset, X_i indicates the i-th input with Y_i as the corresponding label with loss as the multi-class classification loss function. The adversarial perturbations generated using the PGD for each class c are achieved by the given function:

$$\delta_c = PGD(f_\theta, X_i, Y_i, \epsilon, \alpha, J) \tag{7}$$

where X_i is the input with Y_i as the label, ϵ is the size of perturbation, α the step size and J as the number of iterations. The adversarial training loss for each class c is given by:

$$\mathcal{L}_{adv}^c(\theta) = \frac{1}{S_c} \sum_{i=1}^{S_c} \text{loss}(f_\theta(X_i + \delta_c), Y_i) \tag{8}$$

where S_c is the number of samples in class c. The overall adversarial training objective is given as follows:

$$\min_{\theta} \left(\mathcal{L}(\theta) + \lambda \sum_{c=1}^{C} \mathcal{L}_{adv}^c(\theta) \right) \tag{9}$$

Where C is the total number of classes, and λ is a hyper-parameter that balances the importance of the original loss and the adversarial training loss for each class.

3.5   Differential Evolution

This study uses DE as a search strategy in the proposed NAS approach. It finds optimal solutions by iteratively mutating and combining candidate solutions, evaluating their fitness, and replacing less fit solutions. Compared with traditional evolutionary algorithms, DE uses the scaled difference of vectors to produce new candidate solutions in the population [43]. Moreover, DE comprises few parameters and is easy to implement. DE consists of four main stages: initialization, mutation, crossover, and selection. At first, individuals are randomly generated to create the initial population. Then, a mutation operation is performed to generate donor vectors. Most commonly, the DE/best/1 mutation scheme is used.

$$\vec{V}_i^g = \vec{X}_g^{best} + Fx(\vec{X}_{r1}^g - \vec{X}_{r2}^g) \qquad (10)$$

In this equation, $\vec{V}_i^g$ denotes the i-th vector from generation g. $\vec{X}_g^{best}$ is the best vector from generation g, and $\vec{X}_{r1}^g$ and $\vec{X}_{r2}^g$ represent two random vectors from the population g such that $\vec{X}_{r1}^g \neq \vec{X}_{r2}^g$. F is a scaling factor, $F \in (0, 1)$. After the mutation, the crossover operator is performed as follows:

$$\vec{U}_{ij}^g = \begin{cases} \vec{V}_{ij}^g & \text{if } rand_i \leq CR \text{ or } j = ȷ, \\ x_{ij}^g & \text{otherwise.} \end{cases} \qquad (11)$$

Where $X_{ij}'^{(t)}$ defines the j-th dimension of the i-th individual in the g-th generation. In the binomial crossover, a random number $rand_i$ is generated for each individual dimension and compared with the crossover rate CR and $ȷ$ according to the equation to decide whether the crossover operation will occur or not. Finally, the generated vector $\vec{U}_{ij}^g$ is compared with the trial vector $\overrightarrow{U_{ij}^{g'}}$ and the best one is selected for the next generation according to the fitness values. This process is repeated until the stopping criteria or the maximum number of generations is reached.

Algorithm 1: Performance Evaluation

Input: An individual I, Training epochs E, $Dataset_{train}$, $Dataset_{valid}$
Output: Fitness value f_i of an Individual I
(**1**) Decode the individual with the decoding strategy to obtain the model M;
(**2**) **for** $i = 1$ *to* E **do**
(**3**) $\quad$ Train model M on $Dataset_{train}$ using adversarial training; $//$;

(**4**) $f_i \leftarrow$ Test model M on $Dataset_{valid}$ after adversarial attacks (FGSM, PGD, BIM);
(**5**) Return fitness value f_i;

---

4   Experimental Settings and Results

We have performed comprehensive experiments to evaluate the performance of the proposed approach. This section briefly describes the experimental settings used to perform different experiments and the experimental results.

4.1 Datasets Used for Experimentation

The study utilized MedMNIST benchmark datasets, consisting of 2D and 3D images of various organs and modalities [22]. Two datasets, OrganCMNIST and PathMNIST, were used for experiments. The PathMNIST dataset consists of nine different classes containing colorectal cancer histology slides. The OrganCMNIST dataset consists of eleven different classes of abdominal Computed Tomography (CT) scans in the coronal axes. These datasets are described in the Table 1.

Table 1. Datasets description

Dataset Name	Task(Classes)	Data Modality	# Samples	# Training/Validation/Test
PathMNIST	Multi-Class (9)	Colon Pathology	107,180	89,996/10,004/7,180
OrganCMNIST	Multi-Class (11)	Abdominal CT	23,660	13,000/2,392/8,268

4.2 Experimental Settings

The architecture search was conducted on a small network of 8 cells, divided into train, validation, and test sets. The individual was trained on the training set, evaluated on the validation set, and the final architecture was evaluated on the test set. As the parameters of DE play an important role in the quality of the searched solution, we have performed experiments with different CR and F rates, population sizes, and mutation strategies. The best-reported parameters of DE are given in Table 3. Moreover, the best-performing parameter settings of deep learning algorithms and adversarial attack settings are also given in Table 2. For a fair comparison, all the hyperparameters are kept the same. We performed adversarial training using 7-step PGD with a step size of 0.01 and ϵ as 8/255 for 50 epochs.

For training, we used the Stochastic Gradient Descent (SGD) optimizer with a MultiStepLR scheduler and a learning rate of 0.1, a momentum rate of 0.9, and a weight decay of 0.0002. Similarly, the number of epochs for the proposed approach and AdvRush is set to 50 for the final searched architecture. During the search phase of the proposed approach, the number of epochs is 10 to estimate the performance of an individual. The PyTorch library is used for implementing deep learning architectures, and TorchAttacks is used to implement adversarial attacks. All the experiments are performed on NVIDIA A100 Graphical Processing Unit (GPU) on a GPU cluster.

4.3 Experimental Results and Discussion

Medical images exhibit diverse representations and certain textures and can be easily attacked by adversarial perturbations. Studies have shown that hand-crafted architectures are more robust towards adversarial attacks than existing architectures. We have compared our proposed approach with deep learning

Table 2. Parameters settings of different adversarial attacks

Attacks	ϵ	α	Iteration
FGSM	8/255	10/255	–
PGD	8/255	2/255	7
BIM	8/255	2/255	10
FFGSM	8/255	2/255	–
One Attack	–	–	–
EOTPGD	8/255	2/255	–
PGDRS	8/255	2/255	10

Table 3. Parameter settings of differential evolution algorithm

Parameters	Value
Number of Iterations	20
Population Size	15
Crossover Probability (CR)	0.7 (PathMNIST) 0.5 (OrganCMNIST)
Differential Rate (F)	0.5 (PathMNIST) 0.6 (OrganCMNIST)
Chromosome Length	48
Crossover Operator	Simple Crossover
Mutation Operator	DE/rand/2/bin (PathMNIST) DE/best/1/bin (OrganCMNIST)

architectures (VGG16, ResNet-18, ResNet-50) and adversarial NAS approaches for a baseline comparison given in Table 4. In our study, the architecture searched by the NAS is more robust towards multiple attacks due to the search space consisting of multiple attention blocks with different convolution layers and skip connections.

Recent studies have shown that attention layers and skip connections improve the performance and robustness of natural images for classification tasks [30, 45]. These architectures are robust against adversarial attacks and lightweight as we have predefined the number of layers; the NAS algorithm is restricted to searching for robust architecture by trying combinations within the limit. This shows that the learning algorithm and the choice of operations play an essential role in performance enhancement against adversarial attacks. Multiple adversarial NAS approaches exist, but they have not provided their implementation. AdvRush is a famous gradient-based NAS approach that has achieved state-of-the-art results in improving the performance and robustness of the image classification task. Their proposed approach consists of a SuperNet-based NAS approach with a regularization technique that favors a candidate architecture with a smooth loss landscape [15]. Results have shown that AdvRush, a differentiable NAS approach proposed to discover adversarial robust architectures for image classification, has performed better than deep learning architectures. However, the proposed NAS approach has better performance than AdvRush. One potential reason is the search space used; secondly, each architecture is evaluated instead of the weight-sharing approach used by AdvRush.

Table 4. Performance comparison of obtained accuracy scores against different adversarial attacks of the proposed approach with deep learning architectures. These architectures are adversarially trained on the PGD perturbed examples.

Results	PathMNIST						OrganCMNIST					
	Hand Crafted			Robust NAS	Proposed		Hand Crafted			Robust NAS	Proposed	
	ResNet18	ResNet50	VGG16	AdvRush [15]	Proposed (Best Score)	Proposed (Mean)	ResNet18	ResNet50	VGG16	AdvRush [15]	Proposed (Best Score)	Proposed (Mean)
Clean Accuracy	27.14	28.6	29.72	28.76	**35.4**	30.21	11.67	15.29	22.00	28.74	**37.34**	28.33
PGD	15.73	14.13	16.14	26.88	**28.34**	22.81	13.23	19.16	6.80	26.56	**34.05**	25.36
FGSM	14.03	13.58	15.97	28.4	**30.39**	22.39	14.78	19.197	6.19	28.11	**36.31**	22.98
BIM	4.92	7.14	15.6	26.74	**28.18**	21.04	13.22	19.27	6.79	26.67	**34.02**	24.65
One Pixel	13.87	13.19	13.48	11.49	**14.07**	15.43	0.34	1.20	1.81	1.71	**2.58**	3.26
EOTPGD	15.95	21.936	15.77	26.7	**28.04**	17.82	13.22	18.71	6.18	26.64	**34.13**	21.60
PGDRS	5.19	7.35	19.3	28.82	**30.85**	22.07	13.19	18.87	6.78	26.56	**34.21**	23.04
PGD-10	15.7	14.17	16.2	28.37	**28.885**	28.34	13.18	19.85	6.79	26.51	**34.05**	25.25
PGD-20	15.68	14.56	16.17	28.49	**28.59**	22.17	13.14	19.76	6.56	26.64	**33.96**	22.84
PGD-50	15.71	14.25	16.15	28.35	**28.46**	22.03	13.12	19.47	6.43	26.68	**34.12**	22.86
PGD-100	15.41	14.24	16.11	28.33	**28.39**	21.10	13.05	19.21	6.17	26.56	**34.31**	22.88

The limitation of weight-sharing NAS approaches is that they fail to explore diverse architectural components due to constraints on shared weights across diverse architectural components. Therefore, the proposed approach allows greater architectural diversity, allowing exploration of a wide range of model configurations. In the case of standard image classification problems, simple architecture components in weight-sharing based adversarial NAS approach works quite well but perform poorly in case of medical image classification. Medical images exhibit different characteristics apart from standard images.

Table 5. Results comparison of clean accuracy and accuracy after different attacks with and without attention layers in the search space of proposed NAS approach

	OrganCMNIST		PathMNIST	
	Without Attention Layers	With Attention Layers	Without Attention Layers	With Attention Layers
Clean Accuracy	22.19	**37.34**	19.83	**35.4**
PGD	22.51	**34.05**	20.04	**28.34**
FGSM	25.64	**36.31**	19.81	**30.39**
BIM	22.47	**34.02**	17.35	**28.18**
One Pixel	2.3	**3.26**	13.10	**14.7**
EOTPGD	22.52	**34.13**	19.72	**28.04**
PGDRS	26.73	**37.53**	19.94	**30.85**

For performance comparison, the accuracy scores after different adversarial attacks (PGD, FGSM, BIM, FFGSM, PGDRS, EOTPGD, and OnePixel attacks) are computed on the test set. Moreover, to examine if increasing the number of iterations has an impact on the adversarial accuracy, the architectures are evaluated under PGD attack with different numbers of iterations: PGD^{10}, PGD^{20}, PGD^{50}, and PGD^{100}. BIM and FFGSM are extensions of FGSM attacks, in which BIM performs attacks on multiple iterations. FFGSM extends FGSM by applying the gradient sign method on low-dimensional feature representation. PGDRS and EOTPGD represent other extensions of PGD: PGDRS starts with random initial perturbations to avoid getting stuck at the local minimum; EOTPGD considers the performance of attacks when various transformations are applied to the input.

Moreover, One-pixel attack is also considered, which focuses on changing the values of one or a few pixels to induce misclassification. The motivation behind using multiple attacks is to evaluate the effectiveness and robustness of the proposed method against multiple attacks. We reported the best and average results of five runs obtained by our method and compared them with hand-crafted architectures. Similarly, the deep learning architectures are also adversarially trained and evaluated under these attacks.

Furthermore, we have also performed ablation studies to investigate the performance impact of different components (search space). Table 5 describes the results when using attention layers alongside convolution and pooling operations and without attention layers. It is observed that the attention layers exhibit better robustness against different adversarial attacks and better accuracy on clean examples. As attention layers capture relationships between image regions and focus more on salient regions, it also helps to make models robust against adversarial perturbations.

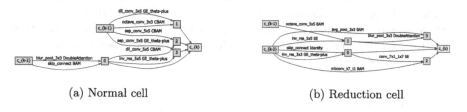

(a) Normal cell   (b) Reduction cell

Fig. 5. The visualization of Normal cell (a) and reduction cell (b) found by the proposed approach for OrganCMNIST dataset

Table 6 demonstrates the number of parameters and model size of the best-performing architecture and the searching time of the proposed NAS approach on PathMNIST and OrganCMNIST. The significant search time is due to adversarial training, in which the model is trained on clean and adversarially perturbed samples. Secondly, the model is tested on perturbed samples with three different types of attacks for fitness evaluation. Compared with deep learning architectures (ResNet, VGG16), the number of parameters and size in MegaBytes (MB) of best-performing architectures are relatively less in numbers. Furthermore, the normal and reduction cells of the best-found architecture on the OrganCMNIST dataset are shown in Figs. 5a and 5b, respectively. Both cells contain a variety of different operations, pooling layers, and different kinds of attention layers.

To compare multiple classifiers and NAS approaches over two MedMNIST datasets, we used the Friedman test to first reject the null hypothesis as given in [44]. Then, a post-hoc analysis using the Wilcoxon signed-rank test is performed with Holm's alpha correction with $\alpha = 0.05$ as the initial value. Then, the average ranks of the classifiers are visualized with a Critical Difference (CD) diagram [44] as shown in Fig. 6. Figure 6 shows the average rank comparison of the proposed approach with deep learning architectures and NAS approaches. The proposed approach has the highest performance over the two datasets from

Fig. 6. Critical difference diagram showing the pairwise statistical comparison of deep learning approaches and AdvRush approach with proposed approach on accuracy after different adversarial attacks.

Table 6. Searching cost (in hours) of the proposed approach on different datasets with model size in MBs and the number of parameters of best-searched architecture

Dataset	Searching Cost	Model Size in MBs	#Parameters
PathMNIST	41 h	2.28	597652
OrganCMNIST	33 h	2.6	703365

the MedMNIST benchmark. The thick line between the average ranks of deep learning architectures indicates no significant difference among these approaches as these architectures share the same architectural components. However, there is no statistical significance between AdvRush and the proposed approach, and AdvRush ranked second.

5   Conclusion

This study proposes an evolutionary NAS approach to search robust architectures for medical image classification. A search space is designed consisting of multiple attention layers and a variety of convolution and pooling operations. Experimental results show that the proposed approach has achieved better results than deep learning architectures and existing NAS approaches. This shows that techniques proposed for image classification would not work in the case of medical images, and a separate approach is very helpful in this case. In the near future, we aim to enhance the proposed approach by incorporating performance estimation strategies, such as surrogate models, to reduce search time.

Acknowledgment. This work was funded by ArtIC project "Artificial Intelligence for Care" (grant ANR-20-THIA-0006-01) and co-funded by IRIMAS Institute/Université de Haute Alsace. The authors would like to thank the Mesocentre of Strasbourg for providing access to the GPU cluster and the providers of all datasets used in this paper.

References

1. Maliamanis, T.V., Papakostas, G.A.: Machine learning vulnerability in medical imaging. In: Machine Learning, Big Data, and IoT for Medical Informatics, pp. 53–70. Academic Press (2021)
2. Apostolidis, K.D., Papakostas, G.A.: A survey on adversarial deep learning robustness in medical image analysis. Electronics 10(17), 2132 (2021)
3. Liu, J., Jin, Y.: Multi-objective search of robust neural architectures against multiple types of adversarial attacks. Neurocomputing 453, 73–84 (2021)
4. Shen, D., Wu, G., Suk, H.I.: Deep learning in medical image analysis. Ann. Rev. Biomed. Eng. 19, 221–248 (2017)
5. Xu, H., et al.: Adversarial attacks and defenses in images, graphs and text: a review. Int. J. Autom. Comput. 17, 151–178 (2020)
6. Goodfellow, I.J., Shlens, J., Szegedy, C.: Explaining and harnessing adversarial examples. arXiv preprint arXiv:1412.6572 (2014)
7. Madry, A., et al.: Towards deep learning models resistant to adversarial attacks. arXiv preprint arXiv:1706.06083 (2017)
8. Dong, J., et al.: Adversarial attack and defense for medical image analysis: methods and applications. arXiv preprint arXiv:2303.14133 (2023)
9. Akhtar, N., Mian, A.: Threat of adversarial attacks on deep learning in computer vision: a survey. IEEE Access 6, 14410–14430 (2018)
10. Qiu, S., et al.: Review of artificial intelligence adversarial attack and defense technologies. Appl. Sci. 9(5), 909 (2019)
11. Ma, L., Liang, L.: Adaptive adversarial training to improve adversarial robustness of DNNs for medical image segmentation and detection. arXiv preprint arXiv:2206.01736 (2022)
12. Xue, F.-F., Peng, J., Wang, R., Zhang, Q., Zheng, W.-S.: Improving robustness of medical image diagnosis with denoising convolutional neural networks. In: Shen, D., et al. (eds.) MICCAI 2019. LNCS, vol. 11769, pp. 846–854. Springer, Cham (2019). https://doi.org/10.1007/978-3-030-32226-7_94
13. Lal, S., et al.: Adversarial attack and defence through adversarial training and feature fusion for diabetic retinopathy recognition. Sensors 21(11), 3922 (2021)
14. Yue, Z., et al.: Effective, efficient and robust neural architecture search. In: 2022 International Joint Conference on Neural Networks (IJCNN). IEEE (2022)
15. Mok, J., et al.: AdvRush: searching for adversarially robust neural architectures. In: Proceedings of the IEEE/CVF International Conference on Computer Vision (2021)
16. Joel, M.Z., et al.: Using adversarial images to assess the robustness of deep learning models trained on diagnostic images in oncology. JCO Clinical Cancer Inf. 6, e2100170 (2022)
17. Rao, C., et al.: A thorough comparison study on adversarial attacks and defenses for common thorax disease classification in chest X-rays. arXiv preprint arXiv:2003.13969 (2020)
18. Li, X., Deng P., Zhu, D.: Defending against adversarial attacks on medical imaging AI system, classification or detection?. In: 2021 IEEE 18th International Symposium on Biomedical Imaging (ISBI). IEEE (2021)
19. Ma, X., et al.: Understanding adversarial attacks on deep learning based medical image analysis systems. Pattern Recogn. 110, 107332 (2021)
20. Kotyan, S., Vargas, D.V.: Towards evolving robust neural architectures to defend from adversarial attacks. In: Proceedings of the 2020 Genetic and Evolutionary Computation Conference Companion (2020)

21. Liu, H., Simonyan, K., Yang, Y.: Darts: differentiable architecture search. arXiv preprint arXiv:1806.09055 (2018)
22. Yang, J., Shi, R., Ni, B.: Medmnist classification decathlon: a lightweight automl benchmark for medical image analysis. In: 2021 IEEE 18th International Symposium on Biomedical Imaging (ISBI). IEEE (2021)
23. Zoph, B., et al.: Learning transferable architectures for scalable image recognition. In: Proceedings of the IEEE Conference on Computer Vision and Pattern Recognition (2018)
24. Guo, M., et al.: When nas meets robustness: in search of robust architectures against adversarial attacks. In: Proceedings of the IEEE/CVF Conference on Computer Vision and Pattern Recognition (2020)
25. Hosseini, R., Yang, X., Xie, P.: DSRNA: differentiable search of robust neural architectures. In: Proceedings of the IEEE/CVF Conference on Computer Vision and Pattern Recognition (2021)
26. Ali, M.J., Akram, M.T., Saleem, H., Raza, B., Shahid, A.R.: Glioma segmentation using ensemble of 2D/3D U-nets and survival prediction using multiple features fusion. In: Crimi, A., Bakas, S. (eds.) BrainLes 2020. LNCS, vol. 12659, pp. 189–199. Springer, Cham (2021). https://doi.org/10.1007/978-3-030-72087-2_17
27. Singh, N.D., Croce, F., Hein, M.: Revisiting adversarial training for imagenet: architectures, training and generalization across threat models. arXiv preprint arXiv:2303.01870 (2023)
28. Athalye, A., Carlini, N., Wagner, D.: Obfuscated gradients give a false sense of security: circumventing defenses to adversarial examples. In: International Conference on Machine Learning. PMLR (2018)
29. Nakai, K., Matsubara, T., Uehara, K.: Att-darts: differentiable neural architecture search for attention. In: 2020 International Joint Conference on Neural Networks (IJCNN). IEEE (2020)
30. Agrawal, P., et al.: Impact of attention on adversarial robustness of image classification models. In: 2021 IEEE International Conference on Big Data (Big Data). IEEE (2021)
31. Papernot, N., et al.: Distillation as a defense to adversarial perturbations against deep neural networks. In: 2016 IEEE Symposium on Security and Privacy (SP). IEEE (2016)
32. Awad, N., Mallik, N., Hutter, F.: Differential evolution for neural architecture search. arXiv preprint arXiv:2012.06400 (2020)
33. Kuş, Z., et al.: Differential evolution-based neural architecture search for brain vessel segmentation. Eng. Sci. Technol. Int. J. **46**, 101502 (2023)
34. Rampavan, M., Ijjina, E.P.: Brake light detection of vehicles using differential evolution based neural architecture search. Appl. Soft Comput. **147**, 110839 (2023)
35. Gülcü, A., Kuş, z.: Neural architecture search using differential evolution in MAML framework for few-shot classification problems. In: Metaheuristics International Conference. Springer, Cham (2022). https://doi.org/10.1007/978-3-031-26504-4_11
36. Yu, C., et al.: EU-Net: automatic U-Net neural architecture search with differential evolutionary algorithm for medical image segmentation. Comput. Biol. Med. **167**, 107579 (2023)
37. Shi, X., et al.: Robust convolutional neural networks against adversarial attacks on medical images. Pattern Recogn. **132**, 108923 (2022)
38. Xu, M., Zhang, T., Zhang, D.: Medrdf: a robust and retrain-less diagnostic framework for medical pretrained models against adversarial attack. IEEE Trans. Med. Imaging 41(8), 2130–2143 (2022)

39. Rodriguez, D., et al.: On the role of deep learning model complexity in adversarial robustness for medical images. BMC Med. Inf. Decis. Mak. **22**(2), 1–15 (2022)
40. Hirano, H., Minagi, A., Takemoto, K.: Universal adversarial attacks on deep neural networks for medical image classification. BMC Med. Imaging **21**, 1–13 (2021)
41. Xu, M., et al.: Towards evaluating the robustness of deep diagnostic models by adversarial attack. Med. Image Anal. **69**, 101977 (2021)
42. Bai, T., Luo, J., Zhao, J., Wen, B., Wang, Q.: Recent advances in adversarial training for adversarial robustness. arXiv preprint arXiv:2102.01356 (2021)
43. Storn, R., Price, K.: Differential evolution-a simple and efficient heuristic for global optimization over continuous spaces. J. Global Optim. **11**, 341–359 (1997)
44. Demšar, J.: Statistical comparisons of classifiers over multiple data sets. J. Mach. Learn. Res. **7**, 1–30 (2006)
45. Huang, S., et al.: Revisiting residual networks for adversarial robustness. In: Proceedings of the IEEE/CVF Conference on Computer Vision and Pattern Recognition (2023)

Progressive Self-supervised Multi-objective NAS for Image Classification

Cosijopii Garcia-Garcia[✉][ID], Alicia Morales-Reyes[ID],
and Hugo Jair Escalante[ID]

Computer Science Department, Instituto Nacional de Astrofísica, Óptica y
Electrónica, Luis Enrique Erro #1, San Andres Cholula, 72840 Puebla, Mexico
{cosijopii,a.morales,hugojair}@inaoep.mx

Abstract. We introduce a novel progressive self-supervised framework for neural architecture search. Our aim is to search for competitive, yet significantly less complex, generic CNN architectures that can be used for multiple tasks (i.e., as a pretrained model). This is achieved through cartesian genetic programming (CGP) for neural architecture search (NAS). Our approach integrates self-supervised learning with a progressive architecture search process. This synergy unfolds within the continuous domain which is tackled via multi-objective evolutionary algorithms (MOEAs). To empirically validate our proposal, we adopted a rigorous evaluation using the non-dominated sorting genetic algorithm II (NSGA-II) for the CIFAR-100, CIFAR-10, SVHN and CINIC-10 datasets. The experimental results showcase the competitiveness of our approach in relation to state-of-the-art proposals concerning both classification performance and model complexity. Additionally, the effectiveness of this method in achieving strong generalization can be inferred.

Keywords: Evolutionary neural architecture search · AutoML · Evolutionary self-supervised learning

1 Introduction

Deep neural networks (DNNs), especially convolutional neural networks (CNNs), have recently gained considerable popularity for tackling diverse challenges [8, 13, 20, 36]. CNN models are highly efficient and have undergone thorough exploration across a wide spectrum of image processing and computer vision tasks. This effectiveness is attributed to factors such as the availability of abundant data, high-performance computing resources, and advancements in machine learning research [21]. On the other hand, neural architecture search (NAS) is a growing field of the automatic design and configuration of CNN models [3] and other types of neural networks, e.g., RNN [21]. NAS methodologies delve into the realm of CNN topologies with the aim of uncovering architectures that meet specific criteria, such as achieving optimal performance or possessing a lightweight parameter structure.

© The Author(s), under exclusive license to Springer Nature Switzerland AG 2024
S. Smith et al. (Eds.): EvoApplications 2024, LNCS 14635, pp. 180–195, 2024.
https://doi.org/10.1007/978-3-031-56855-8_11

Self-supervised learning focuses on architectures able to learn generic (high-level) representations from data by using synthetic labels that can be generated without the need of *human* supervision [9]. An architecture, e.g., a CNN, is trained by minimizing the loss on pretext tasks (e.g., learning to rotate an image). Then, the learned architecture can be used as a starting point (pre-trained model) for approaching other supervised tasks for which labels are available (the downstream task). The advantages of these models are that very often the self-supervised learned model is able to perform decently across several tasks, making this setting attractive to NAS, as the expensive search for a model can *payoff* if it is useful for more than a single downstream task.

In this paper, we introduce a multi-objective NAS method based on self-supervised learning and progressive neural architecture search to design generic CNN architectures for image classification. Our proposed method extends a recently proposed NAS solution designed for supervised learning, CGP-NASV2 [6], by adding a progressive searching mechanism to specialize those high-level blocks used in CGP-NASV2. Moreover, unlike most work on evolutionary NAS, the proposed model targets a self-supervised learning objective (based on RotNet [4]). This objective allows our model to discover generic architectures without the need of labeled data that can be used with other datasets and tasks.

These architectures are linked to a multi-objective searching algorithm that offers a variety of solutions where the complexity or the classification error play a critical role.

The proposed method is evaluated in the context of image classification with CNNs. Experimental results on the CIFAR-100, CIFAR-10, SVHN and CINIC-10 datasets, for transfer learning with the newly found architectures, show that the proposed approach is competitive against several state-of-the-art references. The evolved CNNs architectures have a lower number of parameters as well as lower complexity measured in Multiply-Adds (MAdds) operations in comparison to the state-of-the-art while maintaining a low classification error.

2 Related Work

In this section we review related work to our proposed method, highlighting those where the main role is the evolutionary multi-objective approach and the use of self-supervised learning.

Garcia-Garcia et al. [7] proposed a multi-objective evolutionary NAS method able of finding competitive architectures with low complexity. The search process is performed in the real domain by using two different ways of codifying solutions, allowing it to search for solutions at different levels. As with most work on NAS, this model targets supervised learning; that is, the classification performance of the model is evaluated by using labeled images. We consider this model a starting point for our proposal, as it has shown very competitive results in the design of CNNs for image classification on the CIFAR-100, CIFAR-10, and SVHN datasets.

Nguyen et al. [23] proposed a new method based on sequential model-based optimization (SMBO-TPE) using contrastive self-supervised learning, where the contrastive loss function determines how well the neural architecture has generalized the introduced information. Several types of perturbations are applied to the images, creating pairs of examples (positives and negatives), focusing on mapping these examples when they are similar, and thus being able to discriminate between them. This is a costly process; therefore, an SMBO surrogate is used to reduce the time to evaluate the objective function in the CIFAR-10 and Imagenet datasets, which showed promising results against the state of the art.

Heuillet et al. [11] introduced a novel approach that leverages differentiable NAS to enhance the architecture of Siamese networks. The authors improve the performance of Siamese network frameworks like SimSiam, SimCLR, or MoCo by discovering an encoder-predictor pair using a meta-learner inspired by DARTS, a popular differentiable NAS method. This approach is named "NASiam" (Neural Architecture Search for Siamese Networks), Similar to the previous work, its main approach is based on contrastive self-supervised learning, using the contrastive loss function to guide the search. CIFAR-10, CIFAR-100 and Imagenet datasets were used to validate the proposed model, while achieving a better performance against the other differentiable neural architecture search methods.

Wei et al. [33] proposed a new method for predicting NAS performance metrics. They employ self-supervised learning, specifically contrastive learning algorithms, and introduce a novel architecture encoding scheme. This scheme yields the graph edit distance (GED) metric, which quantifies architectural quality. Consequently, self-supervised learning aids in developing this prediction method, tailored for an evolutionary approach known as the Neural Predictor Guided Evolutionary Neural Architecture Search (NPENAS) algorithm.

In this paper, we propose a multi-objective method that is able to learn generic architectures of competitive performance and moderate complexity. To the best of our knowledge, our proposed model is the first to use evolutionary computation for NAS by using self-supervised learning (that is, our model could find an architecture even if no labeled images are provided). We also emphasize that the critical role of self-supervised learning is when we carry out performance estimation, that is, when architecture is evaluated. Also, we propose a novel progressive search that works in the mating stage of the evolutionary search.

3   Multi-objective Neural Architecture Search

The automated discovery and configuration of CNN architectures has been primarily approached as a single-objective optimization problem aimed at model accuracy. Although effective architectures can be obtained by optimizing model performance, this approach has a series of limitations, the most important of which is perhaps the propensity for solutions to overfit and not having control over their complexity. However, the complexity of the model plays a critical role in the automation of CNN architectures because, in certain environments, it is necessary to have less complex architectures but at the same time with good precision; therefore, this leads to a multi-objective problem.

We propose to automatically design *generic* CNN architectures that maximize the model's accuracy and minimize its complexity. To obtain accurate models based on architectures of moderated complexity we use CGP-NASV2 [7] for the search of models using self-supervised learning. The main idea is to consider scenarios where there is not enough information to train the model and obtain significant performance, but it is necessary to design a customized CNN architecture. Therefore, we rely on self-supervised learning to find a competitive model even when only limited data is available. Our solution is based on a CGP representation that allows us to search in the real domain while using off-the-shell multi-optimization techniques.

The problem can be formulated as a multi-objective optimization one as follows:

$$\text{Minimize } \mathbf{F}(x) = (f_1(x; w^*(x)), f_2(x))^T \tag{1}$$

$$\text{subject to: } w^*(x) \in \text{argmin } \mathcal{L}(w; x) \tag{2}$$

Where f_1 is an objective associated with the classification error of the CNN architecture defined by parameters w^*. Since we adopt a self-supervised learning schema, f_1 can be estimated without labeled images, see Sect. 3.2. f_2 is an objective associated with the model complexity measured in MAdds, in agreement with previous studies [7,16], as they express the total number of operations performed by each architecture. MAdds are a guideline for certain implementation scenarios, such as mobile settings, where the complexity must be less than or equal to 600 MAdds [16,17]. In order to obtain an estimate of the classification error, it is necessary to optimize the weights w of the CNN architecture; thus, an x (solution) depends on this optimization, where normally some training algorithm such as stochastic gradient descent (SDG) is used.

3.1   CGP-NASV2 Solution Representation

Cartesian Genetic Programming (CGP) is a variant of Genetic Programming (GP) based on acyclic graphs to represent solutions that enable forward connections. This representation provides several advantages compared to traditional GP, which employs a tree-based representation. CGP was originally designed for evolving digital circuits. Its name "cartesian" comes from its unique representation of solutions using a two-dimensional grid [7,22].

CGP-NASV2 [7] makes use of CGP and implements block-chained encoding at the top level; it defines a template with some layers as shown in Fig. 1. This representation connects blocks linearly, including those with specific tasks; in CGP-NASV2, the spatial reduction is performed by the pooling layers [19]. At the top-level in the block-chained (see Fig. 1), CGP-NASV2 places a CGP within a "Normal" block while the reduction blocks implement a max pooling layer. In all pooling blocks, the pooling size is fixed as a 2×2 kernel and a stride of 2. Fig. 2 shows an example, after the last block a global average pooling and a fully connected layer are added. Similar guidelines were found in the state of the art

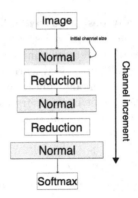

Fig. 1. The solutions representation is structured around chained blocks. Convolution operations are performed in normal block, while pooling occurs in reduction blocks. As the number of blocks increases, the number of channels also grows gradually [7].

review [19,24]. In CGP-NASV2, each "Normal" block from the block-chained representation holds a CGP and it is represented as an integer vector.

In Fig. 2, a CGP-NASV2 solution example is shown. The number of channels and kernel size are defined as parameters associated to each block-chained. The idea is to associate these parameters as weights to each CGP node. Green positions at the integer vector encoding in Fig. 2 represent the assigned hyperparameters within their corresponding CGP block.

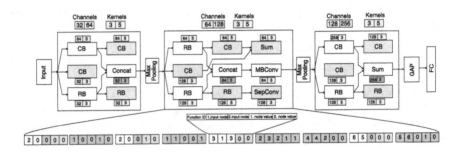

Fig. 2. CGP-NASV2 block-chained representation with the hyperparameters directly encoded.

3.2   Self-supervised Approach for Multi-objective NAS

As explained in Sect. 3.1, CGP-NASV2 uses high-level blocks for solutions representation; in every block, a part of the CNN architecture is searched. In the example of Fig. 2, an architecture of three blocks is represented, from this configuration, the idea is to focalize the search progressively by searching the first block and leaving blocks two and three with a very simple representation; in this case, we use a simple convolution block. After N generations, the searching process continues in the next block leaving the partial representation obtained in the previous block fixed, see Fig. 3. In this preliminary approach, 30 generations were used, where every 10 generations the focalized search moves to the next block to perform the search.

The evaluation is based on a self-supervised approach from the RotNet [4] model using the CIFAR-100 dataset. For this, we created a new dataset from the CIFAR-100 images, but replacing the original classes by rotation levels, thus having four different classes ($0°$, $90°$, $180°$, $270°$) instead of the 100 original classes. Using this data, we proceed to classify the rotations; in this way, we learn a model that solves a related task without the need for large amounts of labeled data. Once a competitive architecture is found, it is trained with labeled data for image classification. One should note that even when an architecture has been learned for a dataset, it can also be applied to other downstream tasks.

The progressive searching in synergy with self-supervised learning allows each block to specialize in learning different features. These sub-architectures are specialized during the evolutionary search; when passing to the next block, the new sub-architectures must adapt to the previous ones, thus maintaining the power of generalization already found. This would occur while targeting the minimization of both, the overall architecture complexity and the classification error. Final solution is a set from which users can select one according to their needs.

To obtain the fitness of each solution, all blocks are assembled together, and the solution is trained with the data from the modified CIFAR-100 dataset, thus obtaining the classification error. To measure the complexity of the model, the sum of all the operations of addition and multiplication in the architecture is performed, thus obtaining the MAdds.

In evolutionary computation, the search is mainly performed by variation operations such as crossover and mutation. To focalize the search, at every generation we performed crossover and mutation only in the currently evolving block, see Fig. 4.

4 Experimental Framework

The proposed experimental framework first assesses the performance of the best evolved architectures in terms of accuracy. After that, a multiple-criteria decision analysis is applied to those models that achieved a good trade-off performance in terms of both accuracy and complexity in MAdds. The experimental settings are defined next, together with the benchmark datasets.

- CGP rows and columns : 10×4
- Mutation probability : $Pm = 0.3$
- Crossover probability : $Pc = 0.9$
- Population : 16

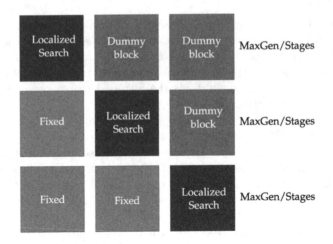

Fig. 3. Proposed progressive search scheme

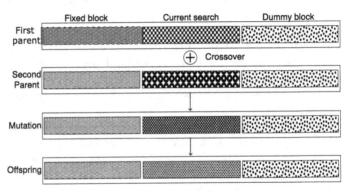

Fig. 4. Example of how the mating stage is carried out, the block where the current search is carried out, changes after each stage defined by a number N of generations

– Generations : 30
– CNN train Epochs: 36

The search was performed on the modified CIFAR-100 dataset. For self-supervised learning, at this stage, only the training partition of the modified CIFAR-100 was used: 80% partition for training and 20% for validation of the evaluated architectures during evolutionary search. After, the evolved architectures were transferred to the CIFAR-100, CIFAR-10, SVHN and CINIC-10 datasets. This means that a *single round of NAS was performed for obtaining a model that was evaluated in four datasets, and the search process did not use any manually labeled image.*

We followed the training mechanisms for the evolved CNN architecture according to CGP-NASV2 [7]: Stochastic Gradient Descent (SGD) as the optimizer together with the cosine annealing learning rate schedule. Our initial learning rate was configured at 0.025, while the momentum was set to 0.9, and the weight decay to 0.0005. The batch size was established at 128, and we conducted a total of 36 training epochs during the evolutionary search.

For both training and testing data, we implemented the following preprocessing steps: a 4-pixel mean subtraction padding on each side and random cropping with a 32 × 32 patch or its horizontally flipped counterpart.

To enhance the training process, we incorporated an auxiliary head classifier [7, 18], which is concatenated after the second reduction block. The loss from this auxiliary head classifier was scaled by a constant factor of 0.4 and added to the loss of the original architecture. This step is applied when we retrain the architectures with a complete dataset. In the final stages, the selected solutions are trained for 600 epochs with the application of the cutout preprocessing technique and a batch size set to 96.

All experiments were conducted on a supercomputing node equipped with 2 Intel Xeon E5-2650 v4 @ 2.20 GHz processors, 6 Nvidia GTX 1080 Ti GPU cards, and 128 GB of RAM. The system operated on the CentOS 7 OS.

5   Experimental Results

Our empirical assessment makes a general performance comparison versus the state of the art approaches. After, the generalization achieved by the self-supervised approach is analyzed using the Grad-CAM [26] method.

5.1   Comparison with the State-of-the-Art

Tables 1, 2, 3 and 4 compare the results between the state-of-the-art and the proposed approach on the CIFAR-100, CIFAR-10, SVHN and CINIC-10 datasets. As explained in Sect. 4, the search was carried out on the modified CIFAR-100 dateset. Therefore, the total time was **5.79** GPU days for the search in the modified CIFAR-100 dataset. And emphasize that the search was not carried out on any of the aforementioned datasets; they were only transferred.

Ten individual experiments were run on the modified CIFAR-100 dataset; the best solution, average, and standard deviation are reported, and two types of solutions are shown: the first named "Best Solution" refers to the best evolved solution found per 10 executions; the solution named "Knee Solution", is the solution determined by the Knee and Boundary Selection Method [5] stating that the solution is the closest to the intersection of the two objectives, i.e., the solution closest to the ideal point, is chosen.

These datasets were selected because they are the most commonly used in the NAS area, starting with CIFAR-100 (see Table 1). Our proposal is fully competent with state-of-the-art solutions, both human-designed and single-objective. In the multi-objective proposals, the finding of solutions with a good trade-off between the number of parameters and low classification error is highlighted. The proposed approach shows a very similar performance to CGP-NASV2 [7], with the advantage of time reduction in terms of GPU-days. When comparing

Table 1. Comparison on CIFAR-100 dataset: Classification error rate, the number of parameters and MAdds are expressed in millions (1×10^6), GPU-days and GPU Hardware. * Transferred architectures

Model	Error rate %	Params	MAdds	GPU-Days	GPU hardware
Human Design					
DenseNet ($k = 12$) [12]	24.42	1.0	–	–	
ResNet ($depth = 101$) [10]	25.16	1.7	–	–	
ResNet ($depth = 1202$) [10]	27.82	10.2	–	–	
VGG [27]	28.05	20.04	–	–	
Single Objective Approaches					
CGP-CNN(ConvSet) [28]	26.7	2.04	–	13	Nvidia 1080Ti
CGP-CNN(ResSet) [28]	25.1	3.43	–	10.9	Nvidia 1080Ti
Large-Scale Evolution [25]	23.0	40.4	–	2750	–
AE-CNN [29]	20.85	5.4	–	36	Nvidia 1080 Ti
Genetic-CNN [34]	29.03	–	–	17	–
(Torabi et al., 2022) [31]	26.03	2.56	–	–	NVIDIA Tesla V100-SXM2
Multi-Objective Approaches					
NSGANetV1 [16]	25.17	0.2	1290	27	Nvidia 2080 Ti
MOGIG-Net [35]	24.71	0.7	–	14	–
EEEA-Net [30]	15.02	3.6	–	0.52	Nvidia RTX 2080 Ti
LF-MOGP [15]	26.37	4.12	–	13	NVIDIA GeForce 3090
CGP-NAS [6]	24.23 (26.41 ± 1.41)	5.43	1581	2.1	Nvidia Titan X
CGP-NASV2 - Best solution [7]	21.18 (22.55 ± 1.24)	4.02 (4 43± 1.57)	457.55 (559.60 ± 476.08)	6.25	Nvidia 1080Ti
CGP-NASV2 - Knee solution [7]	26.35 (29.19 ± 2.20)	0.69 (0.46 ± 0.14)	54.09 (46.96 ± 16.50)	6.25	Nvidia 1080Ti
Ours* - Best solution	22.76 (25.63 ± 2.44)	2.89 (2.67 ± 1.34)	629.11 (545.30 ± 264.25)	5.79	Nvidia 1080Ti
Ours* - Knee solutions	26.57 (30.45 ± 2.67)	0.53 (0.43 ± 0.22)	66.39 (51.89± 28.89)	5.79	Nvidia 1080Ti

with other multi-objective methods, in terms of classification error, the EEEA-NET [30] method shows better performance. In terms of the number of parameters and MAdds, our proposal outperforms the majority of the other methods presented. Our proposal aims to find solutions with a favorable trade-off between classification error and MAdds. In comparison with other proposals that use CGP as a method for architecture representation, such as CGP-CNN [28], Torabi [31], and LF-MOGP [15], our proposal shows superior performance.

In Table 2, we can observe an extensive comparison with different methods on the CIFAR-10 dataset. When performing the comparison with traditional methods designed by humans, our method shows higher performance than the single-objective proposals, both in parameters and classification error. Similar results to those obtained on the CIFAR-100 dataset were achieved with respect to other multi-objective proposals. We should emphasize that our proposal did not perform the search directly in the dataset but rather carried out the transfer of the found architectures using self-supervised learning.

Finally, we assessed the proposed approach on the SVHN and CINIC-10 datasets; corresponding results are shown in Tables 3 and 4. When comparing with methods designed by humans, we can find very close values; however, the solutions found by the proposed method showed a significant reduction in terms of parameters. In comparison to CGP-NASV2, it is observed that, on average, the solutions are less complex and the number of parameters is reduced. On the CINIC-10 dataset, we can see that it is more demanding. We now compare with methods designed by humans; these have a similar performance to those found by our method, highlighting our method in terms of low-complex solutions

Table 2. Comparison on CIFAR-10 dataset. * Transferred architectures

Model	Error rate %	Params	MAdds	GPU-Days	GPU hardware
Human Design					
DenseNet ($k = 12$) [12]	5.24	1.0	–	–	
ResNet ($depth = 101$) [10]	6.43	1.7	–	–	
ResNet ($depth = 1202$) [10]	7.93	10.2	–	–	
VGG [27]	6.66	20.04	–	–	
Single Objective Approaches					
CGP-CNN(ConvSet) [28]	5.92	1.50	–	8	Nvidia 1080Ti
CGP-CNN(ResSet) [28]	5.01	3.52	–	14.7	Nvidia 1080Ti
Large-Scale Evolution [25]	5.4	5.4	–	2750	–
AE-CNN [29]	4.3	2.0	–	27	Nvidia 1080 Ti
Genetic-CNN [34]	7.1	–	–	17	–
(Torabi et al., 2022) [31]	5.69	1.96	–	–	NVIDIA Tesla k80
Multi-Objective Approaches					
NSGANet [16]	3.85	3.3	1290	8	Nvidia 1080 Ti
NSGANetV1 [16]	4.67	0.2	–	27	Nvidia 2080 Ti
MOCNN [32]	4.49	–	–	24	Nvidia 1080 Ti
MOGIG-Net [35]	4.67	0.2	–	14	–
EEEA-Net [30]	2.46	3.6	–	0.52	Nvidia RTX 2080 Ti
EvoApproxNAS [24]	6.80	1.11	458.2	8.8	NVIDIA Tesla V100-SXM2
LF-MOGP [15]	4.13	1.07	–	10	NVIDIA GeForce 3090
CGP-NAS [6]	4.86 (5.42 ± 0.46)	1.40	388.71	1.4	Nvidia Titan X
CGP-NASV2-Best solution [7]	3.70 (4.07± 0.17)	4.04 (5.82 ± 2.70)	636.32 (818.61 ± 372.62)	11.54	Nvidia 1080Ti
CGP-NASV2-Knee solution [7]	4.85 (5.59 ± 0.5)	0.78 (0.71 ± 0.31)	53.99 (79.44 ± 31.96)	11.54	Nvidia 1080Ti
Ours* - Best solution	4.53 (5.30 ± 0.9)	2.11 (2.65 ± 1.34)	684.52 (545.21 ± 264.25)	5.79	Nvidia 1080Ti
Ours* Knee solutions	5.38 (7.43 ± 1.42)	0.82 (0.41 ± 0.22)	87.30 (51.80± 28.88)	5.79	Nvidia 1080Ti

compared to other methods. We have demonstrated that using progressive search in combination with self-supervised learning leads the search to more general architectures. It is worth emphasizing that the search was not carried out on the original dataset. It is a significant achievement to adapt the evolved architectures using our method, which consists of doing the search in a self-supervised way, and thanks to this, the task is simplified because in a conventional search, where for the CIFAR-100 dataset all 100 labels are used, in our approach only 4 are used, which refer to the rotation levels (0°, 90°, 180°, 270°) mentioned above.

5.2   Visual Analysis of the Evolved Architectures

In this section, we visually analyze the architecture found by our proposal using Grad-CAM [26]. Figure 5 shows the architecture trained with the self-supervised approach. Its different network layers are analyzed to see which zones of the images are activated. The initial hypothesis is that when using the self-supervised approach, the CNN architecture will focus on high-level target regions of the images, such as circular or square shapes and more complex shapes like heads or eyes. This architecture was chosen from those found in our proposal. The evolved architecture was trained in two different ways, first in a self-supervised way with the modified CIFAR-100 dataset explained in Sect. 3.2, and in a supervised way on the CINIC-10 dataset.

Two random images from the CINIC-10 dataset were used, which are shown in Fig. 6 and Fig. 7. The first row represents the activation maps using the self-supervised training, and the second row represents the supervised training. Each

Table 3. Comparison on the SVHN dataset. * Transferred architectures

Model	Error rate %	Params	MAdds	GPU-Days	GPU hardware
Human Design					
FractalNet [14]	2.01	38.6	–	–	
Wide ResNet [37]	1.64	2.7	–	–	
ResNet ($depth = 101$) [10]	2.01	1.7	–	–	
DenseNet ($k = 24$) [12]	1.72	15.3	–	–	
Single Objective Approaches					
(Bakhshi et al., 2020) [1]	4.43	19	–	6	
Multi-Objective Approaches					
EvoApproxNAS [24]	3.09	0.90	247.3	8.8	NVIDIA Tesla V100-SXM2
CGP-NASV2 - Best solution [7]	2.70 (2.87± 0.15)	2.21 (4.26 ± 1.88)	399.52 (697.71 ± 210.51)	16.25	Nvidia 1080Ti
CGP-NASV2 - Knee solution [7]	2.88 (3.05 ± 0.18)	0.49 (0.51 ± 0.16)	55.93 (49.31 ± 11.26)	16.25	Nvidia 1080Ti
Ours - Best Solution	2.95 (3.32± 0.43)	2.87 (2.65 ± 1.34)	629.02 (545.21 ± 264.25)	5.79	Nvidia 1080Ti
Ours - Knee Solution	2.82 (3.64± 0.49)	0.51 (0.41 ± 0.22)	66.29 (51.80 ± 28.88)	5.79	Nvidia 1080Ti

Table 4. Comparison on the CINIC-10 dataset. * Transferred architectures

Model	Error rate %	Params	MAdds	GPU-Days	GPU hardware
Human Design					
VGG-16 [2]	12.23	14.7	–	–	
ResNet-18 [2]	9.73	11.2	–	–	
MobileNet [2]	18.00	3.2	–	–	
DenseNet-121 [2]	8.74	7.0	–	–	
GoogLeNet [2]	8.83	6.2	–	–	
Multi-Objective Approaches					
Ours - Best solution	12.16 (13.74 ± 1.4)	3.07 (2.65 ± 1.34)	584.71 (545.21 ± 264.25)	5.79	Nvidia 1080Ti
Ours - Knee solutions	14.33 (17.00 ± 1.92)	0.51 (0.41 ± 0.22)	66.29 (51.80± 28.88)	5.79	Nvidia 1080Ti

image represents an activation map at different depths of the architecture; these are indicated with the layer where the maps were extracted.

We can observe that at first, the architecture found by the self-supervised approach highlights more general details of the input image as well as focuses on sharper edges in it. On the other hand, when the architecture is transferred and trained in a self-supervised way, in the activation maps from the early layers, it focuses on zones of the class to be classified (see Fig. 6). As an illustration, Fig. 7 shows that the bird's head area has a greater impact on the choice made in the end. In contrast, in the self-supervised approach, it seems that the shape and sharp edges of the bird's wings are more important.

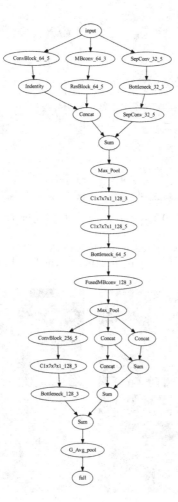

Fig. 5. The evolved architecture for comparison between both approaches using Grad-CAM

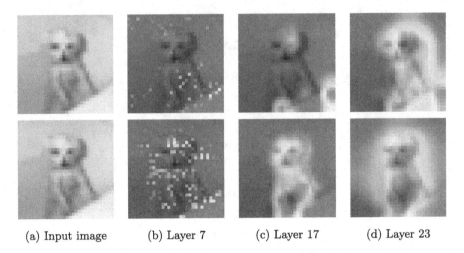

| (a) Input image | (b) Layer 7 | (c) Layer 17 | (d) Layer 23 |

Fig. 6. Activation maps extracted using Grad-CAM. Top: architecture learned with self-supervised learning. Bottom: architecture learned with supervised learning.

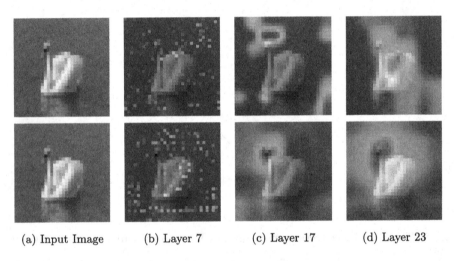

| (a) Input Image | (b) Layer 7 | (c) Layer 17 | (d) Layer 23 |

Fig. 7. Activation maps extracted using Grad-CAM

6   Conclusions

From the results obtained, we can conclude that the architectures found by our method concentrate on operations that can generalize more easily; this is confirmed by the analysis performed with Grad-CAM. Furthermore, from the results obtained, we can say that our proposal has promising results in addition to a considerable reduction in GPU days.

The use of self-supervised learning, in particular the use of RotNet, which focuses on estimating the geometric transformation, allows the network to focus on object shapes, edges, and textures within images. Consequently, the network learns to capture important features from the image. With this idea, the generalization obtained from this method allows for the transfer of the found architectures to datasets with similar data.

In summary, the proposed approach leads a new way to perform the search of neural architectures while relying on the use of self-supervised learning in order to generate models suitable to user-defined characteristics in environments where limited data is present. Moreover, the progressive searching approach opens new research niches after the competitive results achieved in this preliminary study. Finally, our work not only expands the scope of effective neural architecture search but also remarks on the potential of self-supervised learning as a valuable tool for architecture exploration and adaptation.

Acknowledgements. The authors thankfully acknowledge computer resources, technical advice and support provided by Laboratorio Nacional de Supercómputo del Sureste de México (LNS), a member of the CONACYT national laboratories, with project No. 202103083C. This work was supported by CONACyT under grant CB-S-26314.

References

1. Bakhshi, A., Chalup, S., Noman, N.: Fast evolution of CNN architecture for image classification. In: Iba, H., Noman, N. (eds.) Deep Neural Evolution. NCS, pp. 209–229. Springer, Singapore (2020). https://doi.org/10.1007/978-981-15-3685-4_8
2. Darlow, L.N., Crowley, E.J., Antoniou, A., Storkey, A.J.: CINIC-10 is not imagenet or CIFAR-10. CoRR abs/1810.03505 (2018). http://arxiv.org/abs/1810.03505
3. Elsken, T., Metzen, J.H., Hutter, F.: Neural architecture search: a survey. Technical report. (2019). http://jmlr.org/papers/v20/18-598.html
4. Feng, Z., Xu, C., Tao, D.: Self-supervised representation learning by rotation feature decoupling. In: 2019 IEEE/CVF Conference on Computer Vision and Pattern Recognition (CVPR), pp. 10356–10366 (2019). https://doi.org/10.1109/CVPR.2019.01061
5. Fernandes Jr., F.E., Yen, G.G.: Pruning deep convolutional neural networks architectures with evolution strategy. Inf. Sci. **552**, 29–47 (2021). https://doi.org/10.1016/j.ins.2020.11.009. https://linkinghub.elsevier.com/retrieve/pii/S0020025520310951
6. Garcia-Garcia, C., Escalante, H.J., Morales-Reyes, A.: CGP-NAS. In: Proceedings of the Genetic and Evolutionary Computation Conference Companion, vol. 1, pp. 643–646. ACM, New York (2022). https://doi.org/10.1145/3520304.3528963
7. Garcia-Garcia, C., Morales-Reyes, A., Escalante, H.J.: Continuous cartesian genetic programming based representation for multi-objective neural architecture search. Appl. Soft Comput. **147**, 110788 (2023). https://doi.org/10.1016/j.asoc.2023.110788. https://www.sciencedirect.com/science/article/pii/S1568494623008062
8. Grigorescu, S., Trasnea, B., Cocias, T., Macesanu, G.: A survey of deep learning techniques for autonomous driving. J. Field Rob. **37**(3), 362–386 (2020)
9. Gui, J., et al.: A survey on self-supervised learning: algorithms, applications, and future trends (2023)

10. He, K., Zhang, X., Ren, S., Sun, J.: Deep residual learning for image recognition. In: Proceedings of the IEEE Conference on Computer Vision and Pattern Recognition, pp. 770–778 (2016)

11. Heuillet, A., Tabia, H., Arioui, H.: Nasiam: Efficient representation learning using neural architecture search for siamese networks (2023). http://arxiv.org/abs/2302.00059

12. Huang, G., Liu, Z., Van Der Maaten, L., Weinberger, K.Q.: Densely connected convolutional networks. In: Proceedings of the IEEE Conference on Computer Vision and Pattern Recognition, pp. 4700–4708 (2017)

13. Kolbk, M., Tan, Z.H., Jensen, J., Kolbk, M., Tan, Z.H., Jensen, J.: Speech intelligibility potential of general and specialized deep neural network based speech enhancement systems. IEEE/ACM Trans. Audio Speech Lang. Proc. 25(1), 153–167 (2017). https://doi.org/10.1109/TASLP.2016.2628641

14. Larsson, G., Maire, M., Shakhnarovich, G.: Fractalnet: ultra-deep neural networks without residuals. CoRR abs/1605.07648 (2016). http://arxiv.org/abs/1605.07648

15. Liu, Q., Wang, X., Wang, Y., Song, X.: Evolutionary convolutional neural network for image classification based on multi-objective genetic programming with leader-follower mechanism. Complex Intell. Syst. (2022). https://doi.org/10.1007/s40747-022-00919-y

16. Lu, Z., Deb, K., Goodman, E., Banzhaf, W., Boddeti, V.N.: NSGANetV2: evolutionary multi-objective surrogate-assisted neural architecture search. In: Vedaldi, A., Bischof, H., Brox, T., Frahm, J.-M. (eds.) ECCV 2020. LNCS, vol. 12346, pp. 35–51. Springer, Cham (2020). https://doi.org/10.1007/978-3-030-58452-8_3

17. Lu, Z., Sreekumar, G., Goodman, E., Banzhaf, W., Deb, K., Boddeti, V.N.: Neural architecture transfer. IEEE Trans. Pattern Anal. Mach. Intell. (2021). https://doi.org/10.1109/TPAMI.2021.3052758. https://ieeexplore.ieee.org/document/9328602/

18. Lu, Z., et al.: NSGA-Net. In: Proceedings of the Genetic and Evolutionary Computation Conference, pp. 419–427. ACM, New York (2019). https://doi.org/10.1145/3321707.3321729

19. Lu, Z., et al.: Multi-objective evolutionary design of deep convolutional neural networks for image classification. IEEE Trans. Evol. Comput. (2020). https://doi.org/10.1109/TEVC.2020.3024708. https://ieeexplore.ieee.org/document/9201169/

20. Martinez, A.D., et al.: Lights and shadows in evolutionary deep learning: taxonomy, critical methodological analysis, cases of study, learned lessons, recommendations and challenges. Inf. Fusion 67, 161–194 (2021). https://doi.org/10.1016/j.inffus.2020.10.014

21. Miikkulainen, R., et al.: Evolving deep neural networks. In: Artificial Intelligence in the Age of Neural Networks and Brain Computing, pp. 293–312. Elsevier (2019). https://doi.org/10.1016/B978-0-12-815480-9.00015-3. https://linkinghub.elsevier.com/retrieve/pii/B9780128154809000153

22. Miller, J., Thomson, P., Fogarty, T., Ntroduction, I.: Designing electronic circuits using evolutionary algorithms. arithmetic circuits: a case study. Genetic Algor. Evol. Strat. Eng. Comput. Sci. (1999)

23. Nguyen, N., Chang, J.M.: CSNAS: contrastive self-supervised learning neural architecture search via sequential model-based optimization. IEEE Trans. Artif. Intell. 3(4), 609–624 (2022). https://doi.org/10.1109/TAI.2021.3121663

24. Pinos, M., Mrazek, V., Sekanina, L.: Evolutionary approximation and neural architecture search. Genetic Program. Evol. Mach. (2022). https://doi.org/10.1007/s10710-022-09441-z. https://link.springer.com/10.1007/s10710-022-09441-z

25. Real, E., et al.: Large-scale evolution of image classifiers (2017). http://arxiv.org/abs/1703.01041
26. Selvaraju, R.R., Cogswell, M., Das, A., Vedantam, R., Parikh, D., Batra, D.: Gradcam: visual explanations from deep networks via gradient-based localization. In: 2017 IEEE International Conference on Computer Vision (ICCV), pp. 618–626 (2017). https://doi.org/10.1109/ICCV.2017.74
27. Simonyan, K., Zisserman, A.: Very deep convolutional networks for large-scale image recognition. arXiv preprint arXiv:1409.1556 (2014)
28. Suganuma, M., Kobayashi, M., Shirakawa, S., Nagao, T.: Evolution of deep convolutional neural networks using cartesian genetic programming (2020). https://doi.org/10.1162/evco_a_00253
29. Sun, Y., Wang, H., Xue, B., Jin, Y., Yen, G.G., Zhang, M.: Surrogate-assisted evolutionary deep learning using an end-to-end random forest-based performance predictor. IEEE Trans. Evol. Comput. 24(2), 350–364 (2020). https://doi.org/10.1109/TEVC.2019.2924461. https://ieeexplore.ieee.org/document/8744404/
30. Termritthikun, C., Jamtsho, Y., Ieamsaard, J., Muneesawang, P., Lee, I.: EEEA-Net: an early exit evolutionary neural architecture search. Eng. Appl. Artif. Intell. 104, 104397 (2021). https://doi.org/10.1016/j.engappai.2021.104397
31. Torabi, A., Sharifi, A., Teshnehlab, M.: Using cartesian genetic programming approach with new crossover technique to design convolutional neural networks. Neural Process. Lett. (2022). https://doi.org/10.1007/s11063-022-11093-0
32. Wang, B., Xue, B., Zhang, M.: Particle swarm optimization for evolving deep convolutional neural networks for image classification: single- and multi-objective approaches, pp. 155–184 (2020). https://doi.org/10.1007/978-981-15-3685-4
33. Wei, C., Tang, Y., Chuang Niu, C.N., Hu, H., Wang, Y., Liang, J.: Self-supervised representation learning for evolutionary neural architecture search. IEEE Comput. Intell. Mag. 16(3), 33–49 (2021). https://doi.org/10.1109/MCI.2021.3084415
34. Xie, L., Yuille, A.: Genetic CNN. In: 2017 IEEE International Conference on Computer Vision (ICCV), pp. 1388–1397. IEEE (2017). https://doi.org/10.1109/ICCV.2017.154. http://ieeexplore.ieee.org/document/8237416/
35. Xue, Y., Jiang, P., Neri, F., Liang, J.: A Multi-objective evolutionary approach based on graph-in-graph for neural architecture search of convolutional neural networks. Int. J. Neural Syst. 31(9) (2021). https://doi.org/10.1142/S0129065721500350
36. Young, T., Hazarika, D., Poria, S., Cambria, E.: Recent trends in deep learning based natural language processing [review article]. IEEE Comput. Intell. Mag. 13(3), 55–75 (2018). https://doi.org/10.1109/MCI.2018.2840738
37. Zagoruyko, S., Komodakis, N.: Wide residual networks. CoRR abs/1605.07146 (2016). http://arxiv.org/abs/1605.07146

Genetic Programming with Aggregate Channel Features for Flower Localization Using Limited Training Data

Qinyu Wang[1], Ying Bi[2(✉)], Bing Xue[1], and Mengjie Zhang[1]

[1] Victoria University of Wellington, Wellington 6140, New Zealand
{qinyu.wang,bing.xue,mengjie.zhang}@ecs.vuw.ac.nz
[2] Zhengzhou University, Zhengzhou 450001, China
yingbi@zzu.edu.cn

Abstract. Flower localization is a crucial image pre-processing step for subsequent classification/recognition that confronts challenges with diverse flower species, varying imaging conditions, and limited data. Existing flower localization methods face limitations, including reliance on color information, low model interpretability, and a large demand for training data. This paper proposes a new genetic programming (GP) approach called ACFGP with a novel representation to automated flower localization with limited training data. The novel GP representation enables ACFGP to evolve effective programs for generating aggregate channel features and achieving flower localization in diverse scenarios. Comparative evaluations against the baseline benchmark algorithm and YOLOv8 demonstrate ACFGP's superior performance. Further analysis highlights the effectiveness of the aggregate channel features generated by ACFGP programs, demonstrating the superiority of ACFGP in addressing challenging flower localization tasks.

Keywords: Genetic programming · Aggregate channel features · Flower localization

1 Introduction

Flower localization is a computer vision task focused on precisely determining the location of flowers within images. As a common image pre-processing operation, flower localization contributes to the further identification and classification of flower species, supporting botanical research and biodiversity monitoring [12]. The main challenges of flower localization stem from the diversity of flower classes, variations within specific flowers, various imaging conditions, and limited data on some rare flower species.

To address the challenges above, manual segmentation methods, such as GrabCut [17], are typically used to localize the flower and remove the background as image pre-processing in the early stage, which requires labor-intensive efforts. For automatic flower localization, traditional methods typically utilize flower color and shape information to design threshold-based methods [5,11,18]. However, these methods require domain knowledge about flowers for manual selection of feature descriptors and algorithm design. Furthermore, the performance of common threshold-based techniques based on manually crafted image processing programs might deteriorate when confronted with complex background scenes and diverse imaging conditions in flower images. The aggregate channel features (ACF) detector, which extends channels beyond typical color representations to non-color features, has proven effective in enhancing robustness for localizing objects. It has been successfully applied in various object detection tasks, such as localizing pedestrians [3], faces [22], and urine sediments [19]. However, its application in flower localization remains unexplored.

Convolutional neural network (CNN) methods such as Faster R-CNN [16] and YOLO [6] have been widely applied in object detection, including localizing and classifying flowers [15,20]. However, these methods need pre-trained models and still demand a relatively large amount of training data for fine-tuning due to high model complexity. Their large network structures require Graphics processing units (GPUs) for training, resulting in high computational costs. Additionally, the black-box nature of these models reduces the interpretability in explaining object localization processes.

Genetic Programming (GP) is an evolutionary algorithm inspired by natural selection, designed to automatically evolve programs/solutions through the iterative application of evaluation, selection, and genetic operations including mutation and crossover [8]. The tree-based variable-length representation in GP enables the generation of programs with flexible structures and good interpretability. GP has been successfully applied in image classification with limited data [1].

However, existing GP-based methods that involve region detection [2,21] lack flexibility as the positions of the detected regions are randomly selected and remain fixed across all images. Therefore, the region detection mechanism employed in these methods lacks the capability to effectively capture flowers with varying positions within an image.

1.1 Goals

The goal of this paper is to develop a GP approach, named aggregate channel feature based GP (ACFGP), that can automatically generate effective aggregate channel features to localize the flower within an image with limited training data. This paper focuses on the single-object localization task, where the evolved program is expected to produce a bounding box that accurately locates the flower in each image. The specific objectives of this work are summarised as follows:

- Develop a new function set and a new terminal set that include image processing operators to extract and aggregate channel features to enhance the flower, and dynamically compute bounding boxes based on the specific characteristics of each image;
- Design a new GP representation to evolve effective solutions/programs for various flower localization tasks with limited training data;

- Analyze and compare the performance of ACFGP with the baseline benchmark and the YOLOv8 method; and
- Further analyze the trees evolved by ACFGP along with the visualized results to investigate the effectiveness of aggregate channel features.

2   Backgrounds and Related Work

2.1   Existing Methods for Flower Localization

In the early days, manual segmentation algorithms such as GrabCut [17] are used to segment flowers and remove background [14]. GrabCut is a semi-automatic image segmentation algorithm developed based on the color distribution and graph-cut approach, which requires manually framing the bounding box of the flower as a rough segmentation. An automated flower segmentation algorithm [11] is proposed as a pre-processing for automated flower classification [12]. This method consists of two models, one color model for foreground/background segmentation and a generic shape model for petal structure. However, the performance of color-based segmentation might be affected by diverse illumination situations. Threshold-based algorithms are commonly used to localize flowers as pre-processing in traditional flower classification methods [5,18]. A flower is segmented by the threshold calculated based on the intensity values transformed from color information, whose performance might be affected under complex background and lighting conditions.

Flower localization is commonly regarded as part of the flower detection task within CNN methods, often coupled with classification. In [15], an improved Region Proposal Network (RPN) is utilized to generate region proposals for flowers, and a modified Faster R-CNN model is employed for bounding box regression. In [20], flower localization and classification are simultaneously addressed using the YOLOv4 model [6]. Despite the model's simplification via a channel pruning algorithm, the pruned version still contains over two million parameters, resulting in low interpretability. In addition, both methods in [15,20] require pre-trained models, and thousands of flower images are required for fine-tuning.

2.2   Aggregate Channel Features (ACF) for Object Localization

A channel typically refers to a component of an image that represents specific color information, such as red, green, and blue. In this paper, we borrow the concept from [3], where channels are defined as a feature map of the original image, whose pixels are computed from corresponding patches of original pixels. Based on this definition, additional non-color channels are introduced that include a series of feature maps generated through image filters and descriptors. The additional channels enable comprehensive image analysis beyond color information, thus enhancing the capability to comprehend and interpret complex visual data.

ACF methods have been applied to diverse object localization methods. In [3], integral/aggregate channel features are introduced via linear and non-linear transformations. The experimental evaluation demonstrates that ACF enables accurate spatial localization during detection if designed properly. In [22], channel features including color channels, (i.e., Gray-scale, RGB, HSV, and LUV), gradient magnitude, and gradient histograms are used for face detection. Similarly, in [19], ten channel features, i.e., three color channels in LUV space, normalized gradient magnitude, and histogram of oriented gradients (six orientations) are used to localize urine sediments within images. However, these methods rely on a sliding window for localization, which increases the computational cost and limits the ability to localize objects with varying sizes.

In summary, traditional automated flower localization methods primarily rely on thresholding based on color information, making them less effective under complex backgrounds and diverse imaging conditions. CNN-based methods often need fine-tuning with pre-trained models for flower localization, still demanding thousands of images for fine-tuning, and lacking interpretability due to high model complexity. Existing GP-based methods have limitations in flower localization due to randomly selected regions that remain static across all images. Existing ACF methods that rely on sliding windows for localization restrict the adaptability to objects of varying sizes. Therefore, this paper aims to develop a GP approach with ACF to flower localization with limited training data.

3 Proposed Approach

3.1 The New GP Representation

The proposed ACFGP approach has a new GP representation that is based on strongly typed GP (STGP) [10]. In ACFGP, the input for the GP program/tree is an RGB image, and the output is one bounding box identifying the salient flower within the image. ACFGP comprises five layers from input to output, each with a distinct role in the process:

- Input layer: Represents the input RGB image;
- Channel extension layer: Involves color channel selection and channel feature extraction. This layer begins by choosing one from seven color channels (including RGB, LUV, and grayscale) and then applies an image processing filter to the chosen channel, resulting in an extended channel feature.
- Channel aggregation layer: Aggregate the extended channel features from different tree branches to create a saliency mask, highlighting the regions of interest.
- Object localization layer: Detects the salient flower and identifies its bounding box based on the saliency mask.
- Output layer: Represents the bounding box of the flower.

Figure 1 shows how the proposed ACFGP approach constructs a tree using various functions and terminals.

3.2 Terminal Set

Terminals serve as the leaf nodes in a GP tree. There are two types of terminals in ACFGP, i.e., *Img* and *Idx*, as detailed in Table 1. *Img* denotes the RGB image containing salient flowers. It's a three-dimensional array with red, green, and blue pixel values. *Idx* denotes the index used to select a color channel from the RGB image. *Idx* is an integer ranging from 0 to 6, where each index corresponds to one of the seven color channels, i.e., blue, red, green, L, U, V, and grayscale channels.

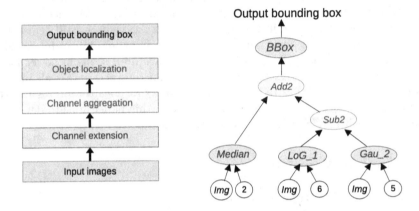

Fig. 1. The program structure and an example program/tree of ACFGP.

Table 1. Terminal Set

Terminal	Type	Value range	Description
Img	array	[0,255]	The 3-channel RGB image
Idx	integer	[0,6]	The index to generate a color channel from an RGB image. Including RGB, LUV, and gray-scale channels

3.3 Function Set

The function set of ACFGP is presented in Table 2 and consists of three categories: channel extension functions, channel aggregation functions, and object localization functions.

Channel extension functions are the image processing filters and descriptors, i.e., *Gau_1*, *Gau_2*, *Median*, *Min_max*, *LoG_1*, *LoG_2*, *Sobel*, *HOG*, and *Saliency*. *Gau_1*, *Gau_2*, and *Median* are used for image smoothing and noise reduction. *Gau_1* and *Gau_2* execute Gaussian filtering with different σ, addressing small/larger scale smoothing. *Min_max*, *LoG_1*, *LoG_2*, and *Sobel* functions are different types of edge detectors. *LoG_1* and *LoG_2* focus on small/larger scale of edge enhancement. *HOG* is a feature descriptor that extracts the distribution and orientation of gradients in an image, capturing important structural information from images, and allowing for robust and reliable pattern recognition. The *Saliency* function is from OpenCV's Saliency module, which detects visually prominent areas in an image based on color, texture, and intensity.

Channel aggregation functions are used to aggregate channel features generated by the channel extension functions into a saliency mask, which is used to localize the salient flower. Since the channel features extracted by the channel extension functions can be considered as a feature map, where pixel values represent feature intensities at specific locations, channel aggregation functions perform addition or subtraction operations on these intensities to enhance features in the region of interest, generating a saliency mask. These functions include *Sub2*, *Add2*, *Add3*, and *Add4*. *Sub2* calculates the absolute difference between two channels, whereas *Add2*, *Add3*, and *Add4* functions perform addition operations on two, three, and four channels, respectively. *Sub2* and *Add2* functions have the flexibility

Table 2. Function Set

Function	Input	Output	Description
Channel extension functions			
Gau_1	Img, Idx	Channel	Perform Gaussian filtering with stand deviation σ 1 on the selected color channel
Gau_2	Img, Idx	Channel	Perform Gaussian filtering with stand deviation σ 2 on the selected color channel
Median	Img, Idx	Channel	Perform 5 × 5 median filtering on the selected color channel
Min_max	Img, Idx	Channel	Perform 3 × 3 min-max filtering on the selected color channel
LoG_1	Img, Idx	Channel	Perform Laplacian of Gaussian filtering with stand deviation σ 1 on the selected color channel
LoG_2	Img, Idx	Channel	Perform Laplacian of Gaussian filtering with stand deviation σ 2 on the selected color channel
Sobel	Img, Idx	Channel	Perform 3 × 3 Sobel filtering on the selected color channel
HOG	Img, Idx	Channel	Extract HOG feature vectors on the selected color channel and transfer them into an image-based representation
Saliency	Img	Channel	Perform static saliency detection and generate a Saliency map
Channel aggregation functions			
Sub2	2 Channels	Channel/Mask	Perform the absolute difference operation between 2 channels
Add2	2 Channels	Channel/Mask	Perform the addition operation on 2 channels
Add3	3 Channels	Mask	Perform the addition operation on 3 channels
Add4	4 Channels	Mask	Perform the addition operation on 4 channels
Object localization functions			
GBBox	Mask	BBox	Calculate the bounding box based on the mask with a 5 × 5 Gaussian filtering
oBBox	Mask	BBox	Calculate the bounding box based on the mask with an opening operation
cBBox	Mask	BBox	Calculate the bounding box based on the mask with a closing operation

to produce either an intermediate channel feature for further aggregation or a saliency mask for object localization.

Object localization functions detect the salient flower in an image using the saliency mask and determine the flower's bounding box. The three functions, *GBBox*, *oBBox*, and *cBBox*, employ OpenCV operations including threshold, morphologyEx, findContours, and boundingRect, to execute the object localization process, as shown in Fig. 2.

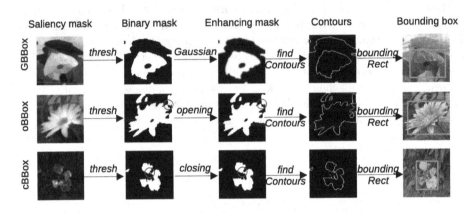

Fig. 2. The pipeline of the *GBBox*, *oBBox*, and *cBBox* functions.

The object localization process starts with the thresholding operation, which converts the saliency mask into a binary mask using a dynamic threshold method [13]. The main differences between *GBBox*, *oBBox*, and *cBBox* functions arise from their approaches to enhancing the binary mask based on different scene characteristics.

1) The opening operation in the *oBBox* function serves to eliminate noise, smooth object boundaries, and separate closely spaced objects.
2) The closing operation in the *cBBox* function aims to fill gaps or holes within objects, connect fragmented structures, and create more compact and complete object representations.
3) The *GBBox* function applies to the case where binary mask quality is satisfied and performs a 5×5 Gaussian filtering operation to smooth the binary mask.

Following the enhancement of the binary mask, the findContours operation identifies potential contours within this mask. The largest contour is considered the salient flower. Finally, the boundingRect operation calculates the bounding box of the largest contour as $[x, y, w, h]$, where $[x, y]$ represents the top-left point of the bounding box, and w and h indicate the width and height of the bounding box, respectively. The output of these object localization functions is represented as $[x, y, x + w, y + h]$, where $[x + w, y + h]$ denotes the bottom-right point of the bounding box. It's worth noting that, unlike existing GP methods, the calculated bounding boxes are different in each image, as they are based on the specific content of the image.

3.4   Fitness Function

The fitness function for ACFGP is the numerical sum of two metrics: detection accuracy and intersection over union (IoU). The formulas for these two metrics are given below.

$$\text{Fitness} = \text{Detection accuracy} + \text{Average IoU} \tag{1}$$

$$\text{Detection accuracy} = \frac{TP}{TP + FP} \tag{2}$$

$$\text{Average IoU} = \frac{1}{N} \sum_{i}^{N} \text{IoU}\left(\text{Prediction}_i, \text{Groundtruth}_i\right) \tag{3}$$

A bounding box is considered a true positive if it captures more than 50% (including 50%) of the target in terms of IoU, while a bounding box with less than 50% IoU is considered a false positive. However, detection accuracy alone cannot differentiate between cases with 50% and 100% IoU. To provide a more comprehensive evaluation for GP programs with the same detection accuracy, the fitness function also incorporates the average of IoU, where N denotes the number of training instances. As a result, the fitness function of ACFGP is a value ranging between 0% and 200%.

3.5   Test Process

In the test process, the best GP tree/program obtained from the evolutionary training process will be used to predict the images in the test set. According to the best GP tree, a set of aggregate channel features will be extended and aggregated to generate a saliency mask, and the bounding box of the salient flower will be calculated in each test image. Similar to fitness evaluation, the detection accuracy and average IoU will be calculated to evaluate the performance of the ACFGP approach but the results are shown separately.

4   Experiment Design

4.1   Datasets

The proposed ACFGP approach is developed for flower localization, which could be applied as image pre-processing for further fine-grained image classification (FGIC). Therefore, the performance of ACFGP is examined on the FGIC dataset: the Oxford_102 dataset. The Oxford_102 dataset contains 102 categories of flower images with 40–258 images per category. The images from the Oxford_102 dataset vary in size, and to ensure uniformity, all images are resized to 224 × 224 for input, which is also a commonly used size in training and testing.

To assess ACFGP's ability to flexibly extend and aggregate different channel features based on the characteristics of different flower species, experiments are conducted on five flower sub-datasets. Each sub-dataset consists of five classes randomly selected from the original Oxford_102 dataset, namely s1-s5 of flower5. To explore ACFGP's performance with limited training data, 10 images from each class are randomly selected as the training set, resulting in a training set of 10×5 images. The remaining images (ranging from 30 to 248 images per class) form the test set for each sub-dataset. For instance, to build flower5-s1, five flower species are randomly selected from the Oxford_102 dataset, with 10 images per class (50 images in total) used for training and the remaining images (279 images in total) assigned as the test set. This process is repeated five times to establish the five sub-datasets. Figure 3 presents example images of the sub-datasets used in the experiments. The flowers in the images vary in color and size, and exist in diverse backgrounds and lighting conditions. Additionally, the positions of the flowers differ in each image.

Fig. 3. Example training images of s1-s5 in flower5.

4.2   Comparison Methods

To assess ACFGP's performance, we compare it with Baseline and YOLOv8. Baseline serves as a benchmark for evaluating aggregate channel features against single-channel features in flower localization. YOLOv8 is chosen as a well-known and latest CNN method in object detection.

Baseline: The Baseline benchmark algorithm employs an exhaustive search across all single-channel features, representing possible programs generated by ACFGP's function set without channel aggregation functions. The search space comprises 171 possible solutions, which include 168 single-channel features extracted by 8 image filters and edge detectors (7 color channels × 8 channel extension functions × 3 object localization functions) and 3 feature maps generated by the *Saliency* function (with 3 object localization functions). For each sub-dataset, Baseline explores all possible solutions, and the program with the highest fitness value on the training set is selected as the best individual. Given its exhaustive nature, only a single-run experiment is conducted on each sub-dataset.

YOLOv8: The YOLOv8 method is a well-known and latest object detection model designed for identifying and localizing objects in images. Due to limited training data, the YOLOv8s (small) model from the YOLOv8 series [7] is used as a comparison method in this paper, which contains 11.2 million parameters.

4.3   Parameter Settings

ACFGP: The population size is 100 and the maximal number of generations is 50. The crossover rate is 0.5, the mutation rate is 0.49, and the elitism rate is 0.01. The Ramped-half-and-half method is used in population initialization and the mutation operation. The minimal and maximal tree depth are set to 2 and 6, respectively. Tournament selection with the size 5 is employed for selection. The ACFGP method is implemented using the DEAP package [4]. 30 independent runs of experiments are conducted with different random seeds for ACFGP on each sub-dataset.

YOLOv8: The YOLOv8s model pre-trained on the COCO dataset [9] is used for fine-tuning for comparison in this paper. The image size input is set to 224×224, the training epoch is set to 100, the batch size is set to 16, the optimizer is SGD, the learning rate is set to 0.01, and the remaining parameters follow the default setting in [7]. Ten independent experimental runs are performed with different random seeds for YOLOv8 on each sub-dataset. In the testing stage, considering that the YOLOv8 method permits multiple bounding box outputs within an image, the final output that localizes the flower is determined by selecting the bounding box with the highest confidence among multiple positive predictions.

5   Results and Discussions

This section compares the object localization performance, including the average IoU and the detection accuracy, of ACFGP, YOLOv8, and Baseline methods over the 30/10/1 runs on the s1-s5 of flower5. The evaluation focuses on two metrics: average IoU, which assesses the quality of object localization, and detection accuracy, which emphasizes the correctness of detections. The testing results, including the maximal (Max) and mean with standard deviation (Mean ± Std) for each metric, are presented in Tables 3 and 4, with the best results highlighted in bold. The Wilcoxon rank-sum test is used for the significance test with the $p\text{-}value = 0.05$. The symbols "+" or "−" in the tables indicate that ACFGP achieves significantly better or worse performance than the compared method. The symbol "=" indicates similar performance between ACFGP and the compared method. The final row in each table summarizes the significance test results on each sub-dataset.

5.1   Average IoU Results

Compared with Baseline: Table 3 demonstrates ACFGP consistently outperforming the Baseline across all comparisons, highlighting its superior ability to extend and aggregate more effective channel features for precise flower localization. This could be attributed to ACFGP's aggregate channel features, which capture diverse and complementary information, enhancing the discrimination of flower features.

Compared with YOLOv8: Table 3 illustrates that ACFGP significantly outperforms YOLOv8 on 3 out of 5 sub-datasets. While YOLOv8 achieves a higher maximal average IoU on flower5-s5, ACFGP maintains competitive performance overall on this sub-dataset. This suggests that ACFGP excels in localizing flowers with limited training data in most comparisons. It's worth noting that in 4 out of 5 comparisons, the standard deviation values of YOLOv8's results are larger than those of ACFGP, indicating that ACFGP achieves a more stable performance than YOLOv8. Despite being pre-trained, YOLOv8 usually demands thousands of flower images to fine-tune its model due to the high model complexity to achieve stable and satisfactory performance. Given the limited training data in this task, capturing discriminative features in flowers becomes challenging for YOLOv8, so it is reasonable that YOLOv8 shows inferior flower localization performance compared to ACFGP.

Table 3. Average IoU (%) of ACFGP, YOLOv8, and Baseline on flower5 sub-datasets

		ACFGP		YOLOv8		Baseline
		Max	Mean ± Std	Max	Mean ± Std	Max/Mean
flower5	s1	**79.12**	**77.53 ± 1.53**	78.97	73.71 ± 3.33+	75.65+
	s2	**81.66**	**77.20 ± 2.96**	79.62	75.35 ± 3.02=	63.88+
	s3	**82.57**	**80.59 ± 0.85**	77.30	73.01 ± 4.95+	78.84+
	s4	**84.03**	**80.90 ± 1.96**	82.22	79.23 ± 1.86+	78.71+
	s5	75.57	72.33 ± 1.79	**77.30**	**73.95 ± 3.18=**	72.21+
Overall					3+, 2=	5+

5.2 Detection Accuracy Results

Compared with Baseline: Table 4 reveals that ACFGP outperforms Baseline significantly in all comparisons. This suggests that ACFGP's aggregate channel features excel in localizing more objects with IoU over 0.5 compared to the single-channel feature in Baseline. Different channel features could highlight different aspects of the flower, such as the shape, texture, or color. Aggregating these channel features improves robustness against variations in lighting conditions and image noise. Consequently, ACFGP achieves better localization results with IoU over 0.5.

Compared with YOLOv8: As shown in Table 4, although YOLOv8 achieves better maximal detection accuracy on flower5-s1 and flower5-s5, ACFGP significantly outperforms YOLOv8 on 4 out of 5 flower sub-datasets in significance tests, and achieves competitive performance on flower5-s5. This indicates that ACFGP outperforms YOLOv8 in localizing more flowers with IoU over 0.5 in most comparisons, aligning with the performance observed in average IoU in Table 3.

Table 4. Detection accuracy (%) of ACFGP, YOLOv8, and Baseline on flower5 sub-datasets

		ACFGP		YOLOv8		Baseline
		Max	Mean ± Std	Max	Mean ± Std	Max/Mean
flower5	s1	89.42	**87.06 ± 1.99**	**89.61**	81.94 ± 4.21+	84.67+
	s2	**91.74**	**88.20 ± 2.58**	88.26	84.00 ± 4.08+	74.35+
	s3	**92.50**	**90.97 ± 0.98**	88.03	80.56 ± 4.76+	89.79+
	s4	**94.50**	**91.88 ± 2.08**	92.98	89.77 ± 1.85+	90.64+
	s5	87.06	82.77 ± 2.21	**88.10**	**82.83 ± 3.37**=	81.03+
Overall					4+, 1=	5+

6   Further Analysis

In this section, we analyze a GP tree/program evolved by ACFGP and compare the object localization performance between ACFGP, Baseline, and YOLOv8 to provide more insight into the effectiveness of the aggregate channel features.

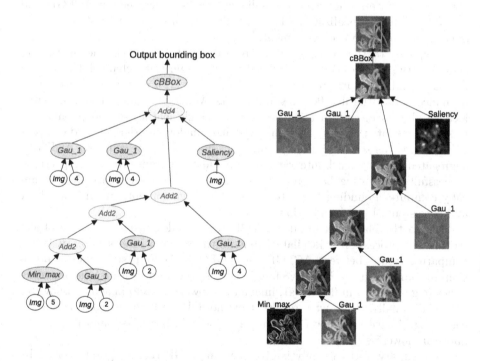

Fig. 4. An example program evolved by ACFGP on flower5-s4 and the example images to show the object localization process using the program.

6.1   Analysis of an Example GP Tree

Figure 4 presents an example GP tree evolved by ACFGP on flower5-s4, which achieves 82.47% average IoU and 93.13% detection accuracy. The program extends and aggregates seven channel features derived from three color channels (U, V, and red channels), incorporating 5 Gaussian features, 1 Min_max feature, and 1 Saliency feature. The corresponding example images reveal that the Gaussian features effectively suppress background noise, and the Min_max and Saliency features which contain edge information of the petal and visually important areas highlight the flower. The generated saliency mask plays a crucial role in reducing background noise and enhancing the flower, which results in a precise bounding box output.

The example GP tree showcases the effectiveness of the ACFGP approach, which autonomously selects informative color channels, extends them into effective channel features using diverse image filters and descriptors, and aggregates these features to generate the saliency mask for precise object localization.

6.2   Visual Comparison Between ACFGP, Baseline, and YOLOv8

Figure 5 displays the results of ACFGP, Baseline, and YOLOv8 across the five flower sub-datasets, alongside the saliency masks generated by ACFGP and Baseline for flower localization. These results are based on the best-run results of each method from the experiment.

Compared to the saliency mask created by the single-channel feature in Baseline, the saliency mask produced by the aggregate channel features in ACFGP exhibits greater robustness, leading to superior object localization performance across the five flower sub-datasets. While Baseline might accurately localize objects in certain instances using a single feature map (e.g., image a in flower5-s3), it struggles with challenges like shadow interference and complex backgrounds (as seen in images c and d in flower5-s3). In contrast, ACFGP's aggregated saliency mask integrates effective features from diverse perspectives, successfully mitigating background noise and shadow interference. It dynamically computes bounding boxes based on the image content, ensuring a robust and consistent object localization performance.

Due to the black-box nature of YOLOv8, detailed insights into its object localization process are unavailable, contributing to poor interpretability. When comparing results between ACFGP and YOLOv8, ACFGP shows better precision. In some instances, YOLOv8 fails to enclose the flower within blue bounding boxes (e.g., image b in flower5-s1, image c in flower5-s2, and image b in flower5-s4). Additionally, YOLOv8's results show notable background noise in several cases (e.g., image e in flower5-s1; image d and e in flower5-s4; and image a, c, and d in flower5-s5).

In summary, further analysis of the example GP tree/program evolved by ACFGP shows its high interpretability. The program provides a clear explanation of how the program extends and aggregates channel features to achieve precise flower localization. In addition, the comparison between ACFGP, Baseline, and YOLOv8 considering saliency masks and the corresponding results, highlights the effectiveness and robustness of aggregate channel features for precise flower localization.

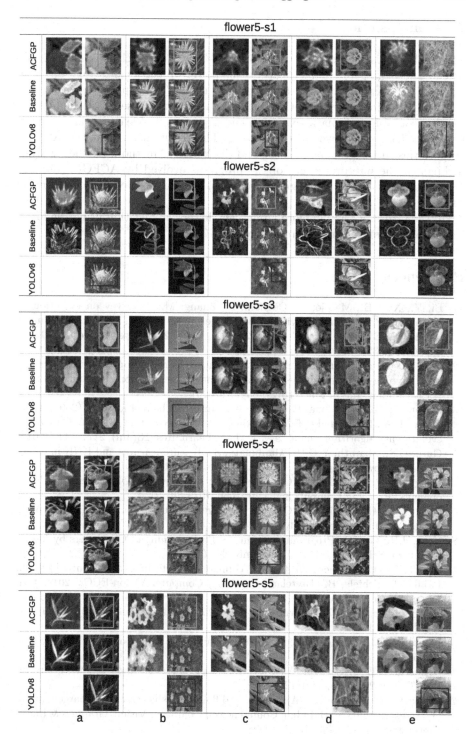

Fig. 5. The saliency masks and corresponding results of ACFGP, Baseline, and YOLOv8 on five flower sub-datasets.

7   Conclusions

The goal of this paper is to develop a GP-based approach to automatically extend and aggregate channel features for flower object localization. The goal has been successfully achieved by developing the ACFGP approach with a new GP representation, a new function set, and a new terminal set. This enables the simultaneous and automatic extension and aggregation of features, and dynamically localizing the flower in each image. The results show that the ACFGP outperforms the baseline benchmark and the YOLOv8 method. Further analysis highlighted the interpretability of the programs evolved by ACFGP, showcasing the effectiveness of the aggregate channel features for flower localization in ACFGP.

In the future, we will explore large-scale flower classification using the localization prediction results obtained through ACFGP.

References

1. Bi, Y., Xue, B., Mesejo, P., Cagnoni, S., Zhang, M.: A survey on evolutionary computation for computer vision and image analysis: past, present, and future trends. IEEE Trans. Evol. Comput. **27**(1), 5–25 (2023). https://doi.org/10.1109/TEVC.2022.3220747
2. Bi, Y., Xue, B., Zhang, M.: An effective feature learning approach using genetic programming with image descriptors for image classification [research frontier]. IEEE Comput. Intell. Mag. **15**(2), 65–77 (2020)
3. Dollár, P., Tu, Z., Perona, P., Belongie, S.: Integral channel features (2009)
4. Fortin, F.A., De Rainville, F.M., Gardner, M.A., Parizeau, M., Gagné, C.: DEAP: evolutionary algorithms made easy. J. Mach. Learn. Res. **13**(Jul), 2171–2175 (2012)
5. Guru, D.S., Sharath, Y.D.H., Manjunath, S.: Texture features and KNN in classification of flower images. Int. J. Comput. Appl. **1**, 21–29 (2010)
6. Jiang, P., Ergu, D., Liu, F., Cai, Y., Ma, B.: A review of YOLO algorithm developments. Procedia Comput. Sci. **199**, 1066–1073 (2022)
7. Jocher, G., Chaurasia, A., Qiu, J.: Ultralytics YOLO, January 2023. https://github.com/ultralytics/ultralytics
8. Koza, J.R.: Genetic Programming: On the Programming of Computers by Means of Natural Selection. MIT Press, Cambridge (1992)
9. Lin, T.Y., et al.: Microsoft COCO: common objects in context. In: Fleet, D., Pajdla, T., Schiele, B., Tuytelaars, T. (eds.) Computer Vision-ECCV 2014: 13th European Conference, Zurich, Switzerland, 6–12 September 2014, Proceedings, Part V 13, pp. 740–755. Springer, Cham (2014). https://doi.org/10.1007/978-3-319-10602-1_48
10. Montana, D.J.: Strongly typed genetic programming. Evol. Comput. **3**(2), 199–230 (1995)
11. Nilsback, M.E., Zisserman, A.: Delving into the whorl of flower segmentation. In: BMVC, vol. 2007, pp. 1–10 (2007)
12. Nilsback, M.E., Zisserman, A.: Automated flower classification over a large number of classes. In: 2008 Sixth Indian Conference on Computer Vision, Graphics & Image Processing, pp. 722–729. IEEE (2008)

13. Ostu, N.: A threshold selection method from gray-histogram. IEEE Trans. Syst. Man Cybern. **9**(1), 62–66 (1979)

14. Pardee, W., Yusungnern, P., Sripian, P.: Flower identification system by image processing. In: 3rd International Conference on Creative Technology CRETECH, vol. 1, pp. 1–4 (2015)

15. Patel, I., Patel, S.: An optimized deep learning model for flower classification using NAS-FPN and faster R-CNN. Int. J. Sci. Technol. Res. **9**(03), 5308–5318 (2020)

16. Ren, S., He, K., Girshick, R., Sun, J.: Faster R-CNN: towards real-time object detection with region proposal networks. In: Advances in Neural Information Processing Systems, vol. 28 (2015)

17. Rother, C., Kolmogorov, V., Blake, A.: " GrabCut" interactive foreground extraction using iterated graph cuts. ACM Trans. Graph. (TOG) **23**(3), 309–314 (2004)

18. Siraj, F., Salahuddin, M.A., Yusof, S.A.M.: Digital image classification for Malaysian blooming flower. In: 2010 Second International Conference on Computational Intelligence, Modelling and Simulation, pp. 33–38. IEEE (2010)

19. Sun, Q., Yang, S., Sun, C., Yang, W.: Exploiting aggregate channel features for urine sediment detection. Multimedia Tools Appl. **78**, 23883–23895 (2019)

20. Wu, D., Lv, S., Jiang, M., Song, H.: Using channel pruning-based YOLO v4 deep learning algorithm for the real-time and accurate detection of apple flowers in natural environments. Comput. Electron. Agric. **178**, 105742 (2020)

21. Yan, Z., Bi, Y., Xue, B., Zhang, M.: Automatically extracting features using genetic programming for low-quality fish image classification. In: Proceedings of 2021 IEEE Congress on Evolutionary Computation (CEC), pp. 2015–2022. IEEE (2021)

22. Yang, B., Yan, J., Lei, Z., Li, S.Z.: Aggregate channel features for multi-view face detection. In: IEEE International Joint Conference on Biometrics, pp. 1–8. IEEE (2014)

Evolutionary Multi-objective Optimization of Large Language Model Prompts for Balancing Sentiments

Jill Baumann[(✉)] and Oliver Kramer

Computational Intelligence Lab, Department of Computing Science,
Carl von Ossietzky Universität Oldenburg, Oldenburg, Germany
{jill.baumann,oliver.kramer}@uni-oldenburg.de

Abstract. The advent of large language models (LLMs) such as Chat-GPT has attracted considerable attention in various domains due to their remarkable performance and versatility. As the use of these models continues to grow, the importance of effective prompt engineering has come to the fore. Prompt optimization emerges as a crucial challenge, as it has a direct impact on model performance and the extraction of relevant information. Recently, evolutionary algorithms (EAs) have shown promise in addressing this issue, paving the way for novel optimization strategies. In this work, we propose a evolutionary multi-objective (EMO) approach specifically tailored for prompt optimization called EMO-Prompts, using sentiment analysis as a case study. We use sentiment analysis capabilities as our experimental targets. Our results demonstrate that EMO-Prompts effectively generates prompts capable of guiding the LLM to produce texts embodying two conflicting emotions simultaneously.

1 Introduction

The rise of ChatGPT [10], Llama 2 [12] and other large language models (LLMs) has revolutionized the field of natural language processing, enabling a wide range of applications from text generation to sentiment analysis. However, the effectiveness of these models is highly dependent on the quality of the input prompts. Prompt optimization stands out as a critical area of research, aiming to refine and tailor prompts to elicit the most accurate and relevant responses from the model.

The organization of this paper is outlined as follows: Sect. 2 provides an overview of related work, laying the groundwork for the subsequent sections. In Sect. 3, we introduce our approach, EMO-Prompts with operators, and detail its integration with the NSGA-II (Non-dominated Sorting Genetic Algorithm II) [2] and the SMS-EMOA (S-metric selection evolutionary multi-objective algorithm) [5]. Section 4 presents experiments conducted with a focus on text writing applications in the context of sentiment analysis, followed by a thorough discussion of the results obtained. Finally, Sect. 5 concludes the paper, summarizing the contributions of this work.

S. Smith et al. (Eds.): EvoApplications 2024, LNCS 14635, pp. 212–224, 2024.
https://doi.org/10.1007/978-3-031-56855-8_13

2   Related Work

Popular prompt engineering techniques, like Chain-of-Thought Prompting [13] or ReAct [14], significantly enhance the reasoning capabilities of LLMs, but often remain sub-optimal. Previous studies have explored various strategies for prompt optimization, highlighting its significance in leveraging the full potential of LLMs. The idea is to find an optimal prompt $p^* \in \mathcal{P}$ in the space $\mathcal{P}$ of prompts w.r.t. an objective function $f(\cdot)$. Examples for typical objective functions are the performance in instruction-induction tasks [3,15], question-answering tasks [3], summarization tasks [4], hate speech recognition [3], or code generation [1,9].

Evolutionary algorithms (EAs) have recently been applied to this domain, showing potential in navigating the vast prompt space for optimal solutions. The Automatic Prompt Engineer (APE) [15] uses LLMs to automatically generate new prompts based on a set of input/output pairs, which is demonstrated to the LLM and select the most promising. For optimization an iterative Monte Carlo search method is applied. APE outperforms human-engineered prompts across two datasets and shows that LLMs can be used as inference models. Meyerson et al. [9] propose a variation operator that is similar to crossover and uses "few-shot" prompting. Its variety is demonstrated through various tasks, like generation of mathematical expressions, English sentences and Python code. Evo-Prompt [4] introduces an evolutionary prompt optimization framework combining LLMs with EAs for automated and efficient prompt optimization. It demonstrates significant improvements over human-engineered prompts and existing methods across various datasets and tasks. The approach showcases substantial advancements, outperforming competitors by up to 25%. EvoPrompting [1] uses the LLM as a mutation and crossover operator to generate convolutional architectures. This method is tested e.g., on MNIST-1D. The results show that EvoPrompting is able to create smaller and more accurate convolutional architectures than manually designed ones. Promptbreeder [3] is a self-referential self-improvement algorithm utilizing an LLM to evolve and adapt prompts across different domains. It not only refines task-prompts for improved performance on benchmarks, but also concurrently optimizes the mutation-prompts used in the evolution process, showcasing its effectiveness on complex challenges such as hate speech classification. In contrast to optimizing discrete prompts, with soft prompting [6–8] only the parameters are tuned. They show effectiveness, but have disadvantages due to their insufficient interpretability and the need to access the parameters of the LLM.

These approaches are designed to optimize prompts to align with a singular objective in the LLM's output, such as ensuring the response is in English. In contrast, EMO-Prompts strives to concurrently fulfill dual objectives in the LLM's response. For instance, not only should the LLM's output be truthful but also informative.

3   EMO-Prompts

Our approach, EMO-Prompts, introduces a evolutionary multi-objective framework for prompt optimization. We employ evolutionary prompt operators to search the space of prompts and NSGA-II [2] as well as SMS-EMOA [5] as selection operators.

An individual is a tuple $(<\text{PROMPT}>, <\text{TEXT}>, (f_1, \ldots, f_n))$ of prompt $<\text{PROMPT}>$, a text $<\text{TEXT}>$ generated by a LLM based on the prompt and n fitness values $f_1, \ldots, f_n$ according to defined objectives. A prompt is the genotype, the generated text the corresponding phenotype.

3.1   Large Language Model

Meta AI's Llama 2 [12] is used as the LLM for our new framework EMO-Prompts. It is open source, can be downloaded and hosted on own infrastructure. Its variants have 7B, 13B or 70B parameters. Compared to Llama 1, Llama 2 was trained with 40% more data and has a twice as big context length.

In consideration of computational intensity, we opted for Llama 2 with 7B parameters. Ollama[1] is used to run Llama 2 with 7B parameters locally and to create customized models with the help of a Modelfile. The Modelfile allows to configure various parameters like temperature and the size of the context window. Apart from defining parameters, a Modelfile offers the option to specify a system prompt and a template, which makes the Modelfile analogous to a blueprint for creating models with Ollama. With a template, "few-shot" prompting can be realized by showing the model a few examples of how the syntax should be. A system prompt, embedded in the template, is used to help the LLM to follow a certain behavior. A exemplary template that can be used to realize "few-shot" prompting is shown in Listing 1.1.

```
 1 TEMPLATE """
 2 ### System:
 3 {{ .System }}
 4 {{- end }}
 5
 6 ### User:
 7 Change the following prompt: provide a 3 sentence story
 8
 9 ### Response:
10 Craft a three-sentence story
11
12 ### User:
13 Modify the following prompt: write a 3 sentence story
14
15 ### Response:
16 Create a three-sentence tale with a twist ending.
17
18 ### User:
19 {{ .Prompt }}
20
21 ### Response:
22 """
```

Listing 1.1. Exemplary Template within a Modelfile

---

[1] https://github.com/jmorganca/ollama.

Langchain[2] is a framework that offers diverse functionalities for developing applications with LLMs. Using Langchain's prompt templates, instructions on how prompts should be generated can be constructed as shown exemplary in Listing 1.2. The prompt template is formatted by inserting the fields in the curly brackets, in this example {mutation_prompt} and {prompt}, into the prompt template.

```
1 """[INST] <<SYS>> Use the following mutation prompt and the following
      prompt, to change the prompt and generate a better prompt. Use one
      sentence maximum, which is a instruction to generate text, and
      keep the answer as concise as possible. <</SYS>>
2 Mutation Prompt: {mutation_prompt}
3 Prompt: {prompt}
4 New Prompt:[/INST]"""
```

Listing 1.2. Prompt Template

3.2 Evolutionary Approach

EMO-Prompts employs the standard EA-loop for prompt optimization, as illustrated in Fig. 1.

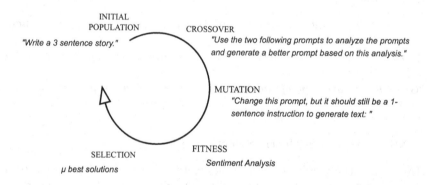

Fig. 1. Evolutionary generation of a new prompt

The initial population is realized by a set of individuals as described above. Ten story generation prompts were manually formulated, prompting the LLM to generate a story. Following this, the fitness of each individual's story was evaluated. To generate a new offspring, two solutions are randomly selected from the population and recombined by our developed crossover operator. A new designed mutation operator, which is also randomly selected from a set of mutation operators is applied to this result. In our EMO-Prompts framework, the LLM operates as a crossover and mutation operator as well as a text generator. For every prompt in the population, the LLM produces the corresponding text, subject to the evaluation through sentiment analysis. Afterwards the μ best

---

[2] https://www.langchain.com.

solutions according to NSGA-II or SMS-EMOA, see next paragraph, are selected for the following generation. This evolutionary process is repeated for a number of generations, or until a satisfactory result is achieved.

To guide the generation of new prompts and define the expected response, it is essential to provide clear instructions that mitigate the risk of hallucinations of the LLM. EMO-Prompts uses the two outlined options, Modelfile and prompt template, to create a customized Llama 7B model for each of its key tasks, including crossover, mutation and text generation.

As can be seen in Fig. 1, new crossover and mutation operator are developed, which are text prompts instructing the LLM to perform crossover or mutation. Either two prompts are taken and a new one is created (crossover) or an existing prompt is changed to a new one (mutation).

Crossover Prompt:

1. *"One prompt is: [...], another prompt is: [...]. Analyze the prompts and generate a better prompt based on this analysis, but it should still be a 1-sentence instruction to generate text."*

Mutation Prompts:

1. *"Change this prompt, but it should still be a 1-sentence instruction to generate text: [...]"*
2. *"Modify this prompt to generate a 1-sentence instruction for text generation: [...]"*
3. *"Generate a variation of the following prompt while keeping the semantic meaning: [...]"*

3.3 NSGA-II and S-Metric Selection

NSGA-II, a multi-objective optimization algorithm, uses non-dominated sorting and crowding distance computation for diverse solution selection. It generates a random population, evaluates them, and sorts them into non-dominated fronts. Solutions within a front are not comparable with each other. The algorithm calculates the crowding distance to maintain diversity, and iterates through crossover and mutation to evolve the population, aiming for Pareto-optimal solutions over several generations. The crowding distance is a measure used to estimate the density of solutions surrounding a particular point in the objective space, favoring less crowded areas to ensure diversity in the solution set.

The S-metric selection algorithm focuses on maximizing the dominated hypervolume in multi-objective optimization, indicating solution quality in convergence and diversity. It selects solutions based on their hypervolume. The hypervolume quantifies the extent of the space encompassed by non-dominated solutions. When the number of non-dominated solutions exceeds μ, this algorithm selects a set of solutions that collectively optimize the overall hypervolume. In contrast, when the count of non-dominated solutions is below μ, the algorithm systematically gives preference to these solutions. The selection process starts by arranging the solutions in ascending order based on their ranking across different fronts, which are essentially tiers of solution quality. Within each front, solutions are further prioritized based on the number of other solutions that dominate them, favoring those with the least domination first.

4   Experiments

4.1   Sentiment Analysis

The sentiment analysis task, facilitated by Hugging Face's tools, serves as the testbed for our approach. In the experiments, we use the 'bhadresh-savani/-distilbert-base-uncased-emotion'[3] model, which serves as an expert text classification tool, specifically designed for the nuanced task of emotion recognition. It uses the DistilBERT [11] architecture, a streamlined variant of BERT that ensures a balance between efficiency and performance; it is 40% smaller in size, yet retains 97% of the original model's language understanding capabilities, thanks to the knowledge distillation process implemented during pre-training. The model is adept at identifying a spectrum of emotions from textual data, including 'sadness', 'anger', 'love', 'surprise', 'joy' and 'fear'. Each emotion is assigned a value between 0 and 1, with all values summing up to 1.

In terms of training, the model was fine-tuned using an emotion dataset and the Hugging Face Trainer, adhering to specific training parameters such as a learning rate of 2^{-5}, a batch size of 64, and a duration of 8 training epochs. The model is conveniently hosted on the Hugging Face model hub and is distributed under the Apache-2.0 License.

4.2   Settings

Based on the emotions of the sentiment analysis, four conflicting emotion pairs are constructed: 'love vs. anger', 'joy vs. fear', 'joy vs. sadness' and 'surprise vs. fear'. The goal is to investigate how prompts generated by EMO-Prompts can cause the LLM to generate texts that contains both emotions of the conflicting emotion pair, e.g., both 'love' and 'anger', which defines the sentence sentiment task. The metric used for evaluation is the hypervolume, introduced in Sect. 3.3. A (10+20) genetic algorithm is performed. The initial population is realized by creating ten initial prompts for text generation. Based on NSGA-II and SMS-EMOA the ten best individuals from the parent and child population are then selected as the new parent population. This is performed for 30 generations and each experiment is repeated ten times.

The genetic algorithm operators are performed by Llama 2. The temperature hyperparameter of Llama 2 is set to 0.7, which was chosen on basis of a few experiments and the size of the context window to 512 due to the token limit of DistilBERT. The rest of the hyper-parameters are default.

---

[3] https://huggingface.co/bhadresh-savani/bert-base-uncased-emotion.

4.3 Results

Table 1 shows the results of the four experiments w.r.t. overall best and worst hypervolume during the optimization process. The mean and standard deviation are reported across ten repetitions. The sentence sentiment task is a maximization problem, i.e., the score of both emotions of an emotion pair should be maximized. Since the emotions are not only semantically conflicting, but also within the sentiment analysis, an ideal hypervolume of 0.44 can be achieved, which corresponds to the area dominated by 10 points equally distributed on the diagonal between (0,1) and (1,0). A higher hypervolume goes hand in hand with an improvement in the quality of the solutions. Due to the way a LLM works, it is not guaranteed that the LLM will provide exactly the same response for the same prompt.

Table 1. Comparison between NSGA-II and SMS-EMOA on four problems measuring hypervolume.

Problem	NSGA-II				SMS-EMOA			
	Best	Worst	Mean	Std Dev	Best	Worst	Mean	Std Dev
Love vs. anger	**0.32**	0.05	0.20	0.08	0.26	0.01	0.16	0.09
Joy vs. fear	0.38	0.32	0.36	0.02	**0.40**	0.30	0.36	0.03
Joy vs. sadness	**0.39**	0.26	0.35	0.04	0.30	0.11	0.23	0.06
Surprise vs. fear	0.43	0.39	0.41	0.01	**0.45**	0.13	0.39	0.09

As illustrated in Table 1, the outcomes of the four experiments are largely similar. Notably, the 'love vs. anger' experiment using SMS-EMOA shows lower values. In comparison, EMO-Prompts utilizing NSGA-II consistently yields higher average fitness function values than SMS-EMOA. Specifically, EMO-Prompts with NSGA-II outperforms in the 'love vs. anger' and 'joy vs. sadness' scenarios, whereas SMS-EMOA excels in the 'joy vs. fear' and 'surprise vs. fear' settings. Remarkably, in the 'surprise vs. fear' experiments, EMO-Prompts attains peak fitness values of 0.45, surpassing the optimal benchmark of 0.44.

Next, the four experiments will be described more detailed.

Love vs. Anger. In the first experiment, we ask the LLM to generate text with the emotions 'love' and 'anger'.

Figure 2 presents a plot comparing the hypervolume progression across generations for the conflicting emotions 'love' and 'anger', using (a) NSGA-II and (b) SMS-EMOA. In this plot, each dashed line signifies an individual repetition, and the solid line depicts the average of all ten repetitions. The hypervolume for EMO-Prompts using NSGA-II shows a consistent increase through generations. In contrast, EMO-Prompts with SMS-EMOA encounters a stagnation in the local optimum, specifically from generations 9 to 17 and again from 21

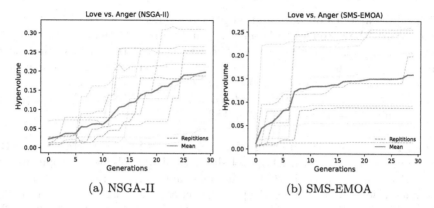

(a) NSGA-II   (b) SMS-EMOA

Fig. 2. 'Love vs. anger': Plots of hypervolume developments for (a) NSGA-II and (b) SMS-EMOA over 30 generations

to 26. This divergence in the progression curves can be attributed to the distinct operational mechanisms of the two algorithms. While EMO-Prompts with SMS-EMOA initially reaches higher fitness values more rapidly, EMO-Prompts utilizing NSGA-II eventually surpasses it in performance.

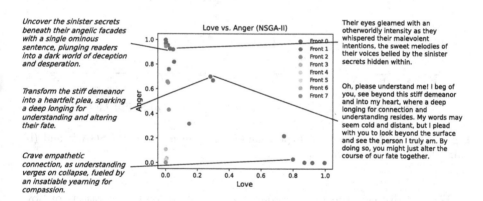

Fig. 3. 'Love vs. anger': Plot of Pareto front (NSGA-II) with examples for prompts (left) and generated text (right)

Figure 3 illustrates the Pareto front approximation for the conflicting emotions of 'love' and 'anger', featuring examples of generated prompts on the left and corresponding texts on the right. The plot showcases the Pareto front approximation from the NSGA-II repetition achieving the best fitness value, indicated by the maximum hypervolume. It includes eight different fronts, with the first front's non-dominated points approximating the Pareto front. In this experiment, the first front is not as tightly clustered as an optimal solution might be, suggesting a challenge for EMO-Prompts in generating prompts that effectively balance the emotions of 'love' and 'anger'. The tendency for solutions to gravitate towards the extremes (0,1) and (1,0) indicates a relative ease in creating prompts that evoke a single emotion. The overall population in this experiment achieves a hypervolume of 0.32. An example of a generated prompt is *"Uncover the sinister secrets beneath their angelic facades with a single ominous sentence, plunging readers into a dark world of deception and desperation."*, which yields a text with sentiment values of (0.05, 0.82).

Joy vs. Fear. In the second experiment we ask the LLM to generate text with the emotions 'joy' and 'fear'. Figure 4 shows the hypervolume development for (a) NSGA-II and (b) SMS-EMOA of the two conflicting emotions 'joy' and 'fear'. The fluctuations of the NSGA-II optimization after a sharp increase in fitness in the first generations lie on average in a certain range between 0.30 and 0.33. On average EMO-Prompts with NSGA-II achieves higher fitness values faster, EMO-Prompts with SMS-EMOA outperforms them afterwards.

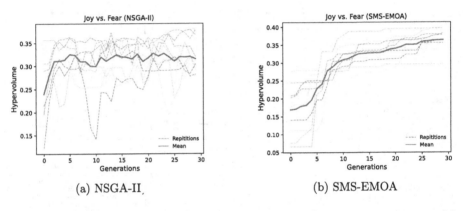

(a) NSGA-II   (b) SMS-EMOA

Fig. 4. 'Joy vs. fear': Plots of hypervolume developments for (a) NSGA-II and (b) SMS-EMOA over 30 generations

Figure 5 displays the Pareto front approximation for the emotions of 'joy' and 'fear', accompanied by examples of generated prompts on the left and their respective texts on the right. This plot focuses on the Pareto front approximation from the SMS-EMOA repetition that achieved the highest fitness value, indicated by the maximum hypervolume. The population features three distinct fronts, with the first front representing a significant portion of the population and covering a hypervolume of 0.40. A notable concentration of points is observed around the point (1,0), suggesting that the LLM tends to generate prompts that predominantly evoke the emotion of 'joy'. For instance, the prompt *"A secret force awakens during an unexpected tempest, exposing a surprising reality about the unlikeliest of saviors."* results in a text with fitness values (0.98, 0.00), illustrating this tendency.

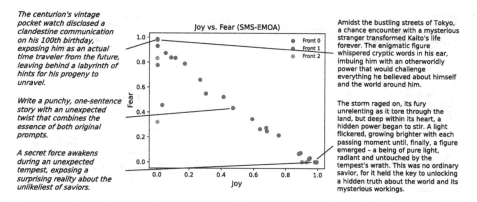

The centurion's vintage pocket watch disclosed a clandestine communication on his 100th birthday, exposing him as an actual time traveler from the future, leaving behind a labyrinth of hints for his progeny to unravel.

Write a punchy, one-sentence story with an unexpected twist that combines the essence of both original prompts.

A secret force awakens during an unexpected tempest, exposing a surprising reality about the unlikeliest of saviors.

Amidst the bustling streets of Tokyo, a chance encounter with a mysterious stranger transformed Kaito's life forever. The enigmatic figure whispered cryptic words in his ear, imbuing him with an otherworldly power that would challenge everything he believed about himself and the world around him.

The storm raged on, its fury unrelenting as it tore through the land, but deep within its heart, a hidden power began to stir. A light flickered, growing brighter with each passing moment until, finally, a figure emerged – a being of pure light, radiant and untouched by the tempest's wrath. This was no ordinary savior, for it held the key to unlocking a hidden truth about the world and its mysterious workings.

Fig. 5. 'Joy vs. fear': Plot of Pareto front (SMS-EMOA) with examples for prompts (left) and generated text (right)

Joy vs. Sadness. In the third experiment, we ask the LLM to generate text with the emotions 'joy' and 'sadness'. Figure 6 shows the hypervolume of (a) NSGA-II and (b) SMS-EMOA for the conflicting emotions 'joy' and 'sadness'. The hypervolume with SMS-EMOA and with NSGA-II increases from generation to generation. Again, on average EMO-Prompts with NSGA-II achieves higher fitness values faster, EMO-Prompts with SMS-EMOA outperforms them afterwards.

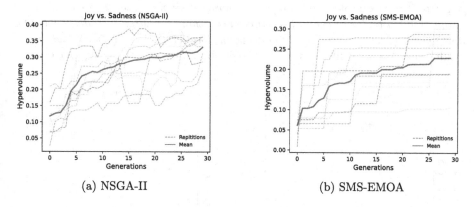

(a) NSGA-II

(b) SMS-EMOA

Fig. 6. 'Joy vs. sadness': Plots of hypervolume developments for (a) NSGA-II and (b) SMS-EMOA over 30 generations

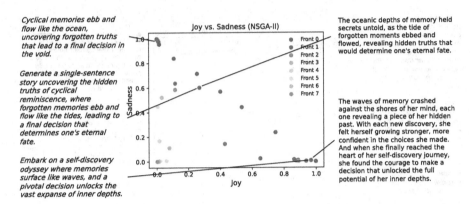

Fig. 7. 'Joy vs. sadness': Plot of Pareto front (NSGA-II) with examples for prompts (left) and generated text (right)

Figure 7 depicts the Pareto front approximation for the emotional dichotomy of 'joy vs. sadness', complete with examples of generated prompts on the left and corresponding texts on the right. The final population includes eight distinct fronts, with the non-dominated points of the first front closely approximating the Pareto front. A significant portion of the population is encompassed within the first front, covering a hypervolume of 0.39. Notably, there is a dense cluster of points near the extreme point (1,0), indicating a tendency of the LLM to produce texts rich in the emotion of 'joy', driven by the nature of the prompts generated. For example, the prompt *"Generate a single-sentence story uncovering the hidden truths of cyclical reminiscence, where forgotten memories ebb and flow like the tides, leading to a final decision that determines one's eternal fate."* results in a text with fitness values of (0.40, 0.57), exemplifying this pattern.

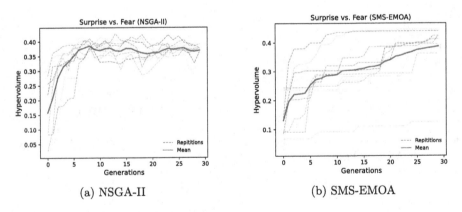

(a) NSGA-II   (b) SMS-EMOA

Fig. 8. 'Surprise vs. fear': Plots of hypervolume developments for (a) NSGA-II and (b) SMS-EMOA over 30 generations

Surprise vs. Fear. The last experiment balances the emotions 'surprise vs. fear'.

Figure 8 presents a comparison of the hypervolume trends for 'surprise vs. fear' using both (a) NSGA-II and (b) SMS-EMOA. In the NSGA-II case, after an initial sharp increase in fitness values during the early generations, the hypervolume fluctuates within a relatively stable range, typically between 0.35 and 0.38. The curve representing the mean indicates that the maximum hypervolume with NSGA-II is achieved around the midpoint of the generations. This difference in the progression patterns between NSGA-II and SMS-EMOA reflects the distinct operational approaches of these algorithms.

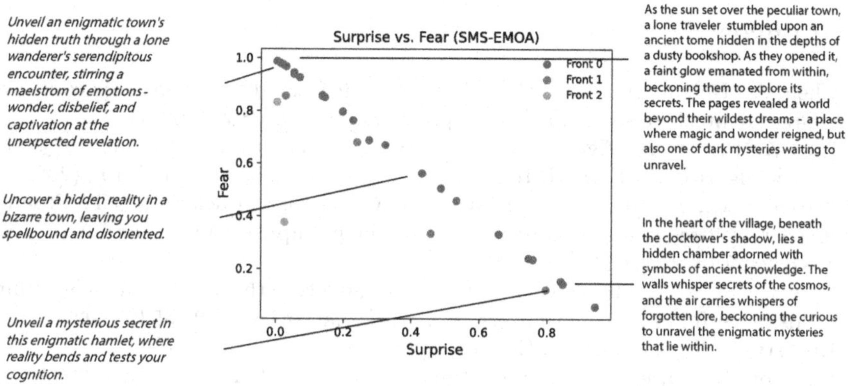

Fig. 9. 'Surprise vs. fear': Plot of Pareto front (SMS-EMOA) with examples for prompts (left) and generated text (right)

Figure 9 illustrates the Pareto front approximation for the emotional contrast of 'surprise vs. fear'. The final population in this plot is divided into three distinct fronts. The overall population achieves a hypervolume of 0.45, marking this as the highest fitness value attained in our experiments and surpassing the ideal target of 0.44. A notable concentration of points is observed around the point $(0,1)$, indicating a prevalence of prompts generated by the LLM that predominantly evoke the emotion of 'surprise' over 'fear'. For instance, the prompt *"Unveil a mysterious secret in this enigmatic hamlet, where reality bends and tests your cognition."* results in a text with fitness values of $(0.85, 0.14)$, exemplifying this trend. Across all experiments, points approximating the sentiment value of $(0.5, 0.5)$ demonstrate the LLM's capability to generate prompts that effectively address both conflicting emotions.

5   Conclusion

In conclusion, our comprehensive experiments have effectively validated the efficiency of the introduced evolutionary operators, in particular the integration of prompt mutation and crossover with NSGA-II and SMS-EMOA, in producing texts with a balanced sentiment.

This research lays a strong foundation for future explorations in prompt optimization and sentiment modulation within text generation. It paves the way for further developments and enhancements in natural language processing. Moving forward, our aim is to broaden the scope of our methodology to include the generation of more extensive texts and those tailored to specific domains. Additionally, we plan to investigate the application of evolutionary prompt techniques to a wider range of tasks involving large language models, with the goal of further pushing the frontiers of text generation technology.

References

1. Chen, A., Dohan, D.M., So, D.R.: EvoPrompting: language models for code-level neural architecture search. arXiv preprint arXiv:2302.14838 (2023)
2. Deb, K., Pratap, A., Agarwal, S., Meyarivan, T.: A fast and elitist multiobjective genetic algorithm: NSGA-II. IEEE Trans. Evol. Comput. **6**(2), 182–197 (2002)
3. Fernando, C., Banarse, D., Michalewski, H., Osindero, S., Rocktäschel, T.: Promptbreeder: self-referential self-improvement via prompt evolution. arXiv preprint arXiv:2309.16797 (2023)
4. Guo, Q., et al.: Connecting large language models with evolutionary algorithms yields powerful prompt optimizers. arXiv preprint arXiv:2309.08532 (2023)
5. Hochstrate, N., Naujoks, B., Emmerich, M.: SMS-EMOA: multiobjective selection based on dominated hypervolume. Eur. J. Oper. Res. **181**, 1653–1669 (2007)
6. Lester, B., Al-Rfou, R., Constant, N.: The power of scale for parameter-efficient prompt tuning. In: Proceedings of the 2021 Conference on Empirical Methods in Natural Language Processing, pp. 3045–3059 (2021)
7. Li, X.L., Liang, P.: Prefix-tuning: optimizing continuous prompts for generation. In: Proceedings of the 59th Annual Meeting of the Association for Computational Linguistics and the 11th International Joint Conference on Natural Language Processing (Volume 1: Long Papers), pp. 4582–4597 (2021)
8. Liu, X., et al.: GPT understands, too. AI Open (2023)
9. Meyerson, E., et al.: Language Model Crossover: Variation through Few-Shot Prompting. arXiv preprint arXiv:2302.12170 (2023)
10. OpenAI. GPT-4 Technical Report. arXiv preprint arXiv:2303.08774 (2023)
11. Sanh, V., Debut, L., Chaumond, J., Wolf, T.: DistilBERT, a distilled version of BERT: smaller, faster, cheaper and lighter. arXiv preprint arXiv:1910.01108 (2019)
12. Touvron, H., et al.: Llama 2: open foundation and fine-tuned chat models. arXiv preprint arXiv:2307.09288 (2023)
13. Wei, J., et al.: Chain of thought prompting elicits reasoning in large language models. In: Advances in Neural Information Processing Systems 35: Annual Conference on Neural Information Processing Systems 2022, NeurIPS 2022, New Orleans, 28 November–9 December 2022 (2022)
14. Yao, S., et al.: ReAct: synergizing reasoning and acting in language models. In: The Eleventh International Conference on Learning Representations, ICLR 2023, Kigali, 1–5 May 2023 (2023)
15. Zhou, Y., et al.: Large language models are human-level prompt engineers. In: The Eleventh International Conference on Learning Representations, ICLR 2023, Kigali, 1–5 May 2023 (2023)

Evolutionary Feature-Binning
with Adaptive Burden Thresholding
for Biomedical Risk Stratification

Harsh Bandhey[1](ID), Sphia Sadek[2], Malek Kamoun[2](ID),
and Ryan Urbanowicz[1(✉)](ID)

[1] Cedars-Sinai Medical Center, Los Angeles, CA 90048, USA
ryan.urbanowicz@cshs.org
[2] University of Pennsylvania, Philadelphia, PA 19104, USA

Abstract. Multivariate associations including additivity, feature interactions, heterogeneous effects, and rare feature states can present significant obstacles in statistical and machine-learning analyses. These
relationships can limit the detection capabilities of many analytical
methodologies when predicting outcomes including risk stratification
in biomedical survival analyses. Feature Inclusion Bin Evolver for Risk
Stratification (FIBERS) was previously proposed using an evolutionary
algorithm to discover groups (i.e. bins) of features wherein the burden
of feature values automatically determined the risk strata of a given
instance in right-censored survival analysis. A key limitation of FIBERS
is that it assumes a fixed threshold for feature burden in stratifying high
vs. low risk, which restricts the flexibility of bin discovery. In the present
work, we extend FIBERS to include different strategies for adaptive
burden thresholding such that feature bins are discovered alongside the
threshold that best separates risk strata. Preliminary comparative performance evaluation was conducted across simulated datasets with different
underlying ideal burden thresholds yielding performance improvements
over the original FIBERS algorithm. This algorithmic feasibility study
lays the groundwork for ongoing application to the real-world problem of
kidney graft failure risk stratification in dealing with the expected population heterogeneity including differences in race, ethnicity, and sex.

Keywords: evolutionary algorithm · feature engineering · binning ·
epistasis · dynamic · risk stratification

1 Introduction

In addition to traditional statistical analyses, a wide variety of machine learning
methodologies are being increasingly used in the domain of biomedical research
to help address the challenge of discovering associations and predictive modeling

---

Supplementary Information The online version contains supplementary material
available at https://doi.org/10.1007/978-3-031-56855-8_14.

in the context of large-scale data and complex multivariate underlying patterns of association [6, 8].

One specific area of biomedical research motivating the present study is the assessment of kidney graft failure (GF) risk in prospective pairs of donors and recipients. The process of donor/recipient matching, integral to the success of kidney transplantation, has traditionally relied on survival analyses as its primary methodology [5, 11]. A pivotal facet of this matching process revolves around the human leukocyte antigen (HLA) loci, which exhibit significant variability within the human population [9]. Conventionally, HLA matching has concentrated on antigen-level (Ag) mismatches (MMs) focusing on the presence or absence of at least one informative MM to classify high or low GF risk and potentially overlooking vital underlying variations that could be indicative of graft failure risk [3]. Emerging research has advocated for the examination of higher resolution amino acid (AA) level MMs, which have been shown to confer substantial incremental risk of kidney graft failure independently of Ag-level MMs [4]. This paradigm shift introduces additional complexities, including (1) a larger feature space, (2) potentially increased heterogeneity of informative AA positions that influence GF risk in different patient subgroups, (3) the presence of infrequent 'rare' MMs in the dataset, and (4) uncertainty regarding the burden (i.e. number) of MMs driving underlying risk, i.e. is a specific MM sufficient to predict GF risk, or, given a set of identified informative MMs, is there a burden of MMs (e.g. 2 or more) over which GF risk increases. HLA allele frequencies are also known to vary between ethnicities [7], further supporting the importance of discovering optimal MM burden thresholds in classifying GF risk.

Evolutionary algorithms, employed as machine learning approaches, have a number of fundamental advantages founded in their flexibility, limited assumptions, and ability to generate interpretable solutions. The "Relevant Association Rare-variant-bin Evolver" (RARE) algorithm, a genetic-algorithm-based approach for feature learning [2], was previously proposed as a pioneering technique for feature construction in data with rare feature variation. RARE avoided a priori assumptions about feature grouping, utilizing evolutionary mechanisms to discover interpretable features sets (i.e bins) that captured rare variant burden to predict a target binary disease outcome.

Recently, the "Feature Inclusion Bin Evolver for Risk Stratification" (FIBERS) algorithm, was proposed as an adaptation of the RARE methodology to address the specific intricacies of 'time-to-event' right-censored survival data for kidney GF analysis [1]. FIBERS adopts a similar genetic algorithm framework but includes a bin fitness based on survival curve differences between high/low risk groups as well as an 'OR' interpretation of bins such that a single MM in any feature of the bin would lead to the instance (i.e. a donor/recipient pair) being categorized as high risk. This means that FIBERS assumes a burden threshold of zero MMs in categorizing instances into either high or low risk groups for subsequent survival curve comparisons, in line with conventional HLA matching assessments. FIBERS was demonstrated to be successful in identifying AA-MM bins that confer kidney graft failure risk beyond the predictive capability of

Ag-level MMs even after adjusting for all covariates. Subsequently, FIBERS was also extended as a scikit-learn compatible software package to promote ease of use and compatibility within a scikit-learn based machine learning analysis pipeline [10].

Despite this success, the likelihood of MM frequency differences and risk-factor heterogeneity in sub-populations, suggests that reliance on a MM burden threshold of 0 limits the efficacy of FIBERS in identifying optimal criteria for stratifying GF risk groups. Therefore, the present study conducts a proof-of-principle methodological expansion of FIBERS to incorporate an adaptive burden threshold that seeks to select the optimal MM burden threshold (separating high vs. low risk groups) for a given problem/dataset. Our goal is to demonstrate that an adaptive burden threshold allows FIBERS to simultaneously discover bins of informative features as well as the corresponding optimal burden threshold in controlled simulations of AA-MM survival data. This algorithmic work is broadly aimed at downstream real-world applications involving survival analyses prediction in data with features that represent matching between pairs of subjects, e.g. organ/tissue compatibility and transplantation research.

In the subsequent sections of this paper, we (1) describe the FIBERS algorithm and proposed adaptive burden threshold implementations to better address diversity in data and bin populations, (2) describe our methodologies for simulating survival data with which to objectively evaluate FIBERS, (3) describe the simulation study datasets and the criteria employed for evaluation, (4) detail our findings, and (5) offer discussion and conclusions while outlining next steps in future work.

2 Methods

2.1 Scikit-FIBERS

The scikit-FIBERS algorithm, detailed in [10], is comprised of three phases as outlined in Fig. 1; (1) bin initialization, (2) bin optimization, via cycles of an evolutionary algorithm, and (3) final bin population evaluation. Bin initialization, randomly generates an initial population of bins where a bin simply specifies a subset of features from the dataset. These features represent HLA AA positions, with a value corresponding to the MM count at that position (0,1,2). Initialized bins are restricted with run parameters to include a defined minimum number of features, and a limit on the number of times a specific feature can occur across the initial population to encourage diversity.

Next the bin population is evolved over a defined number of generations/cycles. Each cycle begins with bin evaluation where each bin assigns instances of the training data to either a high or low risk group by summing any MMs in the data among the features specified within the bin and comparing that sum to a burden threshold (e.g. 0, 1, 2, etc). Scikit-FIBERS previously relied on a fixed burden threshold of zero, meaning that if an instance had no MMs in the features of a bin, it would be placed in the low risk group, otherwise it would be placed in the high risk group. Next, the Logrank test is employed

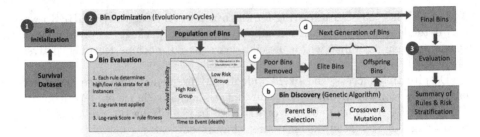

Fig. 1. Overview of the original scikit-FIBERS algorithm.

to assess the divergence between the Kaplan-Meyer survival curves of instances in the high vs. low risk groups. The Logrank score serves as the fitness measure for that bin. If the imbalance between instances in the high vs. low risk groups is too large, the bin is automatically assigned a very low fitness due to potential limitations of the Logrank test. [1].

The next step is bin discovery (i.e. parent selection, crossover, and mutation). Parent bins are selected from the elite bins in the population (i.e. bins with the highest bin fitness), and uniform crossover and mutation mechanisms are applied to generate enough offspring bins (combined with elite bins) to reach the maximum bin population size. Uniform crossover is used over other crossover strategies since FIBERS bins can have variable length and we make the assumption that feature linkage is not important in this context. At this point, bins from the previous generation that have a lower fitness (non-elite set) are removed, with the new generation comprised of elite bins from the previous generation as well as new offspring bins. These steps repeat for a defined number of iterations, with scikit-FIBERS conducting an evaluation of all bins in the final population.

2.2 Algorithmic Expansions for Adaptive Burden Thresholding

In the present study, we have extended scikit-FIBERS to include an adaptive burden threshold, such that the algorithm seeks to automatically discover not only the best features to specify in bins, but what burden threshold to use for separating instances into high/low risk groups. To achieve this, algorithmic modifications were made to both bin evaluation and bin discovery.

Regarding bin evaluation, we implemented two experimental methods for adaptive burden thresholding, i.e. a best-score threshold, and an evolving threshold. These methods are illustrated in Fig. 2. The best-score threshold method checks all defined burden threshold values (e.g. 0, 1, 2, 3) and deterministically selects the one that maximizes survival curve separation between the two risk groups (i.e. fitness). The second strategy evolves the burden threshold for each bin in a manner similar to bin itself (i.e. applying crossover and mutation). Furthermore, users can specify the probability with which the algorithm will utilize the evolving threshold during generations vs. the best-score threshold method, such that a mix of the two strategies can be employed during scikit-FIBERS training. In this study we compare original scikit-FIBERS (with no adaptive burden threshold) to this new implementation where the probability of using the evolving threshold is either zero, i.e. best-score threshold is always employed, 0.5, i.e. an evolving threshold is emplyed half the time, or one, i.e. evolving threshold is always employed.

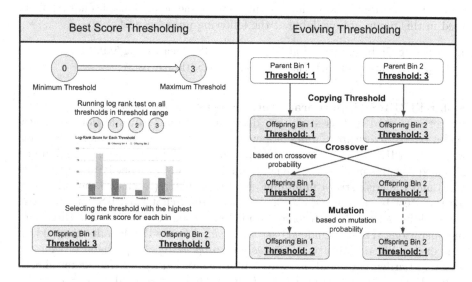

Fig. 2. Illustration of the two experimental adaptive burden thresholding methods implemented and evaluated in this analysis.

Regarding bin discovery, the original scikit-FIBERS imposed a bin length difference restriction, ensuring that each offspring specified no more than twice the number of features as the other. This was done to discourage the generation of bins with a large number of specified features, for simpler, more interpretable bins. When triggered, this mechanism would randomly exchange features between the offspring until the one was no longer twice the size of the other. For the new adaptive threshold FIBERS implementation we removed this mechanism, since bins with a larger number of specified features would be needed in order to optimally solve problems demanding a larger risk group threshold. Furthermore, the original scikit-FIBERS employed a cleanup process after mutation, removing any specified features that were duplicated in offspring bins, and replacing them with random new features from the dataset. The new method simply removes duplicates without replacement. Lastly, during mutation, the original scikit-FIBERS has distinct addition and deletion mechanisms with deletion driven by a fixed mutation probability, and addition probability being dynamically adjusted based on the offspring's bin size compared to the total feature list. The new method simply iterates over all features in the feature list, and for each, decides whether to add or remove it with a fixed probability (mutation probability). All other aspects of the extended scikit-FIBERS algorithm are consistent with the original scikit-FIBERS [10] and FIBERS [2] algorithms.

In this work, our primary focus is evaluating the top-performing bin (i.e., the bin with the highest fitness) as a standalone candidate solution. However, we also examine the top 10 performing bin sets to present a diverse range of potential solutions. The original and adaptive threshold implementations of scikit-FIBERS used in this paper are available at the following links, respectively:

1. https://github.com/UrbsLab/scikit-FIBERS/releases/tag/0.9.3.1
2. https://github.com/UrbsLab/scikit-FIBERS/tree/evostar_24

Scikit-FIBERS Hyperparameters. Experiment independent hyperparameter settings used in this study are detailed in Table 1. In addition to typical evolutionary algorithm parameters such as training iterations and population size, scikit-FIBERS includes the two bin-initialization parameters (a minimum specified features per bin and a maximum number of bins specifying a given feature) and the "risk strata minimum" hyperparameter (to ensure the validity of log-rank test by preventing class imbalance) [10]. Scikit-FIBERS with adaptive thresholding also adds additional parameters to control the dynamic thresholding operations. These include (1) *set threshold*, defining the default threshold for bins during bin initialization, (2) *evolving probability*, the probability that the adaptive burden threshold, is evolved in a generation rather than deterministically selected, (3) *minimum threshold*, the minimum threshold value allowed for bins, (4) *maximum threshold*, the maximum threshold value allowed for bins, and (5) *adaptive thresholding*, a boolean flag activating the use of adaptive thresholds vs. a default threshold of 0.

Table 1. Scikit-FIBERS Hyperparameter Settings

Hyperparameter	Setting
Max Training Iterations	1000
Max Bin Population Size	50
Crossover Probability	0.5
Mutation Probability	0.4
Elitism Proportion	0.8
Min Initial Features Per Bin	2
Max Initial Number Bins With Given Feature	4
Risk Strata Minimum	0.2

2.3 Synthetic Data Simulation

To analyze the capabilities of scikit-FIBERS, we previously designed a methodology for creating synthetic survival data with right-censoring, where a subset of features influences a predefined risk group and time-to-event outcomes [10]. This synthetic data generation draws inspiration from real-world HLA AA-MM data related to kidney transplantation, which was employed in evaluating the original FIBERS algorithm. The underlying assumption in this real HLA data is that the absence of MMs at informative AA positions is predictive of lower GF risk. However in the present simulations we examine scenarios where the count/frequency of MMs across AA positions can variably predict risk (e.g. > 2 MM within a set of AA positions best stratifies risk). Furthermore, real Ag or AA HLA data typically includes position frequency at each position (e.g. 0, 1, or 2). For the sake of simplicity, we simulate AA-MM data here, such that feature positions can have either '0' (no MMs) or '1' (MMs > 0), since we only care about the presence or absence of any MMs within a given feature. Figure 3 summarizes this new bin thresholding concept.

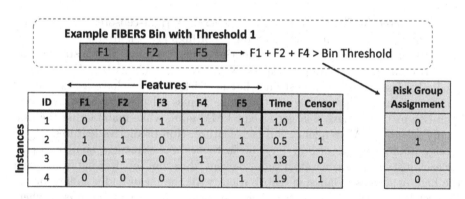

Fig. 3. To illustrate the thresholding concept, we have a small survival dataset where feature values are binary (0 for no MM and 1 for the presence of an MM). The 'Time' variable denotes the time until an event, which could either be the time of death (when 'censor' equals 1) or the time an instance was observed until (when 'censor' equals 0), under the assumption that death occurred at some point after observation. A basic scikit-FIBERS bin can be represented as follows: "IF the sum of MMs in F1, F2, and F5 exceeds the Detected Bin Threshold, THEN categorize the instance as part of the 'high-risk' group."

To test the ability of the adaptive burden threshold to correctly identify the optimal MM burden threshold, we updated this simulation strategy to generate datasets that have different underlying 'ground truth thresholds'. The 'ground truth threshold', is a dataset-wide threshold for the number of MMs that is required to achieve the best separation between high and low-risk survival curves. We take this ground truth threshold as a parameter and generate the datasets in such a way that this would be the burden threshold that would best separate bins into high-risk or low-risk. Ideally, this would be the threshold that is later identified as the optimal bin threshold by this new scikit-FIBERS implementation.

In addition to the ground truth threshold that differentiates low from high-risk instance groups, this simulator also allows users to manipulate various dataset parameters: (1) Total number of instances in the data, (2) proportion of low-risk group instances (fixed at 0.5 in this study to maintain risk group balance), (3) total number of features in the data, (4) number of informative/predictive features in the data (where all other features are noise/non-informative), (5) MM frequency range, which is randomly selected for each feature, (6) Gaussian distribution parameters for simulating time-to-event, including mean and standard deviation of high and low risk groups, respectively (which can be used to tune the signal to noise ratio of the dataset, but fixed in this study), (7), right-censoring frequency, indicating the proportion of instances where the death event was not observed and (8) noise frequency, indicating the proportion of instances where the time-to-event and censoring label of an instance from one risk group is swapped with the other risk group to introduce noise.

As depicted in Fig. 4, the process of generating right-censored survival data begins by categorizing instances into low (0) or high (1) risk groups based on the true risk group (TRG) column, considering the total number of instances and proportion of low-risk-group instances. Next, we randomly assign MM values to predictive features (P_n), ensuring that low-risk instances have a MM sum of no more than the 'ground truth threshold' across all predictive features, and high-risk instances have a larger or equal MM sum, while maintaining the randomly selected MM frequency for each feature. Non-predictive features (R_n) receive random values of 0 or 1 for all instances. Subsequently, similar to [10], for each instance, we generate a time-to-event "duration" by drawing from either the high or low-risk Gaussian distribution, with the mean and standard deviation set by the user. Censoring values are then assigned to the dataset, where 1 signifies that the event (e.g., death) was observed, and 0 indicates that the event occurred after the recorded duration. To simulate censoring, we calculate censoring probabilities for each instance, reflecting the ratio of the instance's survival time to the maximum survival time assigned across instances. We shuffle instances randomly and assign censoring based on these probabilities, stopping when the desired censoring frequency in the data is achieved. Notably, this simulation strategy does not require any specific subset of predictive features to stratify risk, focusing instead on scikit-FIBERS' ability to correctly identify an effective subset of predictive features over non-predictive ones to maximize the separation of risk strata survival curves.

Lastly, noise is generated in these datasets using a user-defined percentage of low-risk instances (noise frequency) and swapping both the time duration and censoring values with corresponding high-risk instances, leading to some low-risk instances having more MMs than the ground truth threshold summed across predictive features, and vice versa for some high-risk instances.

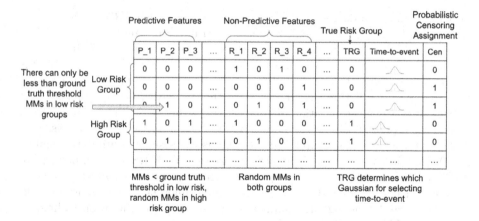

Fig. 4. Example simulation of synthetic right-censored survival data of MM features with a ground truth threshold of ≥ 1.

2.4 Simulation Experiments

In the majority of our experiments, we employ the scikit-FIBERS framework with hyperparameter values as outlined in Table 1, directing our attention towards assessing the performance across top bins instead of just the best bin unless explicitly stated otherwise. Prior research has already substantiated the effectiveness of the original FIBERS algorithm through cross-validation analyses conducted with real-world data [1], as well as an extensive examination involving synthetic positive and negative controls under various noise scenarios [10]. This current study is primarily centered on the systematic evaluation of the new adaptive burden thresholding methodology and with the simplifications added to bin initialization, crossover and mutation. To that end this proof-of-principle study does not require utilizing training and testing partitions, but rather assesses the ability of the algorithm to train on the simulated data and identify the ground truth threshold associated with bins including predictive features, but excluding non-predictive features. In these simulated scenarios, overfitting can easily be identified by the inclusion of non-predictive features, without requiring data partitioning.

Similar to previous work [10], we use an MM frequency range of 0.4–0.5 and a censoring frequency of 0.5, which are employed in all subsequent simulated datasets. We set the Gaussian distribution parameters (Mean, Std.) for selecting time-to-event as (1, 0.2) for high-risk instances and (1.5, 0.2) for low-risk instances, ensuring a distinct separation in the distribution of time-to-events between risk groups. Except for experiments explicitly focused on total feature count, all simulated datasets included 10 predictive and 40 non-predictive features, and a consistent dataset size of 10,000 instances. In contrast, the original FIBERS algorithm was applied to a significantly larger real-world data, with over 100,000 instances and 192 AA-MM features, as detailed in [1]. Smaller simulated datasets were applied in this study for computational efficiency, in this proof-of-principle analysis, but also to demonstrate that FIBERS can be effective even when having a much smaller number of training instances.

We evaluate experiments with three metrics to facilitate a quantitative and qualitative interpretation of the experimental outcomes: (1) *Accuracy*, in the context of knowing the simulated data ground truth, accuracy measures the correctness of predictions as compared to the True Risk Group (2) *Log Rank Score*, evaluates the separation of survival curves between different bin-defined risk groups. (3) *Bin Composition*, referring to the number of predictive vs. non-predictive features ending up in bins as well as the assigned burden threshold.

We investigate the performance of original FIBERS on comparison to FIBERS with adaptive thresholding across simulated datasets featuring different ground truth thresholds (0, 1, 2, 3), employing different thresholding methods (best score and evolving thresholding) and crossover and mutation techniques (regular and simplified). For evolving thresholding we examine probabilities of (0, 0.5, 1.0). Additionally, we examine the impact of total feature count on algorithm performance, including simulations with increasing total feature counts up to 800 while holding the number of predictive features constant at 10. Lastly we examine the resilience of the expanded scikit-FIBERS to increasing noise levels, conducting an analysis with parameters outlined in Table 1 and varying noise generation from 0 to 0.5. Code to generate all simulated datasets and repeat all experiments is included in the github repository for the adaptive threshold implementation scikit-FIBERS specified in Sect. 2.2.

3 Results and Discussion

Figure 5 presents FIBERS accuracy of bin classification (based on the simulated true risk group) with and without adaptive thresholding strategies, as well as with original vs. simplified crossover and mutation strategies. Individual box plots represent the top 10 FIBERS bins (based on Log Rank Score). We observe higher accuracy in predicting the ground truth simulated risk group when employing the adaptive burden threshold in comparison to the original scikit-FIBERS approach for both original and simplified crossover/mutation. Furthermore, the simplified crossover/mutation method consistently yields higher accuracies when coupled with adaptive burden thresholding, in simulated datasets with a ground truth threshold > 0.

As expected, Fig. 5 also indicates that FIBERS with any adaptive burden thresholding consistently outperforms the original FIBERS approach (no adaptive thresholding) in datasets with a ground truth threshold > 0 based on this accuracy metric. This trend is further corroborated by log-rank scores, as evidenced in Table 2 and the box plots of log-rank scores (Supplementary S1). We also observe that in datasets with a larger ground truth threshold (i.e. 2 or 3), applying the adaptive threshold with an evolutionary probability of 0.5 (i.e. a mix of evolution and deterministically picking the best scoring threshold) can yield some better performing bins than the deterministic approach alone. This suggests that the performance of adaptive threshold can benefit from a hyperparameter sweep of this evolution probability, when applied to new datasets.

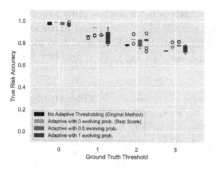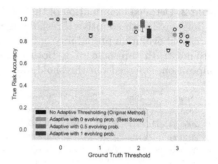

Fig. 5. Box plot of accuracies of the Top 10 FIBERS bins with datasets of different simulated thresholds (0 to 3) using the new scikit-FIBERS adaptive threshold methods and crossover-mutation methods (original - left, simplified - right).

Lastly, we observe a gradual but consistent decrease in true risk accuracy for both original and simplified crossover/mutation, reflecting the increasing difficulty of the underlying binning task, which may be alleviated by increased training iterations and/or maximum bin population size. This will be examine further in future work.

In Table 2, we present the top 10 bins from one experiment conducted within Fig. 5 (i.e. simplified crossover and mutation, a ground truth threshold of 1, and evolving probability of 0), comparing FIBERS with and without adaptive thresholding. We observe higher log rank scores as well as higher accuracies for FIBERS when including the adaptive thresholding, largely consistent across analyses of simulated datasets with a ground truth threshold > 0. Importantly we note that all top 10 bins discovered with FIBERS including adaptive thresholding, were assigned the correct ground truth threshold of 1.

Of note, we observe that bin log-rank score ranking is not perfectly correlated with true risk group accuracy ranking, despite the top performing bins in Table 2 having both the highest log-rank score and accuracy. From this, we recognize that FIBERS log-rank score ranking in real-world data may not always optimally translate into ideal risk group assignment accuracy.

Table 2 also illustrates that FIBERS with adaptive thresholding tended to result in bins with a larger number of specified features. In the given example, all top 10 FIBERS bins correctly only include predictive features and exclude all non-predictive features. Out of all 50 bins discovered by FIBERS in these analyses, non-predictive features did occasionally get included in some lower ranking bins (often alongside a high number of predictive instances), as well as utilizing the incorrect threshold (for the given simulation). For example, see Bin 30 in Supplementary S3: $[P_{10}, P_2, P_8, P_4, P_3, P_7, P_6, R_{73}, P_5, R_{65}]$ with 10 total features including 8 predictive features, an accuracy of 0.864, and a burden threshold of 3 (different from the simulated ground truth threshold of 1). Complete tables detailing all 50 bins for the experiments run in Table 2 can be found in Supplementary S2 and S3.

Table 2. Detailed examination of the top 10 FIBERS bins without adaptive thresholding (left), and with adaptive thresholding using simplified crossover and mutation and a threshold evolution probability of 0 (right). The simulated dataset analyzed included a ground truth threshold of 1. Both bin sets include the log rank score, bin feature composition, assigned bin threshold, and accuracy.

#	Score	Bins	Threshold	Accuracy	#	Score	Bins	Threshold	Accuracy
1	2066.66	$[P_4, P_5, P_6, P_1]$	0	0.8694	1	4652.76	$[P_7, P_4, P_1, P_9, P_5, P_6]$	1	0.9430
2	2050.29	$[P_5, P_2, P_1, P_4]$	0	0.8692	2	4616.55	$[P_9, P_6, P_{10}, P_2, P_8, P_4]$	1	0.9425
3	2046.24	$[P_5, P_1, P_4]$	0	0.8490	3	4569.51	$[P_7, P_5, P_4, P_1, P_2, P_6]$	1	0.9410
4	2040.88	$[P_9, P_2, P_1, P_5]$	0	0.8683	4	4553.11	$[P_5, P_3, P_9, P_6, P_1, P_4]$	1	0.9413
5	2039.54	$[P_5, P_9, P_1, P_8]$	0	0.8658	5	4545.23	$[P_8, P_9, P_1, P_{10}, P_2, P_5]$	1	0.9403
6	2019.14	$[P_5, P_1, P_3, P_4]$	0	0.8685	6	4542.60	$[P_2, P_8, P_7, P_{10}, P_6, P_9]$	1	0.9411
7	2007.44	$[P_5, P_1, P_4, P_{10}]$	0	0.8662	7	4522.94	$[P_3, P_1, P_9, P_1, P_4, P_2]$	1	0.9406
8	2006.21	$[P_9, P_1, P_6, P_5]$	0	0.8672	8	4460.73	$[P_4, P_7, P_1, P_5, P_8, P_3]$	1	0.9398
9	2001.99	$[P_4, P_1, P_9]$	0	0.8478	9	3705.30	$[P_2, P_7, P_1, P_9, P_8]$	1	0.8699
10	2001.47	$[P_9, P_1, P_5, P_{10}]$	0	0.8670	1	3699.89	$[P_2, P_4, P_8, P_6, P_9]$	1	0.8732

Experiments summarized in Fig. 6 examine the impact of an increasing total number of features on FIBERS performance (with and without deterministic adaptive thresholding and original vs. simple crossover and mutation) in simulated data with a ground truth threshold of 2. Here we observe a dramatic improvement in performance when using adaptive thresholding, between 300 and 600 total features, where original FIBERS does no better than random chance. At 700 and 800 total features, both implementations failed to perform, however we expect that a larger number of training iterations, and/or a larger population size would help FIBERS scale up to larger feature spaces. This will be one target of future work.

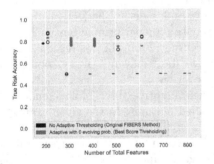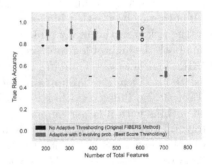

Fig. 6. Performance of FIBERS with varying Total Number of Features: This plot shows a boxplot of accuracies of the top 10 FIBERS bins with increasing total number of features and a fixed ground truth threshold of 2 with 10 predictive features for original (left) and simplified (right) crossover and mutation.

Experiments summarized in Fig. 7 examine the impact of varying noise levels from 0 to 0.5 (i.e. no noise to all noise). For this we again ran FIBERS with and without deterministic adaptive thresholding but limited to simple crossover and mutation in simulated data with a ground truth threshold of 2. Here we observed that FIBERS with adaptive thresholding performs significantly better than original FIBERS (that assumes a fixed bin threshold of 0) in simulated datasets with a noise frequency of 0 to 0.4, and an underlying ground truth threshold greater than 0. This supports the assertion that FIBERS is capable to identify informative feature bins in the presence of significantly noisy signal, as well as further verifying the expected result that FIBERS adaptive binning yields better performance than original FIBERS when the underlying ground truth threshold is greater than 0 (assumed by original FIBERS). Future work applying FIBER with adaptive thresholding to real-world data would thus be expected to converge on top bins that assign a burden threshold that reflects the underlying (unknown) ground truth threshold. This will allow us to discover informative AA-MMs that predict GF risk, without assuming any particular burden threshold.

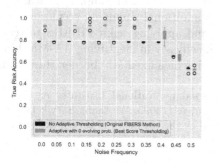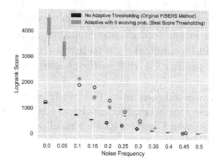

Fig. 7. Increasing noise frequency in the top 10 FIBERS bins with ground truth threshold of 2 using simplified crossover and mutation: Accuracy in predicting true risk group (left) and log-rank scores (right).

Overall, these results support the proof-of-principle efficacy of adaptive thresholding within FIBERS. Future work will more broadly examine scalability, sensitivity, and response to right-censoring in more diverse simulated scenarios. We also aim to further extend the scikit-FIBERS algorithm, to be able to identify more than two risk groups (e.g. high, medium, and low risk), introduce multi-objective pareto-front fitness evaluation based on both log-rank score and bin size, integrate covariate adjustment into bin evolution/discovery to focus FIBERS on risk factors that remain relevant after this adjustment, and leverage ensemble machine learning modeling for risk group assignment. Beyond simulations, we also plan to apply scikit-FIBERS to evaluate kidney GF risk in AA-MM data.

4 Conclusion

In conclusion, this manuscript presents an extension to the scikit-FIBERS framework, introducing adaptive burden thresholding to enhance the flexible identification of feature bins for risk stratification in right-censored survival analyses including donor/recipient matching for tissue/organ transplantation. The incorporation of adaptive thresholds addresses a key limitation of the original FIBERS algorithm, allowing for the simultaneous evolution of feature bins and threshold values. Simulation experiments demonstrated improved performance in terms of accuracy and log-rank scores, especially in scenarios with an increased number of features. The findings suggest that scikit-FIBERS with adaptive burden thresholding holds promise for optimizing risk stratification in complex biomedical survival analyses, such as the HLA AA-MM in kidney graft failure. The algorithm's robustness to noise and scalability further underscore its potential for real-world applications. This work lays the foundation for the ongoing development and deployment of scikit-FIBERS in biomedical research, offering a data-driven and flexible approach to uncovering multifaceted risk factors in complex clinical scenarios. Future work will involve exploring additional simulated scenarios (e.g. with larger feature and instance spaces), and applying scikit-FIBERS to real-world biomedical data to further validate its efficacy in identifying AA-MM bins that predict GF risk, without assuming/relying on a fixed MM burden threshold.

Acknowledgements. The study was supported by National Institutes of Health grants: U01 AI152960 and R01s LM010098, and AI173095. We thank Satvik Dasariraju, Loren Gragert, Keith McCullough as well as the reviewers of this work for their helpful feedback.

References

1. Dasariraju, S., et al.: HLA amino acid mismatch-based risk stratification of kidney allograft failure using a novel machine learning algorithm. J. Biomed. Inform. **142**, 104374 (2023)
2. Dasariraju, S., Urbanowicz, R.J.: Rare: evolutionary feature engineering for rare-variant bin discovery. In: Proceedings of the Genetic and Evolutionary Computation Conference Companion (GECCO 2021), pp. 1335–1343. Association for Computing Machinery, New York (2021). https://doi.org/10.1145/3449726.3463174
3. Egfjord, M., Jakobsen, B., Ladefoged, J.: No impact of cross-reactive group human leucocyte antigen class i matching on long-term kidney graft survival. Scand. J. Immunol. **57**(4), 362–365 (2003)
4. Kamoun, M., et al.: HLA amino acid polymorphisms and kidney allograft survival. Transplantation **101**(5), e170 (2017)
5. Kolonko, A., et al.: Anemia and erythrocytosis after kidney transplantation: a 5-year graft function and survival analysis. In: Transplantation Proceedings, vol. 41, pp. 3046–3051. Elsevier (2009)
6. Kuntzelman, K.M., Williams, J.M., Lim, P.C., Samal, A., Rao, P.K., Johnson, M.R.: Deep-learning-based multivariate pattern analysis (dmvpa): a tutorial and a toolbox. Front. Human Neurosci. **15** (2021). https://doi.org/10.3389/fnhum.2021.638052

7. Luo, Y., et al.: A high-resolution HLA reference panel capturing global population diversity enables multi-ancestry fine-mapping in HIV host response. Nat. Genet. **53**(10), 1504–1516 (2021). https://doi.org/10.1038/s41588-021-00935-7

8. Ng, S., Masarone, S., Watson, D., Barnes, M.R.: The benefits and pitfalls of machine learning for biomarker discovery. Cell Tissue Res. **394**(1), 17–31 (2023). https://doi.org/10.1007/s00441-023-03816-z

9. Robinson, J., Barker, D.J., Georgiou, X., Cooper, M.A., Flicek, P., Marsh, S.G.: Ipd-imgt/hla database. Nucleic Acids Res. **48**(D1), D948–D955 (2020)

10. Urbanowicz, R., Bandhey, H., Kamoun, M., Fogarty, N., Hsieh, Y.A.: Scikit-fibers: an 'or'-rule discovery evolutionary algorithm for risk stratification in right-censored survival analyses. In: Proceedings of the Companion Conference on Genetic and Evolutionary Computation (2023). https://doi.org/10.1145/3583133.3596393

11. Ying, T., Shi, B., Kelly, P.J., Pilmore, H., Clayton, P.A., Chadban, S.J.: Death after kidney transplantation: an analysis by era and time post-transplant. J. Am. Soc. Nephrol. **31**(12), 2887–2899 (2020)

An Evolutionary Deep Learning Approach for Efficient Quantum Algorithms Transpilation

Zakaria Abdelmoiz Dahi[1,2]([✉]), Francisco Chicano[2], and Gabriel Luque[2]

[1] Univ. Lille, Inria, CNRS, Centrale Lille, UMR 9189 CRIStAL, F-59000 Lille, France
abdelmoiz-zakaria.dahi@inria.fr
[2] ITIS Software, University of Malaga, Malaga, Spain
{chicano,gluque}@uma.es

Abstract. Gate-based quantum computation describes algorithms as quantum circuits. These can be seen as a set of quantum gates acting on a set of qubits. To be executable, the circuit requires complex transformations to comply with the physical constraints of the machines. This process is known as transpilation, where qubits' layout initialisation is one of its first and most challenging steps, usually done by considering the device error properties. As the size of the quantum algorithm increases, the transpilation becomes increasingly complex and time-consuming. This constitutes a bottleneck towards agile, fast, and error-robust quantum computation. This work proposes an evolutionary deep neural network that learns the qubits' layout initialisation of the most advanced and complex IBM heuristic used in today's quantum machines. The aim is to progressively replace weakly scalable transpilation heuristics with machine learning models. Previous work using machine learning models for qubits' layout initialisation suffers from some shortcomings in the proposal's correctness and generalisation as well as benchmarks diversity, utility, and availability. The present work solves those flaws by (I) devising a complete Machine Learning pipeline including the ETL component and the evolutionary deep neural model using the linkage learning algorithm P3, (II) a modelling applicable to any quantum algorithm with a special interest to both optimisation and machine learning ones, (III) diverse and fresh benchmarks using calibration data of four real IBM quantum computers collected over 10 months (Dec. 2022 and Oct. 2023) and training dataset built using four types of quantum optimisation and machine learning algorithms, as well as random ones. The proposal has been proven to be more efficient and simple than state-of-the-art deep neural models in the literature.

Keywords: Evolutionary Machine Learning · Deep Neural Architecture Search · Quantum Variational Algorithms · Quantum Transpilation

S. Smith et al. (Eds.): EvoApplications 2024, LNCS 14635, pp. 240–255, 2024.
https://doi.org/10.1007/978-3-031-56855-8_15

1   Introduction

Gate-based quantum computers express computation as quantum circuits, seen as a series of quantum gates acting on a set of quantum bits (or qubits). Current devices are in their Noisy Intermediate Scale Era (NISQ), having a noisy nature and a limited number of qubits (10 to 100 s), not enough to implement efficient quantum error correction [3]. To be executable on quantum machines, such types of quantum algorithms undergo a series of complex transformations, usually known as transpilation, to comply with the hardware constraints of the quantum machine. The transpilation process has a major influence on the efficiency and reliability of quantum computation, especially considering the NISQ nature of today's quantum computers. Qubits' layout initialisation is one of the first, most critical and challenging steps in the transpilation proven to be NP-complete [16]. Due to the complexity of such a task, a good amount of research work can be found in the literature focused on the qubits' initialisation[1], covering several quantum technologies with different approaches. This work focuses on superconducting quantum computers, in particular, IBM quantum computers, considering that they are one of the most investigated and actively deployed in the industry. Most literature (around 68 works in total) and today's transpilers rely on exact [16] or heuristic-based approaches [17]. For small-sized circuits, the transpilation remains decently fast. However, the time increases drastically when the complexity of the algorithms increases (e.g. the authors found that transpiling GHZ circuits on 5627-qubits machines takes around 8.89 h). Clearly, time is a key factor in quantum computation considering that today's quantum hardware is timely limited and affected by noise. Also, tedious and long transpilation might make quantum advantage vanish when an equivalent classical algorithms runs in the same time.

As an alternative to exact and heuristic-based transpilation approaches for qubit initialisation, efforts are made to devise machine-learning-based qubits' initialisers instead. Indeed, heuristics need to run the same costly computation each time a quantum circuit (algorithm) is given (whether used or not before), while Machine Learning (ML) ones would be trained only once profiting from past data. To the best of the authors' knowledge, only three works have explored such an alternative [1,10,11]. Although promising, these former works have shortfalls on several levels, such as incorrect modelling of the qubits' initialisation as a machine learning task which threatens the proposal's reliability. Also, they focus only on the deep learning model, although the entire Extract-Transform-Load (ETL) is quite critical whether for the experiments' reproducibility or their use in on-production transpilers. Moreover, the used benchmarks are not always accessible, they are outdated and limited in terms of complexity, diversity, and utility of the used quantum algorithms and machines. This does not properly reflects the advances in quantum hardware (e.g., noise resilience, error probabilities, etc.) neither the generalisation of the approach to useful quantum algorithms. The present work copes with those shortfalls by

---

[1] For the sake of brevity, we omit the word "layout" in "qubits' layout initialization".

(I) devising a correct and generalisable ML task modelling of the qubits' initialisation problem; (II) devising an entire pipeline including both the ETL components and a deep learning model that is evolved using a linkage learning algorithm known as the Parameter-less Population Pyramid (P3) [7]; and (III) using a diverse benchmark built using continuously-up-to-date noise-calibration data of four real IBM quantum machines over 10 months (Dec. 2022 to Oct. 2023) and a wide range of quantum algorithms including QAOA, quantum classifier, QGANs, and random circuits.

In the rest of the paper we introduce the fundamentals of quantum computation and transpilation in Sect. 2, and we present the neural architecture search in Sect. 3. Section 4 presents the proposed approach, while Sect. 5 experimentally investigates the proposal. Finally, Sect. 6 concludes the paper.

2   Preliminary Concepts

Two quantum computation paradigms exist: adiabatic and gate-based. The first is based on the adiabatic theorem and is dedicated to solving optimisation problems, while the second is a more universal one and has a large range of applications attracting both the interest of Industry and Academia.

We focus in this paper on the gate-based paradigm, which describes computations as quantum circuits (see Fig. 1(a)) composed of a series of *quantum gates* acting on quantum bits (qubits). A qubit state $|\psi\rangle$ is, in general, a superposition of two base states $|0\rangle$ and $|1\rangle$, where $|\psi\rangle = \alpha |0\rangle + \beta |1\rangle$ and $\alpha, \beta \in \mathbb{C}$. They also verify the normalisation condition $|\alpha|^2 + |\beta|^2 = 1$. The quantum state of a multiple-qubits system can be represented by the tensor product of the qubits' quantum state $|\psi\rangle = \bigotimes_{i=1}^{n} |\psi_i\rangle$, where $|\psi\rangle \in \mathbb{C}^{2^n}$. Mathematically, each quantum gate is represented by a *unitary transformation* $U : \mathbb{C}^{2^n} \to \mathbb{C}^{2^n}$, which verifies $U^\dagger U = U U^\dagger = I$, where $U^\dagger$ is the Hermitian adjoint of U and I is the identity matrix. Several quantum gates exist [13] such as the Hadamard H gate, Pauli gates σ_x, σ_y, σ_z, etc.

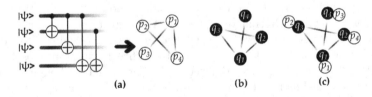

Fig. 1. (a) Quantum circuit, (b) QPU topology and (c) qubits' initialisation

To be executable on gate-based quantum computers, quantum algorithms undergo a *transpilation* process. Transpilation consists in a series of complex and chained tasks to translate the original quantum circuit into another one that fulfills the machines' constraints (e.g., available gates) [17]. Each manufacturer has its own transpilation process with different phases. Indeed, from one work to another (e.g. compilation, synthesis, etc.), and eventually from one manufacturer to another (e.g. IBM: passes, Google: Transformers, etc.), the transpilation process can be named or executed differently. The quality of the transpilation process has a huge impact on the error-robustness, correctness and feasibility of the computation, especially considering the NISQ era of quantum devices. The circuit's depth is a reliable metric of such quality, where for a gate g with a predecessor $P(g)$, the depth(g) is defined by Equation (1). The depth of the circuit is the maximum value of depth(g) over all gates g. In that sense, a special interest is given here to the IBM transpilation process, since it is the most representative and white-boxed one. More particularly, the focus of this work is on the qubits' initialisation problem within IBM quantum machines [5].

$$\texttt{depth}(g) \triangleq \begin{cases} 0 & P(g) = \emptyset \\ 1 + \max_{p \in P(g)} \texttt{depth}(p) & P(g) \neq \emptyset \end{cases} \tag{1}$$

The qubits' initialisation problem is one of the first and most critical transpilation steps. It is proven to be NP-complete [16] and omnipresent in most of today's quantum machines (e.g., IBM, Google, etc.), and spans over several technologies such as superconducting and Ion-trapped qubits [2,21]. Having a quantum circuit C acting on a set of N logical qubits $Q = \{q_1, \ldots, q_N\}$, the qubits' initialisation problem consists in mapping the set of logical qubits Q to a set of M physical qubits $P = \{p_1, \ldots, p_M\}$ on the Quantum Processing Unit (QPU), where $M \geq N$ (see Fig. 1(c)). A solution X to this problem would be one of the possible $_M P_N$ permutations without repetition (partial or complete depending on whether $M = N$ or not) formed by the arrangement of N physical qubits from $\{p_1, \ldots, p_M\}$. In this work, it is admitted that no matter what the arrangement is, the logical qubits $(q_1, \ldots, q_N)$ would be assigned to physical qubits according to a natural order. Meaning, a permutation $(p_1, p_5, p_2, p_4, p_3)$ would represent a solution where q_1 is assigned to p_1, q_2 to p_5, q_3 to p_2, q_4 to p_4, and q_5 to p_3. The quality of a solution has a direct effect on the remaining phases of the transpilation (e.g. qubit routing, optimisation via transitivity and cancellation rules, etc.), the feasibility of computation and its noise-robustness.

3 Deep Learning and Neural Architecture Search

The goal of *supervised learning* is to build a function $f : X \rightarrow Y$ fitting a set of samples $(x, y) \in X \times Y$. There are many techniques for supervised learning. We focus here on *deep learning* models and, in particular, *deep neural networks* (DNNs). A deep neural network is composed of several layers of neurons (Fig. 2(b)). Neurons in the *input layer* represent the input vector x, while the neurons in layer j receive as input the output of layer $j - 1$. This input is weighted and summed before computing an activation function a. The output of the neuron is given by the expression $a(\sum_{i=1}^{n_{(j-1)}} w_i^{(j)} x_i^{(j-1)} + b)$, where $x_i^{(j-1)}$ is the output of the i-th neuron in layer $j - 1$ and $n_{(j-1)}$ is the number of neurons at layer $j - 1$ (see Fig. 2(a)). The goal of the learning process is to find an optimal or sufficiently close to optimal set of connection weights $w_i^{(j)}$. Several factors influence the efficiency of neural networks. For instance, the architecture, the training algorithm, and the choice of features used in training. Training is performed using gradient-decent techniques to minimise a loss function [18].

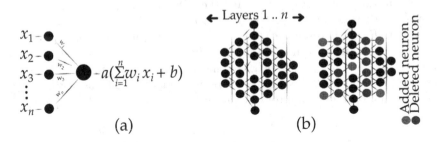

Fig. 2. (a) DNN principle and (b) neural architecture search

Evolutionary computation has been used to enhance the efficiency of machine learning (ML) techniques principally in pre-processing (e.g., feature selection and re-sampling), learning (e.g., parameter setting, membership functions, and neural network topology), and post-processing (e.g., rule optimisation, decision tree/support vectors pruning, etc.). Several well-known EAs exist, although the interest in this paper focuses on *linkage learning* EAs. They represent solutions as d-dimensional vectors and try to learn the relations among the d decision variables to be more efficient during the search. This class of algorithms have been profoundly studied and led to promising advances. One can cite as examples, the LTGA (Linkage Tree Genetic Algorithm) [19], LT-GOMEA (Linkage Tree Gene-pool Optimal Mixing Evolutionary Algorithm) [20], the DSMGA-II (Dependency Structure Matrix Genetic Algorithm II) [9], the 3LOa (Linkage Learning based on Local Optimisation algorithm) [15] and the P3 [7], which we use in this work. We use in this work linkage learning algorithms in Neural Architecture Search (NAS) where the DNN topology is evolved using P3, while the learning process is left to gradient descent approaches [18] (see Fig. 2(b)). Although applying EAs for NAS has been quite investigated, to the best of the authors' knowledge, linkage learning algorithms have never been applied to NAS.

3.1 Deep Learning for Qubits' Initialisation

To the best of the authors' knowledge, there are 68 works that tackle the qubits' initialisation problem whether as a standalone problem [5] or jointly with other transpilation problems [17]. Most of those works approach the problem using exact algorithms with high computational cost as the problem's size increases, or heuristic ones that are meant to be less computational greedy. Unlike previous literature, works in [1,10,11] started exploring the use of deep learning to solve the qubits' initialisation problem. Although, it is worth noting that the works in

[10,11] present the same proposal. The authors in [1] propose a DNN model to tackle the qubits' initialisation as a classification problem. The aim is to learn the initialisation produced by a noise-aware heuristic used in today's IBM quantum machines [12]. That work suffers from some shortfalls. First, the modelling of the qubits' initialisation as a machine learning problem is not completely reliable or generalisable. Actually, it only considers very limited calibration parameters such as T_1, T_2, CNOT error and execution time as well as readout error, while when having a quantum machine with k qubits and n qubits' connectivities, $8k+2\cdot5+2n$ important calibration parameters exist. Also, the circuits modelling does not capture the sequencing of gates execution. Indeed, using the devised modelling, several circuits might be modeled using the same representation, while requiring different intialisation. This turns out to be quite challenging in ML. The benchmark they use is composed of purely random circuits with no known utility and uses only one 5-qubit machine topology called IBM Q Burlington whose calibration data dates back to 2021. This prevents assessing the proposal on quantum algorithms of known use (e.g., optimisation, machine learning, etc.). Another problem is that the machine used has been retired by IBM, so the practical interest of the paper is limited.

The authors in [10,11] combine reinforcement learning with a graph neural network to solve the qubits' initialisation problem. Likewise in [1], the authors used an ML task modelling that is not completely reliable or generalisable. Indeed, calibration data such as the reset gates length are discarded, as well as an important parameter such as the readout length. The circuits modelling is restricted to only four gates, which are not universal. This restricts the generalisation of the approach to other quantum circuits. Most importantly, the circuits feature modelling does not capture the sequencing of the gates which is crucial. Moreover, the authors use only two quantum machines of 7 and 27 qubits (IBM Nairobi and Algiers) and study only 6 quantum circuits. The dataset they use is not made available, which prevents any replication of the results. Last, but not least, all the work done in [1,10,11] focuses only on devising the DNN model and discards other important components in any ML pipeline, especially the ETL part. This prevents the proposal from being constantly up-to-date with the quantum hardware evolution (e.g., noise sensitivity) and therefore any integration in on-production quantum transpilers.

This work attempts to correct all the shortfalls identified in [1,10,11], but it still goes in the same sense as the work done in [1] by learning the qubits' initialisation produced by the IBM heuristic called Noise Adaptive [12].

4   The Proposed Approach

The proposal aims at coping with the shortfalls identified in the previous work: (I) the machine learning task modelling, (II) the proposal's completeness and efficiency, as well as (III) the benchmark diversity/utility and availability. Figure 3 sketches the two main modules of our proposal and their execution flow. The next subsections explain these modules in detail. Also, the full source code of the proposal is made publicly available in [4].

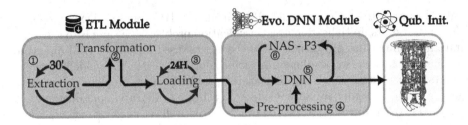

Fig. 3. The modules and components of the proposed approach

4.1   The Extract-Tranform-Load Module

We will start describing the ETL module and its components in this section.

The Extraction Component: This component is a scrapper responsible for the extraction of daily fresh calibration data of all the available IBM quantum machines. It automatically detects which machine(s) is(are) available as well as the quantum transpiler version locally installed. Therefore, it changes automatically the scrapping mechanism as well as handling the servers' failures according to transpiler's versioning. The extraction is performed every 30 min. The calibration data being extracted represent the physical properties (e.g., error, execution time, etc.) of both the qubits and the implementable quantum gates on it. As to the qubits, there are 8 calibration parameters extracted for each qubit: the `decoherence times` (in μs) T_1 and T_2, the frequency and anharmonicity (in GHz), the `readout error probability` and `length` (in ns), the `error probability` of measuring the state $|1\rangle$ when preparing it in $|0\rangle$ and vice versa. As to the quantum gates, two pieces of data are extracted; the `gate error probability` and the `gate length` (in ns) for all possible single (I, R_Z, $S^\dagger$, σ_x, `reset`) and all two-qubits gates (i.e., `CNOT`). In total, having a quantum machine with k qubits and n qubits' connectivities, $8k$ qubit parameters and $2 \cdot 5 + 2n$ gate parameters will be extracted for each of the 4 IBM quantum machines found available: `ibmq_lima`, `ibmq_belem`, `ibmq_quito` and `ibmq_manila`. So, far, data of 11 months have been gathered from the 12th December 2022 to the 2nd October 2023. The scrapper is continuously being executed (24H a day and 7 d a week) to gather new data. The raw calibration data are made available for open public use at [4]. To the authors' knowledge, this is the first time in the literature that such scrapper has been devised and made available together with real world data of this amount and diversity.

The Transformation Component: To correct the modelling previously presented in [1], we propose a modelling based on three components (see Fig. 4): (I) the quantum machine calibration, (II) the quantum machine topology and (III) the quantum circuit.

The first one corresponds to the calibration of the quantum machine. Unlike the previous works [1,10,11], which considers only some calibration data while

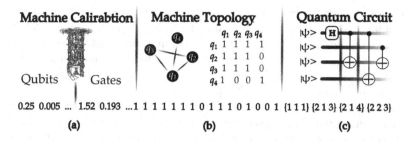

Fig. 4. Representation of ML features generated by transformation component

neglecting others, the presented modelling includes the totality of the data extracted by the extraction component previously introduced. The IBM heuristic Noise Adaptive considers most of those calibration data to calculate the initial qubits' placement (see Fig. 4(a)). This approach allows our proposal to identify uniquely each machine, therefore avoiding that several samples represent multiple machines. The second part of the modelling is dedicated to representing the topology of the qubits' connectivity of the quantum machine. This turns out to be quite important in order to devise a proposal able to learn the Noise Adaptive qubits' intialisation for any quantum machine topology, instead of focusing on one concrete quantum machine (requiring a separate learning process for each machine). The features that represent the topology are a flattened vector of the adjacency matrix of the underlying graph representing the quantum device qubits' connectivity (see Fig. 4(b)).

The third part of the modelling concerns the quantum algorithm itself. Indeed, previous work considers a set of features that does not represent circuits in a unique way: it might map several quantum algorithms to the same representation. In fact, those representation-equivalent algorithms could need a different qubits' initialisation. In the present modelling, the circuit is decomposed into layers, where each layer regroups the gates that can be executed in parallel (see Fig. 4(c)). Therefore a circuit with d gates, would be translated into features by representing each gate in each layer sequentially using a sequence of three digits $\{x_i, y_i, z_i\}$, $i = 1, \ldots, d$, where the first digit x_i identifies the type of gate and the two remaining digits y_i and z_i are dedicated to represent the ID of the qubits taking part in the gate execution. In single-qubit gates, the two digits y_i and z_i will represent the (same) ID of the qubit the gate affects (see Fig. 4).

Unlike the previous literature, which studied either random circuits with no known utility [1], or a set of limited quantum algorithms like in [10,11], in this work the training data is generated by considering four types of algorithms dedicated to optimisation, machine learning and random ones. This way we maximise the generalisation of the devised approach. In particular, we consider the Quantum Approximate Optimisation Algorithm (QAOA) [6], the Quantum Variational Classifier (QVC) and the Quantum Adversarial Neural Network (QGAN) [8]. The QAOA has been used by setting the depth to 1, while the QVC was used with a fully-entangled mapping and classification with depth 1. The QGAN has been built by considering a fully-entangled discriminator and generator with depth 1. Both the QVC and QGAN features' mapping parts have been generated using the R_z, R_y and CR_z quantum gates. The mathematical details of the QAOA, QVC and QGAN go beyond the scope of this work. For further details, please refer to the original works [6,8].

In QAOA, a random instance of the Quadratic Unconstrained Binary Optimisation Problem (QUBO) is solved each time, while the random quantum circuits are generated by considering a universal set of quantum gates: CNOT, H, S and T. The encoding handled by our approach has been thought to go beyond the previous set of gates and can cover a large plethora of quantum gates, such as H, S, T, R_Z, σ_x, U_1, U and CNOT. In all the quantum circuits, a maximum depth of 200 gates is considered. For those quantum algorithms that do not consume all the gates' depth, a padding of zeros is added in their features' vector. So far, the generated dataset contains over 31,109 training samples. The code of the transformation component automatically generates a fresh training dataset (when needed) if new quantum machines/calibration data are available or there is a need for other datasets built using different quantum algorithms.

Moving now to the targets of the training dataset, they have been generated by executing the generated circuits (QAOA, QVC, QGAN or random) using the IBM transpiler by setting it to the third level of optimisation (most sophisticated and costly one) and imposing the `Noise Adaptive` heuristic for qubits' initialisation. The transpiler is executed over 30 executions and the qubits initialisation that produces the circuit with the smallest circuit depth (see Equation (1)) is the one considered as a target for the generated features. Having a quantum machine of n qubits, targets will be organised as tuples of shape $\{x_1, x_2, \ldots x_n\}$, where the rank i of the value in the tuple represents the logical qubit that will be placed in the $x_{i^{th}}$ physical qubit. For instance, the configuration $\{5, 4, 1\}$ indicates that the 1st logical qubit will be placed on the 5th physical one, the 2nd on the 4th, and the 3rd on the 1st.

The Loading Component: The loading component is responsible for the persistence of the training dataset as well as the raw data. This information is stored in ".csv" files considering that the proposal in its current form is meant for direct integration in the IBM transpiler or other CLI standalone use. Indeed, ML libraries such as Scikit-learn, Tensorflow, Pytorch, and Keras, have some readiness to use .csv files via **Pandas** more than other types of files. However, if any API-based use is planned for the proposal, the switch to ".json" files is thought to be quite straightforward and easy to accomplish.

4.2 The Evolutionary Deep Neural Network Module

This section presents the evolutionary deep learning module that will be trained on the data generated by the ETL module (see Sect. 4.1), and learns the qubits' initialisation generated by the IBM transpiling heuristic `Noise Adaptive`.

The Pre-processing Component: Before feeding the training samples generated by the ETL module to the deep neural network, the training dataset is normalised and reduced. Concretely, as for normalisation the Z-score normalisation is applied to convert the dataset samples in such a way that the mean is 0 and the deviation is equal to 1 using the $\frac{x-\mu}{\sigma}$, where μ is the mean of the data and σ is the standard deviation. As a feature reduction, the Principal Component Analysis technique is used with the goal of finding the linear transformation that keeps the variance at maximum using the minimum number of features [14].

The Deep Neural Network Component: The qubits' initialisation is modeled as a multi-class classification problem. Given a quantum algorithm that acts on q qubits and a quantum machine with n qubits, the idea is to determine which of the $1, \ldots, n$ physical qubits (classes) is used for each logical qubit. The original architecture used in this work is the one presented in [1]. The choice was made for this model because it is the one providing the best results so far on this problem (see Fig. 5).

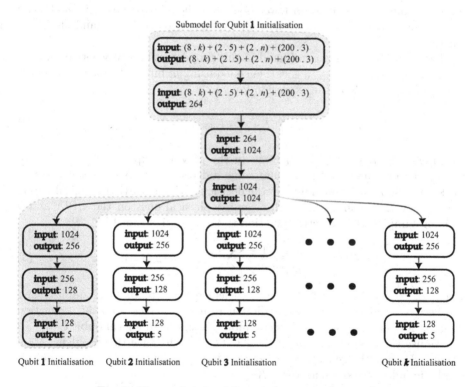

Fig. 5. The used state-of-the-art deep neural network

We can see the DNN model as the union of several sub-DNN models that are trained separately to decide the physical qubit in which a logical qubit is mapped. All neurons in the DNN model are densely connected. Every neural layer uses the ReLU activation function $a(x) = \max(0,x)$ except the dropout and the output layer that uses a softmax function. The output generated by the ETL module is converted into an array of probabilities, where each n variables constitutes the probability of a given logical qubit to be assigned to each physical qubit of the machine.

The Evolutionary Neural Architecture Search Component: The original model presented in [1] is composed of several neural layers of sizes 264, 1024, 256, and 128 (see Fig. 5). The idea in our proposal is to decide the model's architecture as well as some other hyper-parameters using an evolutionary algorithm to reduce to loss function. As loss function, we use the average of the categorical cross-entropy of each sub-DNN model presented in the previous section. The variables to be optimised are the size of each neural layer of the DNN model as well as the number of training epochs and the batch size. In total, 6 optimisation variables $x_{j=1,\ldots,6} \in \mathbb{N}$ are considered (the dropout layer is neglected). Each variable might evolve from the original value used in [1], and $\frac{1}{4}$ of that value, except for the batch size that is allowed to take values up to $\frac{1}{8}$ of the original value. Such a choice of lower bounds is thought to explore if the P3 can reduce the complexity of the model while enhancing (or at least maintaining) its efficiency.

The solution encoding considered in this proposal is binary, where each integer decision variable $y_{j=1,\ldots,6} \in \mathbb{N}$ is encoded using m bits. Let x be the bitstring representing variable y, the expresion for y is:

$$y = l + (1 - 2^{m-1} + u - l)x_{m-1} + \sum_{i=0}^{m-2} 2^i x_i, \tag{2}$$

which ensures that $l \leq y_{j=1,\ldots,6} \leq u$. In the previous expression, l and u represents the minimum and maximum value a decision variable can take. In this work, l is $u/4$ or $u/8$, depending on the decision variable. Having k logical qubits required by the quantum algorithm, and n physical ones on the QPU, and d training samples, the objective function to minimize is defined by

$$f(W) = -\frac{1}{k} \sum_{l=1}^{k} \frac{1}{d} \sum_{i=1}^{d} \sum_{j=1}^{n} t_{(i,l,j)} log(p_{(i,l,j)}), \tag{3}$$

where $t_{(i,l,j)}$ is 1 if sample i assigns logical qubit l to physical qubit j and 0 otherwise, and $p_{(i,l,j)}$ is the prediction returned by the softmax activation function for sample i on the physical qubit j assigned to logical qubit l.

We use the original version of the P3 algorithm to solve this problem. Algorithm 1 sketches the general framework of the P3, while further technical details can be found in the original work [7].

5   Experimental Study

This section presents the experimental design, results, and their discussion in the next three subsections.

Algorithm 1. The P3 pseudo-code

1: Create random solution
2: Apply hill climber
3: **if** solution $\notin$ **hashset then**
4:    Add solution to P_0
5:    Add solution to **hashset**
6: **end if**
7: **while** $\exists\, P_i \in pyramid$ not processed **do**
8:    Mix solution with P_i
9:    **if** solution's fitness has improved **then**
10:       **if** solution $\notin$ **hashset then**
11:          Add solution to P_{i+1}
12:          Add solution to **hashset**
13:       **end if**
14:    **end if**
15: **end while**

5.1 Experiments' Design and Benchmarks

The implementation has been done using both `python` version 3.10.2 and `C++` with a `GCC` version 11.4.0. We use a machine running Ubuntu 22.04.3 LTS 64 bits (7 CPUs and 16 GB of RAM) and Linux Enterprise Server 15 SP4 15.4 OS. The ETL module has been built using IBM quantum kit (Qiskit) version 0.44.2, while the evolutionary deep-learning module has been built using the libraries sckit-learn version 1.3.2, Tensorflow version 2.14.0 and Keras version 2.14.0. The source code of `P3` is the one made publicly available by the authors[2], slightly modified to execute within our framework, including the cluster execution. All the code generated, the datasets and the results are published in a replication package in Zenodo [4]. All the experiments have been run in a cluster with the configuration given in Table 1.

Table 1. Hardware components of the cluster used.

Nodes	CPU/GPU	RAM	InfiniBand	Localscratch
126 × SD530	56 × Intel Xeon Gold 6230R @ 2.10 GHz	200 GB	HDR100	950 GB
24 × Bull R282-Z90	128 × AMD EPYC 7H12 @ 2.6 GHz	2 TB	HDR200	3.5 TB
168 × IBM dx360 M4	16 × Intel E5-2670 @ 2.6 GHz	32 GB	FDR40	400 GB
4 × DGX-A100	8 × A100 Tensor Core	1 TB		14 TB

The comparison basis in this work is composed of the state-of-the-art DNN model devised in [1] (`SOTA-DNN`) and the newly proposed evolutionary DNN using the `P3` algorithm (`P3-DNN`). The original DNN model has been executed using the same parameters as the ones given in [1]. This includes the same neural

---

[2] https://github.com/brianwgoldman/FastEfficientP3.

layers' dimension (i.e., 264, 1024, 256, and 128), the same batch size of 128 and 150 training epochs, and Adam as a learning algorithm with learning rate of 0.0005. Regarding P3, the only parameter it has is the stopping condition, which has been set to 1000 fitness evaluations. Each experiment, whether using the SOTA-DNN or P3-DNN, has been performed over 32 independent executions. Metrics such as the Best, Worst, Median and Median Absolute Deviation (MAD) of the results across all the executions have been recorded.

5.2   Results and Discussion

Table 2 presents the average categorical cross-entropy loss obtained by SOTA-DNN and P3-DNN over all the circuits. The best results are highlighted in bold on the basis of the Median metric. The proposed P3-DNN achieves better results than SOTA-DNN in all the considered metrics.

Table 2. Average categorical cross-entropy loss of SOTA-DNN and P3-DNN over all the quantum circuits.

Model	Best	Worst	Median	MAD
SOTA-DNN	3.1856	9.9488	5.2786	0.9618
P3-DNN	2.6378	5.4681	**3.2379**	0.3076

Interestingly, P3-DNN outperforms SOTA-DNN using a deep neural network that is simpler and easier to train than the original one. While SOTA-DNN is composed of 264, 1024, 256, and 128 neurons in each layer and uses a batch size equal to 128 samples trained over 150 epochs, P3-DNN is composed of 118, 736, 177 and 122 neurons only in each layer and uses batches of size 67 samples trained over 55 epochs. Concretely the SOTA-DNN has 1757929 parameters to train (6864 + 271360 + (5 . 262400) + (5 . 32896) + (5 . 645)), while the P3-DNN has only 854552 parameters to train (3068 + 87584 + (5 . 130449) + (5 . 21716) + (5 . 615)). This indicates that the proposed P3-DNN has reduced the complexity of the model to almost half of the original SOTA-DNN complexity (48.611%).

Figure 6 illustrates a zoom-in on the fitness value evolution of the P3 in one of the best executions, during the first fitness evaluations. It can be seen that the P3 has considerably reduced the loss during the first iterations without consuming yet the 1000 fitness evaluations budget. This might suggest that racing methods combined with P3 could be interesting to investigate to make the P3 even more efficient.

Although finding the adequate DNN architecture using the P3 might take longer than executing directly the original DNN architecture given in [1], the extra time is compensated by the fact that the obtained DNN architecture is much simpler. This means that the optimized DNN architecture will be more efficient and faster to optimise/execute than the original DNN model. In the long run, this advantage will be cumulative and will be specially important

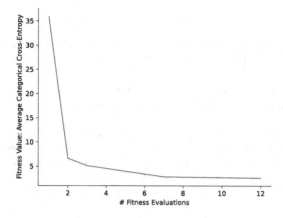

Fig. 6. A zoom-in on a sample of P3 fitness value evolution

given the NISQ nature of the current quantum machines, where the calibration data is continuously changing due to noise and errors. A comparison between the execution time of P3-DNN and SOTA-DNN will not be fair in this work considering that P3-DNN executes the DNN part on python, while P3 is executed in C++ using a continuous communication between both parts.

6   Conclusion and Perspective

This work proposes an evolutionary deep neural network approach to solve the qubits' initialisation problem in IBM/NISQ/gate-based quantum machines. The proposal copes with the shortfalls of previous work, mainly on three aspects: (I) the correctness and completeness of the proposal, (II) a problem modelling that is correct and generalisable, and (III) diversity, usefulness and availability of the benchmarks. The experiments have been made using real calibration data of four IBM quantum machines, circuits for quantum optimisation and machine learning, and we compared our proposal with the state-of-the-art DNN model. The results showed that the proposed approach achieve better results using a simpler DNN model.

As a next step, it is planned to investigate the use of other linkage learning techniques such as LT-GOMEA, tackle the problem as a regression task, explore other ML techniques, include other quantum machines of different size, and build training datasets based on different metrics in addition to the circuits' depth.

Acknowledgments. The corresponding author declares that this work was made and initially submitted while at the University of Malaga. This research is partially funded by the PID 2020-116727RB-I00 (HUmove) funded by MCIN/AEI/ 10.13039/501100011033; and TAILOR ICT-48 Network (No 952215) funded by EU Horizon 2020 research and innovation programme. This work is also partially funded by the Junta de Andalucia, Spain, under contract QUAL21 010UMA.

References

1. Acampora, G., Schiattarella, R.: Deep neural networks for quantum circuit mapping. Neural Comput. Appl. **33**(20), 13723–13743 (2021). https://doi.org/10.1007/s00521-021-06009-3

2. Bahreini, T., Mohammadzadeh, N.: An minlp model for scheduling and placement of quantum circuits with a heuristic solution approach. J. Emerg. Technol. Comput. Syst. **12**(3) (2015). https://doi.org/10.1145/2766452

3. Brooks, M.: Beyond quantum supremacy: the hunt for useful quantum computers. Nature **574**(7776), 19–21 (2019). https://doi.org/10.1038/d41586-019-02936-3

4. DAHI, Z.A., Chicano, F., Luque, G.: Replication package for the paper "An evolutionary deep learning approach for efficient quantum algorithms transpilation" (2023). https://doi.org/10.5281/zenodo.10141872

5. Dahi, Z.A., Chicano, F., Luque, G., Alba, E.: Genetic algorithm for qubits initialisation in noisy intermediate-scale quantum machines: the IBM case study. In: Proceedings of the Genetic and Evolutionary Computation Conference (GECCO 2022), pp. 1164–1172. Association for Computing Machinery, New York (2022). https://doi.org/10.1145/3512290.3528830

6. Farhi, E., Goldstone, J., Gutmann, S.: A quantum approximate optimization algorithm (2014). https://doi.org/10.48550/arXiv.1411.4028

7. Goldman, B.W., Punch, W.F.: Parameter-less population pyramid. In: Proceedings of the 2014 Annual Conference on Genetic and Evolutionary Computation (GECCO 2014), pp. 785–792. Association for Computing Machinery, New York (2014). https://doi.org/10.1145/2576768.2598350

8. Havlíček, V., et al.: Supervised learning with quantum-enhanced feature spaces. Nature **567**(7747), 209–212 (2019). https://doi.org/10.1038/s41586-019-0980-2

9. Hsu, S.H., Yu, T.L.: Optimization by pairwise linkage detection, incremental linkage set, and restricted/back mixing: DSMGA-II. In: Proceedings of the 2015 Annual Conference on Genetic and Evolutionary Computation (GECCO 2015), pp. 519–526. Association for Computing Machinery, New York (2015). https://doi.org/10.1145/2739480.2754737

10. LeCompte, T., Qi, F., Yuan, X., Tzeng, N.F., Najafi, M.H., Peng, L.: Graph neural network assisted quantum compilation for qubit allocation. In: Proceedings of the Great Lakes Symposium on VLSI 2023 (GLSVLSI 2023), pp. 415–419. Association for Computing Machinery, New York (2023). https://doi.org/10.1145/3583781.3590300

11. LeCompte, T., Qi, F., Yuan, X., Tzeng, N.F., Najafi, M.H., Peng, L.: Machine-learning-based qubit allocation for error reduction in quantum circuits. IEEE Trans. Quant. Eng. **4**, 1–14 (2023). https://doi.org/10.1109/TQE.2023.3301899

12. Murali, P., Baker, J.M., Abhari, A.J., Chong, F.T., Martonosi, M.: Noise-adaptive compiler mappings for noisy intermediate-scale quantum computers (2019). https://doi.org/10.48550/ARXIV.1901.11054

13. Nielsen, M.A., Chuang, I.L.: Quantum Computation and Quantum Information. Cambridge University Press (2000)

14. Pearson, K.: LIII on lines and planes of closest fit to systems of points in space. London, Edinburgh, and Dublin Philos. Magaz. J. Sci. **2**(11), 559–572 (1901). https://doi.org/10.1080/14786440109462720

15. Przewozniczek, M.W., Komarnicki, M.M.: Empirical linkage learning. IEEE Trans. Evol. Comput. **24**(6), 1097–1111 (2020). https://doi.org/10.1109/TEVC.2020.2985497

16. Siraichi, M.Y., Santos, V.F.d., Collange, C., Pereira, F.M.Q.: Qubit allocation. In: Proceedings of the 2018 International Symposium on Code Generation and Optimization (CGO 2018), pp. 113–125. Association for Computing Machinery, New York (2018). https://doi.org/10.1145/3168822

17. Sivarajah, S., Dilkes, S., Cowtan, A., Simmons, W., Edgington, A., Duncan, R.: t|ket⟩: a retargetable compiler for nisq devices. Quant. Sci. Technol. **6**(1), 014003 (2020). https://doi.org/10.1088/2058-9565/ab8e92

18. Telikani, A., Tahmassebi, A., Banzhaf, W., Gandomi, A.H.: Evolutionary machine learning: a survey. ACM Comput. Surv. **54**(8), 1–35 (2021). https://doi.org/10.1145/3467477

19. Thierens, D.: The linkage tree genetic algorithm. In: Schaefer, R., Cotta, C., Kołodziej, J., Rudolph, G. (eds.) Parallel Problem Solving from Nature, PPSN XI, pp. 264–273. Springer, Heidelberg (2010). https://doi.org/10.1007/978-3-642-15844-5_27

20. Thierens, D., Bosman, P.A.: Optimal mixing evolutionary algorithms. In: Proceedings of the 13th Annual Conference on Genetic and Evolutionary Computation (GECCO 2011), pp. 617–624. Association for Computing Machinery, New York (2011). https://doi.org/10.1145/2001576.2001661

21. Zulehner, A., Wille, R.: Compiling SU(4) quantum circuits to IBM QX architectures. In: Proceedings of the 24th Asia and South Pacific Design Automation Conference (ASPDAC 2019), pp. 185–190. Association for Computing Machinery, New York (2019). https://doi.org/10.1145/3287624.3287704

Measuring Similarities in Model Structure of Metaheuristic Rule Set Learners

David Pätzel[1]([✉]) [iD], Richard Nordsieck[2] [iD], and Jörg Hähner[1] [iD]

[1] University of Augsburg, Augsburg, Germany
david.paetzel@uni-a.de
[2] XITASO GmbH IT & Software Solutions, Augsburg, Germany

Abstract. We present a way to measure similarity between sets of rules for regression tasks. This was identified to be an important but missing tool to investigate Metaheuristic Rule Set Learners (MRSLs), a class of algorithms that utilize metaheuristics such as Genetic Algorithms to solve learning tasks: The commonly-used predictive performance-based metrics such as mean absolute error do not capture most users' actual preferences when they choose these kinds of models since they typically aim for model interpretability (i. e. low number of rules, meaningful rule placement etc.) and not low error alone. Our similarity measure is based on a form of metaheuristic-agnostic edit distance. It is meant to be used—in conjunction with a certain class of benchmark problems—for analysing and improving an as-of-yet underresearched part of MRSL algorithms: The metaheuristic that optimizes the model's structure (i. e. the set of rule conditions). We discuss the measure's most important properties and demonstrate its applicability by performing experiments on the best-known MRSL, XCSF, comparing it with two non-metaheuristic Rule Set Learners, Decision Trees and Random Forests.

Keywords: Benchmarking · Metaheuristic Rule Learning · Model Similarity · Learning Classifier Systems · Rule Learning

1 Introduction

This paper presents a novel tool for evaluating *Metaheuristic Rule Set Learners* (MRSL) for *regression tasks*. MRSLs are a subclass of *Rule Set Learners* (RSLs) which are machine learning (ML) algorithms that learn *sets of rules*. While non-metaheuristic RSLs like the well-known C4.5 algorithm [34] typically use local heuristics to generate sets of rules, MRSLs use metaheuristics such as Genetic Algorithms (GAs) to perform some form of global search for well-performing rule sets. Examples for MRSL systems are Learning Classifier Systems such as XCS [41], Fuzzy Rule-based Systems [7] or Ant-Miner [26]. While there are MRSL approaches for unsupervised learning (e. g. [38]), reinforcement learning (e. g. [5]) and classification (e. g. [3,26]) as well, the present paper focusses on *regression tasks*. MRSL systems solving regression tasks include [2,17,42].

S. Smith et al. (Eds.): EvoApplications 2024, LNCS 14635, pp. 256–272, 2024.
https://doi.org/10.1007/978-3-031-56855-8_16

In the regression tasks that we are concerned with, the goal is to find an in some sense optimal model $\hat{f} : \mathcal{X} \to \mathcal{Y}$ that maps inputs $x \in \mathcal{X} = \mathbb{R}^{\mathcal{D}_X}$ to outputs $y \in \mathcal{Y} = \mathbb{R}$. At that, the only guidance given in order to find that optimal model is a *training set* consisting of N inputs $(x_n)_{n=1}^{N} \subsetneq \mathcal{X}$ and outputs $(y_n)_{n=1}^{N} \subsetneq \mathcal{Y}$. A common way to measure and improve model optimality is to compute predictive error measures such as the *mean absolute error* (MAE) or the *mean squared error*. While MRSLs are commonly evaluated using these kinds of metrics as well, Pätzel et al. [27] as well as Kovacs and Kerber [20,21] argue that during their development, MRSLs should actually be handled differently due to the fact that these algorithms not only optimize some parametric model's fixed set of *parameters* (like, for example, fitting neural network connection weights) but actually optimize the number, the conditions *and* the model parameters of a set of rules. This means that MRSLs actually perform both parameter optimization and *model structure* optimization (which in the neural network example translates to fitting connection weights *and* optimizing the network's architecture). This dual nature of MRSL algorithms entails that predictive error measures alone can typically not capture a user's actual preferences: They chose MRSL algorithms over other high-predictive-performance options such as Neural Networks for their increased interpretability and therefore also require the created model to have a low number of rules, meaningfully placed rules, little rule overlap and similar properties. That being said, the optimization target of supervised MRSL algorithms is far from clear and that may be one of the reasons why most MRSL research has focussed on improving these systems' predictive performance alone.

Pätzel et al. [27] present a concept to resolve that mismatch between MRSL research targets and MRSL user preferences. That concept is based on generating certain data-generating processes that serve as a new form of benchmark learning tasks for MRSL algorithms (this is summarized in Sect. 3). The advantage of these processes over other approaches is that they are of the same form as the models built by MRSL algorithms: Each process is a set of rules. The goal is to enable the following workflow for investigating MRSL algorithms:

1. Generate a random data-generating process (or, rather, many of them).
2. Generate training data using that process.
3. Apply an MRSL algorithm to the generated training data.
4. Compare the model (a set of rules) created by the MRSL algorithm with the set of rules of the original data-generating process.

At that, the very last step is what Pätzel et al.'s proposal is all about: The data-generating process being a set of rules allows to not only consider predictive performance but also *whether the MRSL algorithm was able to reconstruct the original data-generating process*. This enables directly measuring the progress made by the MRSL algorithm's *metaheuristic* since the model structures of the learnt model and of the data-generating process can be compared.

While Pätzel et al. [27] explained the overall concept of these principled benchmarks such as how data-generating processes can be generated, their paper fell short of proposing an actual way to compare model structures consisting of sets of rules. This is where the present paper comes in: We identify and discuss the desired properties of dissimilarity measures that could be used for this task and then *present a novel dissimilarity measure between sets of rules* that fulfills them. In combination with above-mentioned benchmark tasks, the measure allows improving characteristics of MRSL algorithms other than predictive performance. Specifically, in the context of XAI, this allows to play to the strengths of MRSL (inherent explainability) and quantify the tradeoff between raw predictive performance and comprehensible model structures. We demonstrate our measure's applicability by applying it to the analysis of differences between several parametrizations of one of the best-known MRSLs, XCSF, and two non-metaheuristic RSLs, namely DTs and *Random Forests* (RFs).

2   Metaheuristic Rule Set Learners

In this section, we try to give a, due to space restrictions very rough, idea of the models built and assumptions made by *Metaheuristic Rule Set Learners* (MRSLs; for more details see [27]). This is necessary in order to be able to describe in Sect. 3 how the benchmark learning tasks look like and in Sect. 4 the dissimilarity measure between model structures.

As was already said above, this paper focusses on regression tasks (i. e. learning a mapping $\mathcal{X} \to \mathcal{Y}$ with $\mathcal{X} = \mathbb{R}^{\mathcal{D}_{\mathcal{X}}}$ and $\mathcal{Y} = \mathbb{R}$, $\mathcal{D}_{\mathcal{X}} \in \mathbb{N}$). Common and well-established MRSLs (e. g. XCSF [42]) as well as more recently developed systems (e. g. SupRB [17]) solve regression tasks by building discriminative models of the following form [27]:

$$\widehat{f}_{\mathcal{M}}(\theta, x) = \sum_{k=1}^{K} m(\psi_k; x)\, \gamma_k\, \widehat{f}_k(\theta_k; x) \tag{1}$$

At that,

- $m(\psi_k; \cdot) : \mathcal{X} \to \{0, 1\}$ is the *condition* or *matching function* of rule k (parametrized by ψ_k) which states for any $x \in \mathcal{X}$ whether rule k applies or not and correspondingly whether it will influence the overall prediction for that input (if $m(\psi_k; x) = 1$, we say $m(\psi_k; \cdot)$ *matches* data point x),
- γ_k is the *mixing weight* of rule k which allows to weigh rules against each other in areas of overlap,
- $\widehat{f}(\theta_k; \cdot) : \mathcal{X} \to \mathcal{Y}$ is the *local model* of rule k (parametrized by θ_k and fitted on the subset of the training data that $m(\psi_k; \cdot)$ matches) which gives the rule's output for any input $x \in \mathcal{X}$,
- the model's *parameters* form $\theta = \left((\gamma_k)_{k=1}^{K}, (\theta_k)_{k=1}^{K} \right)$,
- the model's *model structure* is $\mathcal{M} = \left(K, (\psi_k)_{k=1}^{K} \right)$.

Given a certain fixed model structure, a model's parameters' optimization is often straightforward: Each local model k's parameters θ_k only have to be optimized on a subset of the training data (i. e. the data points where the local model's condition is fulfilled) and local model families are typically simple such as linear regression models or just constants [27]. While optimal mixing weights γ_k are often computationally expensive to obtain, there exist well-performing heuristics which are often-used in MRSL algorithms [9].

Other than model parameter optimization, optimization of the model structure is a difficult task and, correspondingly, most of the compute of (M)RSLs goes into doing so. While non-evolutionary RSLs such as *Decision Trees* (DTs) choose the model structure based on (often, local) heuristics, MRSLs use metaheuristics and often some form of global search (e. g. Genetic Algorithms [41]); a recent overview of techniques used was given by Heider et al. [14]. How exactly model structure optimization is done strongly depends on the condition family used. For real-valued tasks such as the regression tasks considered in the present paper, a common choice are interval-based conditions where

$$m(\psi_k, x) = m((l_k, u_k), x) = \begin{cases} 1, & x \in [l_k, u_k) \\ 0, & \text{otherwise.} \end{cases} \tag{2}$$

Since intervals are easily comprehensible greater-/less-than statements, interval-based conditions often yield models with higher interpretability than more sophisticated condition families (e. g. ellipsoid-based matching [37]) and are thus often preferred [15]. At the same time, they are reasonably expressive and share a lot of similarity with other axis-parallel ways to subdivide the input space (e. g. the axis-parallel cuts made by DT algorithms); the latter being especially important if comparisons with such algorithms are being conducted—which we do in this paper. Overall, this led us to use interval-based conditions for the purposes of this paper as well and as a result, the $(m(\psi_k, \cdot))_{k=1}^{K}$ correspond to (and are sometimes called) a *set of ($\mathcal{D}_\mathcal{X}$-dimensional) intervals*.

3   Generating Benchmark Tasks

As was already said in the introduction, Pätzel et al.'s framework [27] is based on generating a set of benchmark learning tasks that have the same form as the common MRSL models introduced in Sect. 2 (cf. Eq. (1)). This means that each benchmark learning task is a data-generating process which corresponds to, for each input $x \in \mathcal{D}_\mathcal{X}$, a random variable of the following form:

$$Y = \sum_{k=1}^{K} m(\psi_k; x)\, \gamma_k\, (f_k(\theta_k; x) + \epsilon_k) \tag{3}$$

where the $\{\epsilon_k\}_{k=1}^{K}$ are normally distributed random variables corresponding to the respective local model's noise. These processes can be generated randomly by drawing all of the required parameters from suitable random distributions.

Compared to the original work [27,33], we introduced a minimum coverage parameter that allows us to ensure that the generated rules properly cover the input space. Since the exact way how the learning tasks are generated is not relevant for the discussion of our dissimilarity measure, we refer to our code [31] for details on this. We further changed the local model family to constant models since they are less expensive to compute and the algorithms we use for our demonstration in Sect. 5 use constant local models as well.

4   Measuring Dissimilarity of Sets of Rules

This section describes our proposal for measuring dissimilarity between the model structures of models generated by MRSL algorithms. There are many existing ways (e. g. the Hausdorff distance [10] or intersection over union) to measure the dissimilarity between sets and most can be adapted to work on the sets corresponding to MRSL model structures (i. e. sets of rule conditions). However, due to the nature of MRSL model structures, the dissimilarity measure should fulfill certain properties which we found difficult or impossible to fulfill using the available dissimilarity measures. We start this section with a discussion of said properties and then introduce our dissimilarity measure.

4.1   Desired Properties

A dissimilarity measure $d(\cdot, \cdot)$ for MRSL model structures should have the following properties:

Property 1 (Symmetry). Since we may not only want to examine dissimilarities to the data-generating process model but also between two models generated by two different (or even the same) algorithm, we want the dissimilarity measure to be symmetric in order to avoid having to choose which of the two models should be used as a reference. Symmetry can be expressed formally as

$$d(\mathcal{M}_1, \mathcal{M}_2) = d(\mathcal{M}_2, \mathcal{M}_1) \tag{4}$$

Property 2 (Same training match sets[1] should yield minimal dissimilarity). While the training data and with it, the input space $\mathcal{X}$, is typically scaled (e. g. min-max normalized) before it is fed into an ML algorithm, we still would like the dissimilarity measure to be agnostic of input-space-based distances between interval bounds. The reason for this is that we do not want to punish cases where a metaheuristic has *bad luck*: Any predictive-performance-based part of a fitness function of a metaheuristic in an MRSL algorithms can only ever distinguish model structures if there is a change in at least one rule's training match set (See footnote 1). If we did not require this property, then a model structure $\mathcal{M}_1$ may be considered less or more similar to a reference model structure $\mathcal{M}_0$ than another model structure $\mathcal{M}_2$ despite $\mathcal{M}_1$ and $\mathcal{M}_2$ resulting in the exact same overall model with respect to the training data (i. e. despite differing ψ_k, all match sets are equal due to where in input space the training data points

---

[1] A *rule k's training match set* is the set of training data points that $m(\psi_k; \cdot)$ is fulfilled for, i. e. $\{x \in X \mid m(\psi_k; x) = 1\}$.

lie). Since most metaheuristics are non-deterministic, we conjecture that this occurring is not merely pathological. In short: If this property is fulfilled then two model structures that induce the same training data set for each local model have zero dissimilarity—even if the model structure parameters are actually different. Formally,[2]

$$(m_k(\psi_{1k}; X))_{k=1}^{K_1} = (m_k(\psi_{2k}; X))_{k=1}^{K_2} \quad \Leftrightarrow \quad d(\mathcal{M}_1, \mathcal{M}_2) = 0 \qquad (5)$$

Property 3 (Different training match sets should yield non-minimal dissimilarity). This is somewhat of an inverse of Property 2. Whenever two model structures differ enough that the training data of at least one of the local models changes, the two model structures should not be considered maximally similar:

$$(m_k(\psi_{1k}; X))_{k=1}^{K_1} \neq (m_k(\psi_{2k}; X))_{k=1}^{K_2} \quad \Leftrightarrow \quad d(\mathcal{M}_1, \mathcal{M}_2) > 0 \qquad (6)$$

Property 4 (Sensible behaviour even if conditions do not overlap). Even if two conditions (or sets of conditions) do *not* overlap, they can still be more or less similar to each other from both a metaheuristics point of view and also from a human user's perspective. For example, a single condition (corresponding to a set of rules of size one) should be seen as more similar to a set of conditions that lies closer to it than to a set that lies further away—even if it does not overlap with either of both sets. We therefore want the dissimilarity measure to be able to sensibly differentiate between pairs of conditions even if these conditions do not overlap. Since the concept of *sensibility* is rather vague, we deliberately do not try to formalize this property.

4.2 Dissimilarity Measure

Before we define our dissimilarity measure formally, we try to provide some intuition about it. The dissimilarity measure is ultimately meant to allow investigating the progress and behaviour of metaheuristics which optimize the model structures of MRSL models. As previously mentioned we consider MRSLs with interval-based conditions and hard binary matching (i.e. either a training data point is matched or it is not). For these MRSLs, a finite data supervised learning setting effectively induces a form of quantization of the training data signal when the model structure is changed: The model's output for the training data will only be different if at least one of the conditions is changed enough that at least one of the local models is fitted on a different subset of the training data. This gave us the idea of developing a measure similar to *edit distances* which are often used in discrete spaces (e.g. in graphs [12]): We compute an upper bound on the maximum number of training data prediction-changing edits required to transform rules into each other. This can be seen as an edit operator-agnostic

---

[2] We slightly abuse notation here and overload the matching function m to be able to pass the training data input $N \times \mathcal{D}_{\mathcal{X}}$ matrix X consisting of N vectors $x_n \in \mathcal{X}$ to a single condition $m(\psi; \cdot)$ to get an N-vector, i.e. $m(\psi; X) = (m(\psi; x_n))_{n=1}^{N} \in \{0, 1\}^N$.

(and thus metaheuristic operator-agnostic) edit distance; we count an edit only if it would change behaviour on the training data.

This can be defined formally using two measures at two levels: $\delta_X(\cdot, \cdot)$ measures the dissimilarity between *two particular rule conditions* whereas $d_X(\cdot, \cdot)$ combines these condition-wise dissimilarities into a dissimilarity measure over *sets* of conditions (i.e. over model structures). We start by explaining $\delta_X(\cdot, \cdot)$ which is parametrized with a given training set's input data points X. As was already explained above, we use interval-based conditions and a condition's $m(\psi; \cdot)$ parameter vector ψ therefore induces a ($\mathcal{D}_X$-dimensional) interval $[l, u]$. In order to measure the dissimilarity between two such intervals, we *count the number of edits required to transform each of the bounds into the corresponding bound of the other interval*. At that, a single edit corresponds to moving one bound so that one more (or one less) training data point is included in the interval. We consider each dimension independently of the others (which is why this is an upper bound on the worst-case edit distance) because computing optimal sequences of edits is computationally infeasible. We can compute the edit count by considering each training data point independently and counting which interval bounds could *traverse* it. This *traversal count* can be formalized as (a visualization of this for two two-dimensional intervals is given in Fig. 1):

$$\delta_X(\psi_1, \psi_2) = \delta_X([l_1, u_1], [l_2, u_2]) \tag{7}$$

$$= \sum_{d=1}^{\mathcal{D}_X} |\{x \in X \mid x \in H(\psi_1, \psi_2), L_{\min}(d) \le x_d \le L_{\max}(d)\}|^{\frac{1}{\mathcal{D}_X}} \tag{8}$$

$$+ |\{x \in X \mid x \in H(\psi_1, \psi_2), U_{\min}(d) \le x_d \le U_{\max}(d)\}|^{\frac{1}{\mathcal{D}_X}}$$

where $H(\psi_1, \psi_2) = \text{Hull}([l_1, u_1], [l_2, u_2])$ is the convex hull (this arises naturally, see Fig. 1) of the two intervals and

$$L_{\min}(d) = \min(l_{1d}, l_{2d}), \qquad L_{\max}(d) = \max(l_{1d}, l_{2d}), \tag{9}$$
$$U_{\min}(d) = \min(u_{1d}, u_{2d}), \qquad U_{\max}(d) = \max(u_{1d}, u_{2d}). \tag{10}$$

Equation (8) is a sum that iterates over all input space dimensions and for each of them computes and sums the cardinality of two sets. The first set contains all data points $x \in X$ that lie in the interval's convex hull *and* between the lower of the two lower bounds of the two intervals in that dimension and the upper of the two lower bounds of the two intervals in that dimension. The second set analogously treats the intervals' upper bounds. The traversal count counts each $x \in X$ multiple times since it computes each dimension independently of the others which yields the aforementioned maximum number of traversals per data point. We further take the $\mathcal{D}_X$th root of the set cardinalities in order to treat each dimension independently and compute a *per-dimension* instead of a per-volume value; without the $\mathcal{D}_X$th root, conditions that match more training data points (i.e. larger intervals) would have a generally higher dissimilarity score just due to their higher volume.

This dissimilarity measure fulfills at the conditon-level (i.e. for model structures of size $K = 1$) the properties that we introduced above: Symmetry

(Property 1) is rather straightforward to show, Properties 2 and 3 follow directly from how the measure considers training data points being matched and we argue that Property 4 is fulfilled as well since the measure will continue to count training data points between conditions even if they do not overlap.

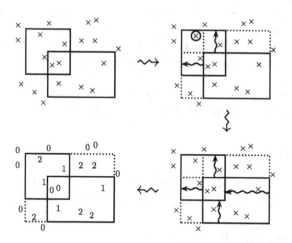

Fig. 1. Condition-wise dissimilarity measurement for two two-dimensional interval-based conditions. *Top left:* Data points as crosses, intervals as solid line rectangles (we call the square interval on the left ψ_1 and the non-square one on the right ψ_2). *Top right:* One data point x and the edit movements of the bounds that are relevant for possible traversals of x marked. If ψ_2's y-axis upper bound is transformed into ψ_1's y-axis upper bound first, then x is traversed when ψ_2's x-axis lower bound is transformed into ψ_1's x-axis lower bound. If ψ_2's x-axis lower bound is transformed into ψ_1's x-axis lower bound first, then x is traversed when ψ_2's y-axis upper bound is transformed into ψ_1's y-axis upper bound. This yields two possible traversals for x. *Bottom right:* All edit movements of this two-dimensional example. *Bottom left:* Data point crosses replaced by their respective number of traversals; as can be seen, only data points in the convex hull (dotted line) of the two intervals can ever be traversed and data points within the intersection of the intervals are never traversed.

In order to compute the dissimilarity between two model structures $\mathcal{M}_1$ and $\mathcal{M}_2$ (i.e. two *sets* of conditions), we compute the (mean) *sum of minimum distances*, which is a known set-wise dissimilarity measure (cf. e.g. [10]):

$$d_X(\mathcal{M}_1, \mathcal{M}_2) = \frac{1}{2} \left(\sum_{\psi_1 \in \mathcal{M}_1} \min_{\psi_2 \in \mathcal{M}_2} \delta_X(\psi_1, \psi_2) + \sum_{\psi_2 \in \mathcal{M}_2} \min_{\psi_1 \in \mathcal{M}_1} \delta_X(\psi_1, \psi_2) \right) \quad (11)$$

Note that only taking one of the sums (instead of their mean) would not yield a symmetric measure since minimizing dissimilarity may yield different results depending on which set of conditions is minimized over. Further, there are other options for combining the condition-wise dissimilarities into a set-wise dissimilarity; an overview of several is given by Eiter and Mannila [10].

The properties given in Sect. 4.1 are fulfilled for $d_X(\cdot, \cdot)$ as well, they carry over naturally from $\delta_X(\cdot, \cdot)$. As an aside, be aware that proving whether or not d is a *metric* (it is not yet clear whether it fulfills the triangle equality—we conjecture that it does not) is out of the scope of this paper.

Finally, it should be noted that computing $d_X(\mathcal{M}_1, \mathcal{M}_2)$ can be computationally expensive, especially if both $\mathcal{M}_1$ and $\mathcal{M}_2$ contain many rules. This is due to having to compute $\delta_X(\psi_1, \psi_2)$ for *all possible pairings* of $\psi_1 \in \mathcal{M}_1$ and $\psi_2 \in \mathcal{M}_2$ in order to compute the two summands in Eq. (11). At that, evaluating $\delta_X(\psi_1, \psi_2)$ can in itself be expensive for higher dimensions and more training data points.[3] In many cases, this is not that much of a problem, though, since the dissimilarity measure is meant mainly for post-hoc analysis and *not* for being evaluated during the investigated algorithm's runtimes.

5   Demonstration

To show the applicability of our dissimilarity measure, we apply it to the comparison of three different parametrizations each of DTs, RFs and XCSF.

5.1   Data-Generating Processes

We first generated a set of data-generating processes. Since the processes themselves are not relevant for the following demonstration of the dissimilarity measure, we will not go into detail here and refer the reader to our code [31]. We generated a range of learning tasks for dimensionalities $\mathcal{D}_\mathcal{X} \in \{3, 5, 8\}$, rule counts $K \in \{2, 4, 8, 10, 14, 18\}$ and two minimum coverage rates $\kappa_{\min} \in \{0.75, 0.9\}$. Sampling for these parameters produced a total of 132 learning tasks with an average of 3.77 tasks per configuration (and correspondingly 7.54 tasks if pooling over minimum coverage rates). From each learning task we generated training and test data sets by uniformly sampling inputs from the respective input space and based on those then sampling the random variable corresponding to outputs (see Eq. (3)). The number of training and test data points depended on the learning task's dimensionality: We generated $200 \cdot 10^{\mathcal{D}_\mathcal{X}/5}$ training data points and ten times as many test data points for each task, this yielded 796, 2000 and 7962 training data points for the three dimensionalities considered.

5.2   Evaluation of Repeated Runs

Next, we performed a set of experiments for which we used the DT and RF implementations provided by the *Scikit-learn* Python library [28] and the XCSF implementation provided by Preen's *XCSF* Python library [30]. We use three

---

[3] For $N = 768$ training data points, our own (not at all optimized) code took around $0.0005\,\text{s}$ per computation of δ_X (mean over *all* computations of d_X with $N = 768$ performed for Fig. 2) and correspondingly around $0.2\,\text{s}$ for computing d_X for two model structures of size 20. For $N = 2000$, we measured $0.002\,\text{s}$ per δ_X computation (and correspondingly $0.8\,\text{s}$ for size 20 model structures).

different parametrizations for each of the algorithm families by allowing a certain range for the maximum number of rules in the final model: up to 50, up to 100 and up to 200 rules (and 20 times as many for RFs). Further, all three algorithms are set to use constant models (for DTs and RFs this is already the default) and XCSF is set to use interval-based conditions.

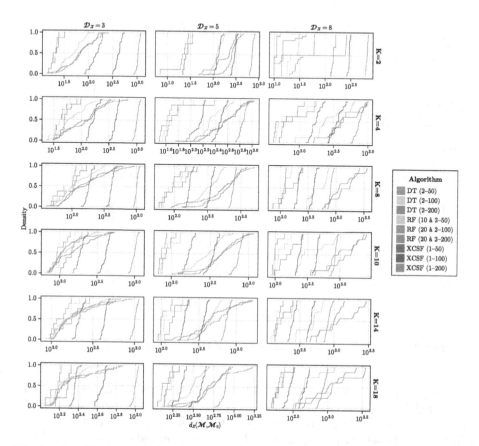

Fig. 2. Empirical cumulative distribution of $d(\cdot, \mathcal{M}_0)$ when pooling all the runs of each algorithm (i. e. pool over repetitions and minimum coverage rate values) for each combination of dimensionality $\mathcal{D}_{\mathcal{X}}$ (columns) and number of rules K (rows). Number ranges in the legend show the respective variant's range of allowed rule set sizes (in case of RF, a fixed number of used DTs and a range of allowed sizes for each).

We performed hyperparameter optimization on each of the 132 generated training data sets for each of the 9 algorithms used; we used the Optuna framework [1] to do so. Due to this paper's space constraints, we have to refer the reader to our code [32] for the parameter ranges and configurations used.

For each of the nine algorithm variants listed above, we then performed ten repetitions (ten consecutive random seeds) on each of the learning tasks

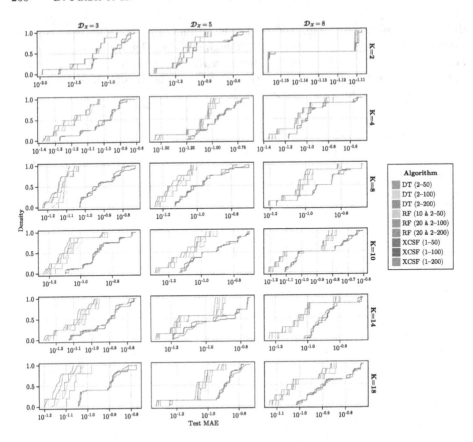

Fig. 3. Empirical cumulative distribution of test MAE. See Fig. 2 for details.

yielding an overall number $132 \times 9 \times 10 = 11880$ trials. Each final model is then evaluated for its predictive performance (MAE) using the holdout test set which was generated for the respective learning task. Further, we extract from each final model the set of rules $\mathcal{M}$ in interval representation and compute $d(\mathcal{M}, \mathcal{M}_0)$, its dissimilarity (see Eq. (11)) to the respective learning task's data-generating process's model structure $\mathcal{M}_0$. It should be noted at this point that a DT creates a set of non-overlapping rules that fully cover the input space, XCSF's rules are *allowed* to overlap, and an RF algorithm generates a fixed number d of DTs which means that there are at any one point in input space d rules.

Figure 2 shows the empirical cumulative distributions of $d(\cdot, \mathcal{M}_0)$ when pooling all the runs of each algorithm (i. e. pool over repetitions and minimum coverage rate values) for each combination of dimensionality $\mathcal{D}_\mathcal{X}$ (columns) and number of rules K (rows). In this figure, each graph in each diagram corresponds to one of the nine algorithm variants. This figure first of all shows that XCSF with a population size of 200 performs worse in terms of $d(\cdot, \mathcal{M}_0)$ than most of the other algorithms in most cases. In fact it only outperforms RFs containing at least 20 DT for $\mathcal{D}_{\mathcal{X}} = 8$ and $K \geq 4$. Suprisingly, the three XCSF variants are spaced roughly proportional to the their maximum population sizes; upon closer investigation, this effect can be explained by the fact that the variants with maximum population sizes 50 and 100 exhaust that maximum fully in all cases and the variant with maximum population size 200 is only able to slightly reduce the number of rules in some cases using the subsumption mechanism. The DTs better performance is mainly due to creating models with an overall much smaller number of rules (mean±std, rounded to one decimal, 10.3±4.9, 13.4±7.4 and 16.3±9.8 rules for the three DT variants) which means that there are much fewer summands in Eq. (11). The RFs, on the other hand, perform worse than the XCSF variants on a subset of the learning tasks because they have a larger number of rules than XCSF (yielding more summands in Eq. (11)).

When comparing Fig. 2 which shows $d(\cdot, \mathcal{M}_0)$ with Fig. 3 which shows MAE we can see that evaluating for MAE only separates the tree-based algorithms from the XCSF variants while $d(\cdot, \mathcal{M}_0)$ provides more nuance. It is interesting to note here that the RFs seem to perform better relative to the other algorithms with respect to test MAE but not with respect to $d(\cdot, \mathcal{M}_0)$. One possible explanation is the performance-interpretability tradeoff: RFs use more rules than DTs and XCSF (worse interpretability) but at the same time the increase in parameters results in them being able to model the data better.

6   Related Work

This section discusses some more related work that has not yet been mentioned in the preceding text.

In some sense the present work can be seen as being in the same spirit as the work that Kovacs and Kerber did in the early 2000s where they questioned the common practice that performance of Learning Classifier Systems (LCSs, which are a prominent member of the MRSL family), most notably XCS, was measured using accuracy (or, more generally, error) alone [20,21]. They argued that this does not fully capture the actual target one has in mind when thinking about the performance of these systems.

To circumvent the weaknesses of only measuring predictive performance and more directly assess progress of the metaheuristic, Kovacs defined several alternative measures [19,20] based on comparing the sets of rules to some form of known optimal solution. However, Kovacs's work was restricted to binary classification tasks with binary vector inputs and transfer to regression tasks does not seem possible. That aside, the used notion of optimal rule sets was informed by already observed behaviour of the systems to be investigated [20] whereas we try hard to stay algorithm-agnostic with our approach. Another weakness of Kovacs's measures of closeness to the data-generating process is that they only check *how many* of the rules in the optimal rule set the final rule set contains. However, *extracting* optimal rules from a rule set that contains them is not solved yet in general (there are merely some heuristics available, each with strengths and weaknesses, that try to do that [25]). Other authors agree with us on that,

e. g. Drugowitsch [9] writes in 2008 "[studies by Kovacs et al.] aimed at defining the optimal set for limited classifier representations [but there] was still no general definition available" and then continues to derive a probabilistic framework for defining rule set fitness from first principles. However, that framework is, as of now, computationally infeasible even for low input space dimensionalities.

Tan et al. look at different rule compaction approaches and in order to understand their difference, they compute similarity scores of the rule sets that these different approaches generate [39]. Their score is computed as the *ratio of rules preserved* and is thus not symmetric (Property 1).

Heider et al. [16] investigate novelty search for discovering rules in an MRSL algorithm called SupRB. In order to measure a candidate rule's novelty (i. e. a form of dissimilarity), they use the Hamming distance to compare its match set to the matching vectors of its nearest neighbours. However, this measure assigns to two rules that do not overlap a dissimilarity value that does not change with spatial distance but only with the rules' volumes (Property 4).

Setnes et al. [36] consider compacting models generated by fuzzy rule-based systems. They define similarity between two rules using the fuzzy set-equivalent to the rules' match sets and therefore there is, again, no differentiation between rule positions as soon as rules do not overlap anymore (Property 4).

Kharbat et al. [18] perform for binary classification with binary vector inputs a form of rule set compaction based on clustering rules. They define rule similarity using the Jaccard binary similarity coefficient which also shares the problem of not differentiating between non-overlapping rules (Property 4).

[11] propose a dissimilarity measure between *data sets* which is based on training DTs on the data sets and then computing a dissimilarity between the resulting models. In order to compare two DTs, they compute a third DT (they call this the *greatest common refinement*) which contains all the feature splits from the two base DTs and allows to compare the DTs' predictions region-by-region. However, comparing predictions is not the target of our work.

Serpen and Sabhnani [35] compare two rule sets like the ones we consider by computing their respective volumes as well as the volume of their overlap. However, volume alone is too weak a signal to investigate MRSL metaheuristics; in particular, this does not fulfill Property 4.

Earlier RSL literature explored heuristical detection of changes in data distributions. Some of these approaches (e. g. [22–24, 29, 40]) fit models to the data repeatedly and then compare these models in order to detect changes in the data distributions. However, since these approaches naturally assume the rate of change is small, their dissimilarity measures are not suitable for models that may be fundamentally different (e. g. Property 4). Aside from that, they handle only classification tasks most of the time.

Edit distance (sometimes also called *mutation distance*) has been studied for improving Genetic Programming algorithms; for example, Gustafson and Vanneschi [13] give an overview of approaches and propose a crossover-based edit distance measure. However, distances between trees are fundamentally different from distances between sets of intervals that are allowed to overlap.

Finally, while there have been proposed for metaheuristics many similarity measures (cf. [4,6]) as well as diversity measures (e. g. [8]) that often also measure similarity, to our knowledge, none of the currently available ones consider phenotypes that correspond to sets of intervals and even less so intervals that correspond to the conditions of rules.

7   Future Work

The main segment of future work will be to *use* our proposed dissimilarity metric to benchmark different variants of existing MRSL algorithms, analyse their models and based on the findings develop for these systems new and better-performing metaheuristic operators (or combinations thereof). Aside from that, we conjecture that the measure is suited for a hyperparameter tuning regime: Given an unknown learning task, one could capture one's knowledge of and beliefs about it (e. g. input space dimensionality, suspected number of rules that can approximate the task well, distribution over condition volumes etc.) as a set of *helper tasks*—a set of rule-set-based data-generating processes [27] with matching properties. One could then minimize on the helper tasks an MRSL algorithm's hyperparameters with respect to the dissimilarity measure in order to obtain a hyperparametrization for the unknown learning task.

Apart from measuring dissimilarity to a known data-generating process, the measure may prove useful to be included *within* population-based MRSLs as a measure of similarity between solutions (e. g. to build mechanisms that encourage population diversity, increase exploration and reduce preliminary convergence).

We chose the sum of minimum distances in Eq. (11) for now for its simplicity and the ease with which one can reason about it but intend to implement and test Eiter and Mannila's alternative, the link distance [10], as well.

As noted above, computing the dissimilarity measure can be expensive; however, large parts of that computation can be parallelized which should yield significant speedups on now-typical hardware.

8   Conclusion

We proposed a way to measure dissimilarity between rule sets created by regression *Metaheuristic Rule Set Learners* (MRSLs). Dissimilarity between two rules is defined as a lower bound on the maximum number of prediction-changing edits required to transform one rule into the other; this is extended to sets of rules using a well-known set distance measure. The measure tries to capture better the preferences of a user using MRSL algorithms in scenarios where interpretability is aimed at (i. e. a few well-placed rules are preferred over the highest possible performance). Our dissimilarity measure provides an additional target for MRSL development that allows to measure metaheuristic progress more directly than would be possible using predictive error-based measures such as mean absolute error. We expect that the presented tooling helps in overcoming the percieved decline in research activity on MRSLs by allowing to better analyse the metaheuristic's effects on model structure.

References

1. Akiba, T., Sano, S., Yanase, T., Ohta, T., Koyama, M.: Optuna: a next-generation hyperparameter optimization framework. In: Proceedings of the 25th ACM SIGKDD International Conference on Knowledge Discovery and Data Mining (2019)
2. Alcala, R., Gacto, M.J., Herrera, F.: A fast and scalable multiobjective genetic fuzzy system for linguistic fuzzy modeling in high-dimensional regression problems. IEEE Trans. Fuzzy Syst. **19**(4), 666–681 (2011). https://doi.org/10.1109/TFUZZ.2011.2131657
3. Bernadó-Mansilla, E., Garrell-Guiu, J.M.: Accuracy-based learning classifier systems: models, analysis and applications to classification tasks. Evolut. Comput. **11**(3), 209–238 (2003). https://doi.org/10.1162/106365603322365289
4. Brusco, M., Cradit, J.D., Steinley, D.: A comparison of 71 binary similarity coefficients: the effect of base rates. Plos One **16**(4) (2021)
5. Butz, M.V., Stolzmann, W.: An algorithmic description of ACS2. In: Lanzi, P.L., Stolzmann, W., Wilson, S.W. (eds.) IWLCS 2001. LNCS (LNAI), vol. 2321, pp. 211–229. Springer, Heidelberg (2002). https://doi.org/10.1007/3-540-48104-4_13
6. Choi, S.S., Cha, S.H., Tappert, C.C.: A survey of binary similarity and distance measures. J. Syst. Cybernet. Inform. **8**(1), 43–48 (2010)
7. Cordón, O.: A historical review of evolutionary learning methods for mamdani-type fuzzy rule-based systems: designing interpretable genetic fuzzy systems. Int. J. Approximate Reasoning **52**(6), 894–913 (2011). https://doi.org/10.1016/j.ijar.2011.03.004
8. Corriveau, G., Guilbault, R., Tahan, A., Sabourin, R.: Review and study of genotypic diversity measures for real-coded representations. IEEE Trans. Evol. Comput. **16**(5), 695–710 (2012). https://doi.org/10.1109/TEVC.2011.2170075
9. Drugowitsch, J.: Design and Analysis of Learning Classifier Systems - A Probabilistic Approach. SCI, vol. 139. Springer, Berlin (2008). https://doi.org/10.1007/978-3-540-79866-8
10. Eiter, T., Mannila, H.: Distance measures for point sets and their computation. Acta Informatica **34**(2), 109–133 (1997). https://doi.org/10.1007/S002360050075
11. Ganti, V., Gehrke, J., Ramakrishnan, R.: A framework for measuring changes in data characteristics. In: Proceedings of the Eighteenth ACM SIGMOD-SIGACT-SIGART Symposium on Principles of Database Systems, PODS 1999 pp. 126–137. Association for Computing Machinery, New York (1999). https://doi.org/10.1145/303976.303989
12. Gao, X., Xiao, B., Tao, D., Li, X.: A survey of graph edit distance. Pattern Anal. Appl. **13**(1), 113–129 (2010). https://doi.org/10.1007/S10044-008-0141-Y
13. Gustafson, S., Vanneschi, L.: Crossover-based tree distance in genetic programming. IEEE Trans. Evol. Comput. **12**(4), 506–524 (2008). https://doi.org/10.1109/TEVC.2008.915993
14. Heider, M., Pätzel, D., Stegherr, H., Hähner, J.: A Metaheuristic Perspective on Learning Classifier Systems, pp. 73–98. Springer Nature Singapore, Singapore (2023). https://doi.org/10.1007/978-981-19-3888-7_3
15. Heider, M., Stegherr, H., Nordsieck, R., Hähner, J.: Learning classifier systems for self-explaining socio-technical-systems (2022)
16. Heider, M., et al.: Discovering rules for rule-based machine learning with the help of novelty search. SN Comput. Sci. **4**(6), 778 (2023). https://doi.org/10.1007/s42979-023-02198-x

17. Heider, M., Stegherr, H., Wurth, J., Sraj, R., Hähner, J.: Separating rule discovery and global solution composition in a learning classifier system. In: Proceedings of the Genetic and Evolutionary Computation Conference Companion, GECCO 2022, pp. 248–251. Association for Computing Machinery, New York(2022). https://doi.org/10.1145/3520304.3529014

18. Kharbat, F., Odeh, M., Bull, L.: New approach for extracting knowledge from the XCS learning classifier system. Inter. J. Hybrid Intell. Syst. **4**, 49–62 (2007). https://doi.org/10.3233/HIS-2007-4201

19. Kovacs, T.: Deletion schemes for classifier systems. In: Proceedings of the 1st Annual Conference on Genetic and Evolutionary Computation, pp. 329–336 (1999)

20. Kovacs, T.: What should a classifier system learn and how should we measure it? Soft. Comput. **6**(3), 171–182 (2002)

21. Kovacs, T., Kerber, M.: High classification accuracy does not imply effective genetic search. In: Deb, K. (ed.) GECCO 2004. LNCS, vol. 3103, pp. 785–796. Springer, Heidelberg (2004). https://doi.org/10.1007/978-3-540-24855-2_93

22. Liu, B., Hsu, W., Han, H.-S., Xia, Y.: Mining changes for real-life applications. In: Kambayashi, Y., Mohania, M., Tjoa, A.M. (eds.) DaWaK 2000. LNCS, vol. 1874, pp. 337–346. Springer, Heidelberg (2000). https://doi.org/10.1007/3-540-44466-1_34

23. Liu, B., Hsu, W., Ma, Y.: Discovering the set of fundamental rule changes. In: Proceedings of the Seventh ACM SIGKDD International Conference on Knowledge Discovery and Data Mining, KDD 2001, pp. 335–340. Association for Computing Machinery, New York (2001). https://doi.org/10.1145/502512.502561

24. Liu, B., Ma, Y., Lee, R.: Analyzing the interestingness of association rules from the temporal dimension. In: Proceedings 2001 IEEE International Conference on Data Mining, pp. 377–384 (2001). https://doi.org/10.1109/ICDM.2001.989542

25. Liu, Y., Browne, W.N., Xue, B.: A comparison of learning classifier systems' rule compaction algorithms for knowledge visualization. ACM Trans. Evol. Learn. Optim. **1**(3) (2021). https://doi.org/10.1145/3468166

26. Parpinelli, R.S., Lopes, H.S., Freitas, A.A.: An ant colony algorithm for classification rule discovery. In: Data Mining, pp. 191–208. IGI Global (2002). https://doi.org/10.4018/978-1-930708-25-9.ch010

27. Pätzel, D., Heider, M., Hähner, J.: Towards principled synthetic benchmarks for explainable rule set learning algorithms. In: Proceedings of the Companion Conference on Genetic and Evolutionary Computation, GECCO 2023 Companion, pp. 1657–1662. Association for Computing Machinery, New York (2023). https://doi.org/10.1145/3583133.3596416

28. Pedregosa, F., et al.: Scikit-learn: machine learning in Python. J. Mach. Learn. Res. **12**, 2825–2830 (2011)

29. Pekerskaya, I., Pei, J., Wang, K.: Mining changing regions from access-constrained snapshots: a cluster-embedded decision tree approach. J. Intell. Inf. Syst. **27**(3), 215–242 (2006). https://doi.org/10.1007/S10844-006-9951-9

30. Preen, R.J., Pätzel, D.: Xcsf (2023). https://doi.org/10.5281/zenodo.8193688

31. Pätzel, D.: dpaetzel/rslmodels.jl: v0.1.1. https://doi.org/10.5281/zenodo.10557400

32. Pätzel, D.: dpaetzel/run-rsl-bench: v1.1.0. https://doi.org/10.5281/zenodo.10550923

33. Pätzel, D.: dpaetzel/syn-rsl-benchs: v1.0.0 (May 2023). https://doi.org/10.5281/zenodo.7919420

34. Quinlan, J.R.: C4.5: Programs for Machine Learning. Morgan Kaufmann Publishers Inc., San Francisco, CA, USA (1993)

35. Serpen, G., Sabhnani, M.: Measuring similarity in feature space of knowledge entailed by two separate rule sets. Knowl.-Based Syst. **19**(1), 67–76 (2006). https://doi.org/10.1016/j.knosys.2003.11.001

36. Setnes, M., Babuska, R., Kaymak, U., van Nauta Lemke, H.: Similarity measures in fuzzy rule base simplification. IEEE Trans. Syst. Man Cybernet. Part B (Cybernetics) **28**(3), 376–386 (1998). https://doi.org/10.1109/3477.678632

37. Stalph, P.O., Butz, M.V.: Guided evolution in XCSF. In: Proceedings of the 14th Annual Conference on Genetic and Evolutionary Computation, GECCO 2012, pp. 911–918. Association for Computing Machinery, New York (2012). https://doi.org/10.1145/2330163.2330289

38. Tamee, K., Bull, L., Pinngern, O.: Towards clustering with XCS. In: Proceedings of the 9th Annual Conference on Genetic and Evolutionary Computation, GECCO 2007, pp. 1854–1860. Association for Computing Machinery, New York (2007). https://doi.org/10.1145/1276958.1277326

39. Tan, J., Moore, J.H., Urbanowicz, R.J.: Rapid rule compaction strategies for global knowledge discovery in a supervised learning classifier system. In: Liò, P., Miglino, O., Nicosia, G., Nolfi, S., Pavone, M. (eds.) Proceedings of the Twelfth European Conference on the Synthesis and Simulation of Living Systems: Advances in Artificial Life, ECAL 2013, Sicily, Italy, 2–6 September 2013, pp. 110–117. MIT Press (2013). https://doi.org/10.7551/978-0-262-31709-2-CH017

40. Wang, K., Zhou, S., Fu, C.A., Yu, J.X.: Mining Changes of Classification by Correspondence Tracing, pp. 95–106. https://doi.org/10.1137/1.9781611972733.9

41. Wilson, S.W.: Classifier fitness based on accuracy. Evol. Comput. **3**(2), 149–175 (1995)

42. Wilson, S.W.: Classifiers that approximate functions. Nat. Comput. **1**(2), 211–234 (2002). https://doi.org/10.1023/A:1016535925043

Machine Learning and AI in Digital Healthcare and Personalized Medicine

Incremental Growth on Compositional Pattern Producing Networks Based Optimization of Biohybrid Actuators

Michail-Antisthenis Tsompanas[✉][iD]

School of Computing and Creative Technologies, University of the West of England, Bristol, UK
Antisthenis.Tsompanas@uwe.ac.uk

Abstract. One of the training methods of Artificial Neural Networks is Neuroevolution (NE) or the application of Evolutionary Optimization on the architecture and weights of networks to fit the target behaviour. In order to provide competitive results, three key concepts of the NE methods require more attention, i.e., the crossover operator, the niching capacity and the incremental growth of the solutions' complexity. Here we study an appropriate implementation of the incremental growth for an application of NE on Compositional Pattern Producing Networks (CPPNs) that encode the morphologies of biohybrid actuators. The target for these actuators is to enable the efficient angular movement of a drug-delivering catheter in order to reach difficult areas in the human body. As a result, the methods presented here can be a part of a modular software pipeline that will enable the automatic design of Biohybrid Machines (BHMs) for a variety of applications. The proposed initialization with minimal complexity of these networks resulted in faster computation for the predefined computational budget in terms of number of generations, notwithstanding that the emerged champions have achieved similar fitness values with the ones that emerged from the baseline method. Here, fitness was defined as the maximum deflection of the biohybrid actuator from its initial position after 10 s of simulated time on an open-source physics simulator. Since, the implementation of niching was already employed in the existing baseline version of the methodology, future work will focus on the application of crossover operators.

Keywords: Biohybrid machines · Compositional Pattern Producing Networks · optimization · evolutionary algorithms · machine learning

1 Introduction

Machine learning and, particularly, Artificial Neural Networks (ANNs) have become an increasingly prominent method for building accurate and efficient models with minimal required effort and background knowledge of the under study system. The widespread acceptance of ANNs is attributed to well established computational methods of training these networks, i.e., backpropagation,

© The Author(s), under exclusive license to Springer Nature Switzerland AG 2024
S. Smith et al. (Eds.): EvoApplications 2024, LNCS 14635, pp. 275–289, 2024.
https://doi.org/10.1007/978-3-031-56855-8_17

alongside other factors, such as big data availability. Another noteworthy and interesting training method of ANNs is Neuroevolution (NE), which proved to produce equally robust models [4].

NE is referring to the methodology of applying principles of Evolutionary Algorithms (EAs) to the process of training and optimizing ANNs [14]. The inspiration behind this technique was drawn from natural evolution, in order to realize a bio-inspired method of optimization. Populations of possible solutions or network instances (i.e. network architectures and connection weights) are produced as the result of simulated evolution and tested against a predefined fitness function. The fittest solutions are selected to mutate and reproduce, in order to provide more possible solutions that are then injected in the following generations, and so forth, until the computational budget is spent or a target efficiency is reached.

The implementation of NE can be employed in the evolution of different kinds of networks, like Compositional Pattern Producing Networks (CPPNs) [13]. Since CPPNs are formalized in a similar way as ANNs, there is no need for extensive changes in well-established methodologies applied on the latter, while favourable results are expected. The main difference between the two types of networks is the activation functions of their nodes; whereas, CPPNs are not restricted in any way around this area, ANNs are mainly employing monotonic functions. As a result, CPPNs are better suited for applications that require the production of complex patterns and structures [3].

Here, the use of CPPNs is studied as a tool for the primary discovery of morphologies of biohybrid machines (BHMs), in a similar way as previous works that delivered promising results [1,2,7]. By adopting the open source code developed previously [7], we aim to empower the integration of advanced algorithms during the initial stages of the BHM development, helping identify crude designs of efficient morphologies. The primary objective of our project is to establish a BHM design process and employ this framework to pioneer the development of a ground-breaking medical device, namely a biohybrid catheter capable of delivering pharmaceuticals to challenging to reach areas of the human physiology. In specific, the main goal of the developed software module is to produce BHM actuators that will facilitate robust angular movement of a catheter in the labyrinth-like environment of the circulatory system of humans.

Some NE algorithms, i.e., NeuroEvolution of Augmenting Topologies (NEAT) [15], have been proved to be more efficient than others, because of three critical factors described in the following. These methodologies (i) employ means that enable crossover during evolution without complicated topology analysis, (ii) include niching capacity that is able to protect innovative individuals from premature exclusion and (iii) encourage the incremental growth of complexity in solutions, on account of the initial populations being structures of minimal complexity. Reviewing the algorithmic approach implemented in [7], we could locate a variant method for niching, but there were no provisions for the other two characteristic factors. While the crossover factor was not included intentionally for simplicity reasons based on the authors' reasoning, we could not pinpoint the motivation behind initialization of population with networks that were not of minimal complexity.

As a result, this work takes into consideration the initialization of the populations with minimal structure networks and compares the outputs with the method followed by the original work [7]. To test the appropriateness of starting at minimal dimensions and, as a result, allowing incremental growth of complexity, the open source code was altered towards including that characteristic. Moreover, a comparative analysis of the champions discovered was performed, in order to justify previous findings [15], i.e., that this characteristic enables higher effectiveness. The results show that starting at minimal dimension provides solution candidates with similar effectiveness in terms of fitness, however a significant acceleration of the computation for the same target of total generations is achieved. This can be attributed to the lower complexity of the networks being managed throughout the computational process.

The rest of the manuscript is organized as in the following. Section 2 provides some background on NE and the aspects that render it a suitable surrogate of other training methods, along with basic characteristics of CPPNs. Section 3 describes the methodology used in this study, i.e., details of the simulators, algorithms and the proposed initialization method. Then, Sect. 4 presents the results of the tests for both initialization methods and Sect. 5 concludes this study.

2 Background

Some typical paradigms of NE methods [5,9] assumed a fixed structure for the networks that were studied and their dimensionality was manually set before evolution began. One hidden layer was included with neurons fully connected with the input and output neurons, while the evolution was assessing the weights of the connections. Because of the fixed topology of networks, crossover and mutation operators were trivially applied to the weights of the connections and optimization enabled the training of networks towards a desired behaviour.

Nevertheless, the weights of network connections are not a sole indicator of how neural networks function. The structure that defines the number of nodes and how they are connected, plays a significant role as well. Thus, enhanced NE methods were proposed under the term Topology and Weight Evolving Artificial Neural Networks (TWEANNs) employing evolution of both topologies and connection weights [8,11]. These techniques take advantage of increasing structural complexity through mutations. Although the addition of randomly formed nodes may cause a decrease in fitness initially, the modulation of connection weights during subsequent evolution steps can result in an ultimately higher fitness.

An innovative NE method, named NEAT [15] has motivated a large range of variants relevant till this day [10] and managed to outperform previous methods, as it was more thoroughly designed, in order to exploit the fact that smaller dimensionality networks can be optimized faster. It proved to be a superior methodology, because, according to the authors [15], (i) it would include a crossover operator, while previous versions did not, (ii) it would safeguard the innovation in network architecture with initial low fitness against premature exclusion of these promising architectures and (iii) it would allow for incremental growth on the complexity of networks by initializing populations at minimal complexities. The authors tested what each of these three aspects contributed to the overall efficiency and concluded that all aspects and their combinations were significant for providing even better efficiency.

In a similar setting, the conclusions in [16] argue that the robustness in evolutionary methods is achieved by an initial population of minimal and non-complicated genomes. As generations lapse these genomes undergo the introduction of additional genes that serve as enablers to the expansion of the search space. Therefore, novel dimensions are introduced and evolutionary exploration is initially exploring a relatively small and manageable space, before moving to additional dimensions that are included only if necessary, i.e., after the search to the given search space dimensionality stagnates. This incremental process is called complexification and is a technique used to partially tackle the curse of dimensionality. Moreover, complexification is not limited to enhance the results of NE methods, but, also, the efficiency of more typical evolutionary algorithms. In our previous works on optimization of individuals with variable genome lengths [17,18], the ability to optimize and complexify the genome were both included in the methodology to produce fitter solutions, while the initial populations were of minimal genome lengths.

CPPNs are similar to ANNs, with the main difference being the relaxation of rules on the activation functions of nodes of the former type of networks. Namely, CPPNs are better suited for generating complex patterns and structures [3], since the graph that represents them define associations between a variety of functions (or activation functions) that are depicted as nodes (as depicted in an example of CPPN in Fig. 1). Connections are characterised by weights that determine the impact of each node output to the input of the next layer node. In cases where multiple connections terminate to the same node, all weighted outputs of the previous nodes are aggregated and used as inputs to the current node. An additional difference, that is also essential for CPPNs' functionality is that the topology of the graph is not restricted in any way, thus, enabling higher levels of representation liberty that achieves more complex patterns.

Another, more semantic difference between these networks is that while ANNs emulate the functionality of human brain in learning, CPPNs simulate a completely different biological process, namely, the developmental process [13]. Consequently, another attractive feature of CPPNs in applications of producing patterns is that when they are queried on an absolute coordinate frame (i.e. x, y in a two-dimensional space), there is no need for the specific definition of local interactions within the representation. When using each specific point in a Cartesian coordinate system as an input of CPPN, the outputs will formulate a pattern without the phenotype (i.e. the CPPN) requiring local interactions or temporal sequencing. The network will use the coordinates of all points in a

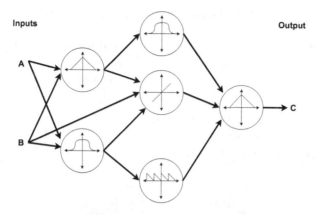

Fig. 1. An example of a CPPN with two inputs and one output. Each node represents a specific function, while connections are weighted and represent the sum of weighted intakes of each function.

space as inputs and the output provided will precisely specify the entities and characteristics of the specific location in space that was used as input every time.

3   Methods

To evaluate some primary morphologies of BHM catheter actuators with no detailed investigation of all the possible components of the underlying mechanisms and no biotechnology laboratory overheads, the *in silico* investigation is preferred. Thus, a simulator that would be capable of mimicking behaviours of truly heterogeneous materials is required. Thus, Voxelyze [6] was employed as the test-bed of morphologies composed of different materials with diverse physical properties, such as Poisson's ratio, stiffness, density and friction coefficients. Moreover, Voxelyze has the capacity to simulate external forces along with volumetric actuation of entities; a characteristic that enables the representation of novel architectures like the ones found in BHMs, namely accommodating contracting muscle cells. In Voxelyze each elementary volume, designated as a voxel, can encode a different material and the distance between neighboring voxels is modeled as Euler-Bernoulli beams. Moreover, additional environmental settings can be defined to illustrate specific scenarios, such as gravitational acceleration, collision rules and friction between the range of different voxels and a static floor. Here, to follow the scenarios investigated in previous studies [7], two types of voxels were outlined with the parameters depicted in Table 1. Specifically, one type is an active voxel that can contract and provide the energy required for movement; whereas, the other type is a passive voxel with similar physical properties, but, no motion capacity included.

Voxelyze acts as a test-bed for the fitness function, namely, morphologies of $8 \times 7 \times 7$ voxels in a Cartesian grid are evaluated based on their simulated behaviours. The morphologies of a maximum of 392 voxels are constructed in

Table 1. Parameters of active and passive voxel.

Parameters	Active voxel	Passive voxel
Elastic modulus (MPa)	5	5
Density (kg/m^3)	1,000,000	1,000,000
Poisson's ratio	0.35	0.35
Coefficient of Thermal Expansion $(1/°C)$	0.01	0
Coefficient of static friction	1	1
Coefficient of dynamic friction	0.5	0.5

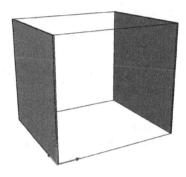

Fig. 2. Boundaries in Voxelyze representing a fixed end and a free end of a catheter actuator.

the virtual environment of Voxelyze, the simulation starts and after 1 s of initial simulation time, the morphology will be settled from possible gravitational motion into the starting point for the evaluation. Following this, a further 10 s are simulated, in order to record the final displacement of the whole morphology and calculate the deflection achieved by the simulated actuator. Note here, that in order to better represent the scenario of a catheter actuator, one end of the morphology is fixed (the YZ plane for $x = 0$, depicted as the green plane in Fig. 2), whereas the other end is free to perform translational and rotational motion based on the global behavior derived from all the individual active voxels' activity (the YZ plane for $x = 8$, depicted as the purple plane in Fig. 2). The fitness for each candidate morphology is provided by the total deflection at $t = 10s$ of simulated time, i.e., the distance of the projection on any YZ plane of one of the top and outer voxels that are adjacent to the free end of the morphology.

The indirect encoding concept that exploits CPPNs is utilized to symbolize candidate morphologies in the evolutionary optimization process. To decode the individuals, the Cartesian coordinates (x, y and z) are used as inputs for the network (in addition with the distance from the center of the available space d and a bias b) and the output represents the voxel type for the respective combination of coordinates (as illustrated in Fig. 3). After querying the network

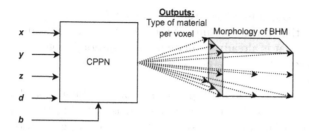

Fig. 3. Decoding of the CPPN into a morphology by using coordinates in a three-dimensional space as inputs and material types as outputs. The functionality of the CPPN is illustrated in Fig. 1

with all the possible combinations on the aforementioned $8 \times 7 \times 7$ grid, the types of all the 392 voxels are provided, thus, the morphology is decoded into the 3D space and can be inserted to Voxelyze to calculate its fitness. Note here, that the CPPN output can denote active or passive voxels, but, also empty space in order to permit more elaborate morphologies.

The nodes in the hidden layers of the CPPNs can represent any of the predefined mathematical functions (i.e., sine, absolute value, negative absolute value, square, negative square, square root and negative square root). The weight of the connection between two nodes represent the multiplication factor of the outgoing result. When several connections terminate to the same node, then, the addition of the weighted in-going results is used as input to the node's activation function.

For the evolutionary algorithm, the genotype of the individuals is in a form of a CPPN, whereas the phenotype is in a form of a 3D morphology of the BHM actuator, which is derived by querying the CPPN genotype. Following the concept of the open source code [7] and for a clearer implementation, no crossover operator was implemented, however, the Age-Fitness Pareto Optimization (AFPO) algorithm [12] was utilized. Particularly, a population of 50 randomly generated CPPNs was produced through an intricate initialization process. Afterwards, these 50 individuals were decoded into BHM morphologies and evaluated through Voxelyze. Then, 50 additional individuals were created through mutation operations over the initial population. These 50 additional individuals with the inclusion of one more randomly generated individual were evaluated and from the total of 101 available individuals the 50 fittest were selected to comprise the next generation. This would complete one evolution cycle and the new generation would go again through the mutation operator and so forth, until 2000 generations were evaluated.

The mutation process for each individual involves the application of one type out of six possible alternations in the CPPN genotype with a 0.167 probability. The possible alternations are the addition of a node or connection, the modification of a node or connection and the removal of a node or connection. As a result, both the weights of the CPPN and its architecture can be modified to permit training and complexification (or even simplification) of the network. It is noteworthy, that if the decoding of the CPPN genotype produces a morphology phenotype that is already evaluated, the mutation is considered neutral and the process is repeated until a non-neutral mutation is found, for a maximum of 1500 attempts.

The utilization of AFPO [12] provides a basic niching capacity. On the condition that this multi-objective optimization is pursuing the dominance over fitness and age for individuals to survive for consequent generations, premature convergence is avoided. In specific, the individuals are selected based on their higher fitness value and lower genotypic age, in a multi-objective Pareto front optimization. As a result, individuals that have emerged latter in evolution can coexist in the same population with older and fitter individuals, because they are not dominated on the age dimension of the Pareto front.

The final aspect that needs to be clarified is the initialization process, as it was determined one of the three key factors for NE effectiveness. The original initialization process, used in [7] and denoted here as the baseline, begins with building the minimal possible network, namely connecting input and output nodes with edges of weight zero. Then, the mutation operator is executed multiple times, i.e., 10 times for random node addition, 10 times for random connection addition, 5 times for random connection removal, 100 times for random node modification and 100 times for random connection weight modification. After that, the network is pruned to remove any erroneous nodes and connections. The notable operations here are the 10 node additions, the 10 connection additions and the pruning of the network, which can result to a network with a maximum of 10 hidden layer nodes. However, there is no guarantee that this is the minimal or, even, close to the minimal possible structure.

To study the potential of incremental growth in the NE of CPPNs, we altered the aforementioned initialization process to build a population with lower amount of hidden layer nodes. Specifically, we kept the same methodology as described previously, however, we altered the amount of random node additions during initialization from 10 to 2. This small number was enough to allow the production of an initial population with acceptable diversity and a range of hidden layer nodes from zero to 2. This is obviously an initial population of simpler network structures that will more probably permit complexification alongside optimization.

4   Results

In the following, the comparison of the outputs for 10 runs with the same random generator seeds are presented for both initialization processes. Execution times of the two variants are presented in Fig. 4. It is apparent that the original initialization stems a significantly slower evaluation of the 2000 generations, when compared with the minimal structure initialization proposed here (Wilcoxon rank-sum test, $p < 0.001$). In specific, execution times of the original initialization have a mean of 19.71 h (samples = 10, minmax = (18.69, 20.88), variance

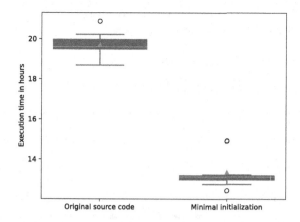

Fig. 4. Execution times for the two variants of the population initialization.

$= 0.35$, skewness $= 0.35$, kurtosis$= 0.0998$), while for the proposed minimal initialization a mean of 13.37 h (samples $= 10$, minmax $= (12.44, 14.93)$, variance $= 0.72$, skewness $= 1.19$, kurtosis $= -0.0620$) was observed. So, the minimal initialization economizes 6 h of computational time per run of 2000 generations.

This acceleration can be attributed to the bottleneck in the whole process of finding non-neutral mutated networks. As mentioned previously, after each mutation operation the phenotype (3D morphology) produced from the genotype (CPPN) was compared with the phenotype of the pre-mutated genotype. If the new CPPN would produce the same morphology, a new mutation would be attempted, unless 1500 unsuccessful attempts are executed. We realised that this technique introduces significant overheads to the whole evolutionary computation process. Moreover, it is evident that this procedure requires more computational resources when assessing networks with higher numbers of nodes in the hidden layer. On the contrary, when assessing less complex networks the discovery of non-neutral mutations would happen faster, as realised from the execution times in Fig. 4.

In order to compare the complexity of the networks within the initial population for both variants of the initialization process, Fig. 5 illustrates the number of nodes in the hidden layer. These data are collected for all 50 initial individuals for each of the 10 runs of both variants (i.e. 500 individuals). As expected the original source code initialization produces more complex networks, with a mean of 7.734 nodes (variance $= 2.436$), whereas, the proposed minimal initialization produces simplified networks with a mean of 1.9 nodes (variance $= 0.110$) in the hidden layer.

Moreover, Fig. 6 depicts the distribution of the nodes in the networks within the population after 2000 generations of evolutionary optimization. Note that the data collected include all 50 final individuals for each of the 10 runs of both

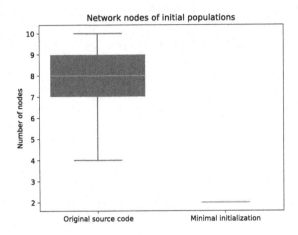

Fig. 5. Distributions of nodes of networks in the initial populations.

variants (i.e. 500 individuals). While the final populations of the baseline method have higher mean (=8.49) and variance (=10.35) in terms of network nodes, the proposed methodology illustrates sufficient diversity in the final populations as well (mean= 3.45, variance= 6.80). Nonetheless, by comparing the means of the initial and final populations, it can be seen that the complexification of the networks is more prominent in the minimal initialization process. Figure 7 outlines the distribution of the complexity (or number of nodes) of the fittest individuals at the end of the 2000 generations. It can be concluded here that champions are found in the higher network complexity available in the populations for both variations. Moreover, the minimal initialization methodology manages to discover champions encoded by networks with up to c. 10 hidden layer nodes, despite the starting point of only 2 nodes.

To investigate the impact of incremental growth that clearly manifested in the aforementioned, Table 2 describes the fitness of the champions (i.e., fittest individuals) after 2000 generations for both variants. Each line illustrates the fitness for both runs with the same random generator seed. The original initialization method is producing fitter champions in half of the runs (samples = 10, minmax = (0.253, 0.433), mean = 0.346, variance = 0.00345, skewness = −0.115, kurtosis = −0.8651). On the other hand, the minimum complexity initialization is not falling far behind and, despite the minimal networks present

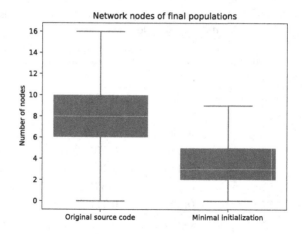

Fig. 6. Distributions of nodes of networks in the final populations.

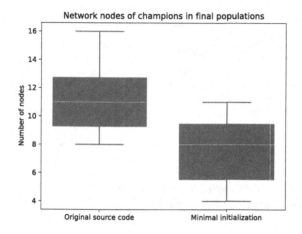

Fig. 7. Distributions of nodes of networks of the champions in the final populations.

in the first populations, it manages to produce comparable champions (samples = 10, minmax = (0.176, 0.469), mean = 0.333, variance = 0.00679, skewness = −0.275, kurtosis = −0.2162). Moreover, comparing the distributions of the fitness of both sets of champions, we can not reject the null hypothesis that the two sets of fitnesses are drawn from the same distribution (Wilcoxon rank-sum test, $p = 0.879$). Thus, no advantage is apparent for starting from a minimal or complex population, other than the acceleration in computations.

To demonstrate the incremental growth of network complexities throughout evolution, the range of the amount of nodes for whole populations, the median and the champions are provided for both variations in runs with seeds 52 and 58 (the fittest champions of both variations) in Figs. 8 and 9 respectively. For illustration reasons, the morphology of the highest performing champion of all

Table 2. Fitness of the champions after 2000 generations for both initialization procedures (higher is better and indicated by bold fonts).

Random seed	Original init.	Minimum init.
50	**0.382**	0.352
51	0.253	**0.255**
52	0.328	**0.469**
53	0.266	**0.417**
54	**0.325**	0.176
55	**0.419**	0.293
56	0.341	**0.375**
57	**0.375**	0.344
58	**0.433**	0.303
59	0.337	**0.349**

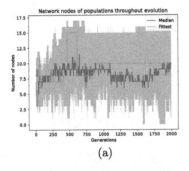 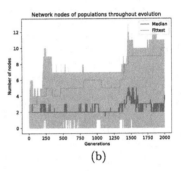

(a)   (b)

Fig. 8. Evolution of CPPN node numbers for runs with seed 52 for (a) the original initialization process and (b) the proposed minimal initialization.

the runs (found in the minimal initialization variant with seed 52) is illustrated in Fig. 10 (a) at its initial position and in Fig. 10 (b) at its final position (after 10 s of simulation time). Moreover, the CPPN that was evolved from the algorithm to encode this morphology is depicted in Fig. 11, where the input and output nodes, activation functions and connection weights are defined.

5   Discussion

The implementation of CPPNs, as an indirect representation of individuals, in evolutionary optimization has proved to be quite efficient. However, the initialization of a previously published work did not follow the incremental growth concept that is of paramount importance to the efficiency of NE methods. To prove the effects of starting at minimal complexity and, thus, allowing incremental growth through evolution (or complexification) along optimization, we altered the initialization methodology to apply this concept. The software framework

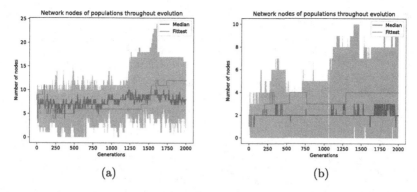

Fig. 9. Evolution of CPPN node numbers for runs with seed 58 for (a) the original initialization process and (b) the proposed minimal initialization.

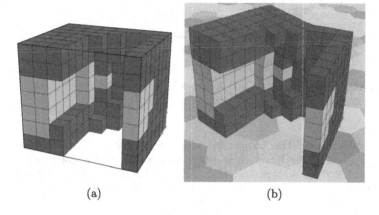

Fig. 10. The highest performing champion of this study simulated in the Voxelyze environment at its (a) initial and (b) final position.

was an application of NE optimization on the first module of a design pipeline for a BHM catheter actuator.

The results show that there is no significant advantage in the fitness of the emerging champions for the populations that started from higher complexity for the baseline implementation. On the contrary, the methodology that employed minimal initialization performed at the same degree of efficiency, maintained high diversity in the final populations and required less computational resources to reach the same degree of efficiency. As a result, aspects of future work will be the comparison of the two methods with the same computational budget, but in terms of wall-time and not amount of generations. Moreover, the effect of the crossover will be studied as it has been proved [15] that enhances the capabilities of NE in well-established benchmarks.

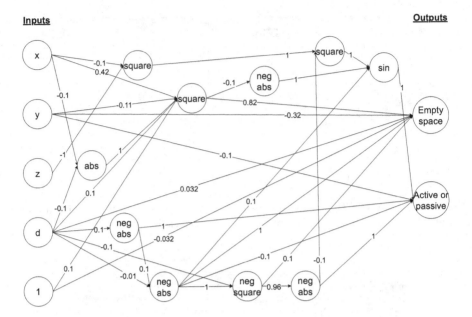

Fig. 11. The CPPN structure that decodes into the morphology of the highest performing champion of this study illustrated in Fig. 10

Acknowledgement. This project has received funding from the European Union's Horizon Europe research and innovation programme under grant agreement No. 101070328. UWE researchers were funded by the UK Research and Innovation grant No. 10044516.

Code availability. The code to reproduce the results can be found here: https:// github.com/Antisthenis/reconfigurable_organisms/tree/biomeld_dev2.

References

1. Cheney, N., Bongard, J., Lipson, H.: Evolving soft robots in tight spaces. In: Proceedings of the 2015 Annual Conference on Genetic and Evolutionary Computation, pp. 935–942 (2015)
2. Cheney, N., MacCurdy, R., Clune, J., Lipson, H.: Unshackling evolution: evolving soft robots with multiple materials and a powerful generative encoding. ACM SIGEVOlution **7**(1), 11–23 (2014)
3. Clune, J., Lipson, H.: Evolving 3d objects with a generative encoding inspired by developmental biology. ACM SIGEVOlution **5**(4), 2–12 (2011)
4. Galván, E., Mooney, P.: Neuroevolution in deep neural networks: Current trends and future challenges. IEEE Trans. Artif. Intell. **2**(6), 476–493 (2021)
5. Gomez, F.J., Miikkulainen, R., et al.: Solving non-Markovian control tasks with neuroevolution. In: IJCAI, vol. 99, pp. 1356–1361. CiteSeer (1999)
6. Hiller, J., Lipson, H.: Dynamic simulation of soft multimaterial 3d-printed objects. Soft Rob. **1**(1), 88–101 (2014)

7. Kriegman, S., Blackiston, D., Levin, M., Bongard, J.: A scalable pipeline for designing reconfigurable organisms. Proc. Natl. Acad. Sci. **117**(4), 1853–1859 (2020)
8. Lee, C.H., Kim, J.H.: Evolutionary ordered neural network with a linked-list encoding scheme. In: Proceedings of IEEE International Conference on Evolutionary Computation, pp. 665–669. IEEE (1996)
9. Moriarty, D.E., Mikkulainen, R.: Efficient reinforcement learning through symbiotic evolution. Mach. Learn. **22**, 11–32 (1996)
10. Papavasileiou, E., Cornelis, J., Jansen, B.: A systematic literature review of the successors of "neuroevolution of augmenting topologies". Evol. Comput. **29**(1), 1–73 (2021)
11. Pujol, J.C.F., Poli, R.: Evolving the topology and the weights of neural networks using a dual representation. Appl. Intell. **8**, 73–84 (1998)
12. Schmidt, M.D., Lipson, H.: Age-fitness pareto optimization. In: Proceedings of the 12th annual conference on Genetic and evolutionary computation, pp. 543–544 (2010)
13. Stanley, K.O.: Compositional pattern producing networks: a novel abstraction of development. Genet. Program Evolvable Mach. **8**, 131–162 (2007)
14. Stanley, K.O., Clune, J., Lehman, J., Miikkulainen, R.: Designing neural networks through neuroevolution. Nat. Mach. Intell. **1**(1), 24–35 (2019)
15. Stanley, K.O., Miikkulainen, R.: Evolving neural networks through augmenting topologies. Evol. Comput. **10**(2), 99–127 (2002)
16. Stanley, K.O., Miikkulainen, R.: Competitive coevolution through evolutionary complexification. J. Artif. Intell. Res. **21**, 63–100 (2004)
17. Tsompanas, M.A., Bull, L., Adamatzky, A., Balaz, I.: Evolutionary algorithms designing nanoparticle cancer treatments with multiple particle types [application notes]. IEEE Comput. Intell. Mag. **16**(4), 85–99 (2021)
18. Tsompanas, M.A., Bull, L., Adamatzky, A., Balaz, I.: Metameric representations on optimization of nano particle cancer treatment. Biocybern. Biomed. Eng. **41**(2), 352–361 (2021)

Problem Landscape Analysis for Efficient Optimization

Hilbert Curves for Efficient Exploratory Landscape Analysis Neighbourhood Sampling

Johannes J. Pienaar[1], Anna S. Boman[1] , and Katherine M. Malan[2]([✉])

[1] University of Pretoria, Pretoria, South Africa
anna.bosman@up.ac.za
[2] University of South Africa, Pretoria, South Africa
malankm@unisa.ac.za

Abstract. Landscape analysis aims to characterise optimisation problems based on their objective (or fitness) function landscape properties. The problem search space is typically sampled, and various landscape features are estimated based on the samples. One particularly salient set of features is *information content*, which requires the samples to be sequences of neighbouring solutions, such that the local relationships between consecutive sample points are preserved. Generating such spatially correlated samples that also provide good search space coverage is challenging. It is therefore common to first obtain an unordered sample with good search space coverage, and then apply an ordering algorithm such as the nearest neighbour to minimise the distance between consecutive points in the sample. However, the nearest neighbour algorithm becomes computationally prohibitive in higher dimensions, thus there is a need for more efficient alternatives. In this study, Hilbert space-filling curves are proposed as a method to efficiently obtain high-quality ordered samples. Hilbert curves are a special case of fractal curves, and guarantee uniform coverage of a bounded search space while providing a spatially correlated sample. We study the effectiveness of Hilbert curves as samplers, and discover that they are capable of extracting salient features at a fraction of the computational cost compared to Latin hypercube sampling with post-factum ordering. Further, we investigate the use of Hilbert curves as an ordering strategy, and find that they order the sample significantly faster than the nearest neighbour ordering, without sacrificing the saliency of the extracted features.

Keywords: Fitness landscape analysis · sampling · Hilbert curve

1 Introduction

Search landscape analysis has established itself as a useful approach for understanding complex optimisation problems and analysing evolutionary algorithm behaviour [18]. Many landscape analysis techniques have been developed over

Supported by the National Research Foundation of South Africa Thuthuka Grant Number 138194/TTK210316590115.

the years, with the most widely used techniques including fitness distance correlation [11], local optima networks [26], and exploratory landscape analysis (ELA) [24]. When landscape analysis produces numeric outputs, the resulting feature vectors can be used as abstract representations of problem instances, where instances with similar feature vectors are assumed to fall into similar problem classes. If these feature vectors effectively capture the important characteristics of problems, they can be used as the feature component for automated algorithm design (AAD) – specifically, automated algorithm configuration and selection. A number of recent studies have achieved different aspects of AAD using landscape analysis in specific contexts [2, 13, 14, 16, 17, 32].

Landscape analysis approaches differ in terms of what they measure or predict (e.g. ruggedness, modality, presence of funnels, and so on), and also on what they produce (e.g. numerical results or a visualisation of a phenomenon) [21]. They can also be distinguished based on the scale of the analysis [27] – a global approach attempts to characterise the features of the search space as a whole, while a local approach will consider the features of the landscape in the neighbourhood of solutions. Fitness distance correlation [11] and local optima networks [26] are both examples of global approaches, whereas the average length of an adaptive walk [37] is an example of a local landscape feature.

Many landscape analysis techniques that measure local features are based on samples that are *spatially correlated*, i.e. sequences of neighbouring solutions, as opposed to a sample of independent solutions from the whole search space. Techniques that require such sequences of sampled solutions include correlation length for measuring ruggedness [38], entropic profiles of ruggedness and smoothness with respect to neutrality [35, 36] – adapted as a single measure of ruggedness for continuous spaces [19], approximations of gradient [20], information content features [25], and measures of neutrality [34].

In the context of numerical optimisation, a wide range of landscape analysis approaches have been implemented in the R package `flacco` [12], which has also been re-implemented in Python (`pflacco`[1]). Over 300 landscape metrics are included in `flacco`, organised into 17 sets, which include the original ELA metrics [24] covering six of the feature sets, and the set of information content features [25] consisting of five metrics. In a study of the subset of the "cheap" feature sets in `flacco`, Renau et al. [29] found that the two most salient (in terms of distinguishing between problems) and robust feature metrics were from the information content and ELA meta-model feature sets. Unlike the other feature sets in `flacco`, the information content metrics require spatially correlated samples. The saliency of the information content feature metrics and the additional requirement of neighbourhood ordering is the motivation for this study.

To date, the two main strategies proposed for generating spatially correlated samples in continuous search spaces are random walks [22] and post-factum ordering of global samples [25]. Desirable properties of a sampling strategy for estimating local features are that the solutions provide good coverage of the search space [15], successive points are positioned close to each other (compared

---

[1] https://pypi.org/project/pflacco/.

to other solutions in the sample) to capture landscape changes in the neighbour-hood, and that the process has low computational expense. The most commonly used approaches to neighbourhood sampling for numerical optimisation are uniform random sampling or Latin hypercube sampling (LHS) [28], followed by either random or nearest-neighbour ordering. These are the approaches implemented in `flacco` and `pflacco`.

In this paper, we investigate the use of Hilbert curves [10] as a new sampling strategy. Hilbert curves are fractal space-filling curves with two desirable properties: (1) they guarantee uniform search space coverage, and (2) neighbouring points on the curve are located close to one another. We show that Hilbert curves are comparable to ordered LHS in terms of the saliency of the extracted features, but are significantly cheaper to compute. Additionally, we show that a Hilbert curve can be used to spatially order an LHS sample, resulting in significant computational gains compared to the commonly used nearest neighbour ordering method.

2   Hilbert Curves

A space-filling curve is a surjective continuous function from the unit interval $[0, 1]$ to a unit hypercube $[0, 1]^d$. Surjectivity implies that every point in the hypercube maps to at least one point in the interval, and continuity ensures that no areas in $[0, 1]^d$ are missed [5]. Space-filling curves are a special case of fractal curves, and are guaranteed to fill a continuous space in the limit.

Our interest in space-filling curves derives from two useful properties that they offer, namely (1) uniform coverage of a bounded d-dimensional space, and (2) the ability to provide a unique mapping between points in the d-dimensional space and points on the 1-dimensional curve [31]. Specifically, this study considers the Hilbert space-filling curve, first proposed by D. Hilbert in 1891 [10]. The Hilbert curve is defined recursively, and can be constructed through a limit process of iteration. Each successive iteration creates an approximation of the true Hilbert curve that passes through more points in the d-dimensional unit hypercube. For practical purposes, the number of iterations is chosen to be finite, and is further referred to as the *order* of the Hilbert curve.

Consider the construction of a Hilbert curve in $2D$. For the first iteration, a line on the closed interval $[0, 1]$ and a square $[0, 1]^2$ are taken. Four equidistant points are selected on the line, where the starting point is at 0 and the end point is at 1. The square is subdivided into four equal parts. Intervals of the line connecting each pair of points are then mapped onto the square such that the intervals adjacent on the line share a common edge on the square. This results in a simple U-shape, illustrated in Fig. 1(a). At each subsequent iteration, the curve from the previous iteration is divided into four equal parts. Each part is then shrunk by a factor of 1/2, rotated, and repositioned such that the four curves connect at their endpoints in a U-shaped or reverse U-shaped pattern. Figure 1 shows $2D$ Hilbert curves of order 1 to 3.

The Hilbert curve is bijective, i.e. the mapping between points on the curve and a d-dimensional space is reversible. Bijectivity of space-filling curves has been

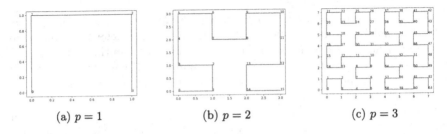

(a) $p = 1$ (b) $p = 2$ (c) $p = 3$

Fig. 1. Visualisation of a $2D$ Hilbert curve of order $p = \{1, 2, 3\}$.

useful in organising multi-dimensional data storage and retrieval systems, since it allows a linear index to the data to be constructed and searched [1,30]. Ideally, closely related data in the d-dimensional space would also be close together on the 1-dimensional curve. Faloutsos and Roseman [6] investigated the clustering properties of Hilbert curves compared to other space-filling curves, and found Hilbert curves to be superior when measuring the average distance (along the curve) to the nearest neighbours of a given point. This finding suggests that Hilbert curves may serve as a viable alternative to nearest neighbour sorting of a sample.

For the Hilbert curve implementation in this study, we use the `hilbertcurve` Python package (https://pypi.org/project/hilbertcurve/), based on the algorithm presented by Skilling [33]. All code used for the experiments conducted in this study can be found at https://github.com/jpienaar-tuks/MIT807.

3   Hilbert Curves as Samplers

Given a Hilbert curve of a particular order, the vertices can be used as a basis for a sample in the associated multidimensional space. Table 1 shows how the number of vertices grows with the order of the curve and the dimension of the search space. When sampling for ELA, it is common practice to use sample sizes of $10^2 \times d$ to $10^3 \times d$ [25]. Table 1 highlights in bold the order and dimension combinations where the number of vertices exceeds a common sample budget of $10^3 \times d$. To remain within a sampling budget, we sub-sample randomly from the set of vertices on a Hilbert curve with a minimum order of 3.

3.1   Stochastic Sampling Using Hilbert Curves

Since the generation of a Hilbert curve is deterministic, the following two strategies were evaluated to introduce stochasticity:

- Selecting random points along the edges of the Hilbert curve, illustrated in Fig. 2(a): Given any two sequential vertices, P_i and P_{i+1}, a new point is selected $P_j = rP_i + (1 - r)P_{i+1}$, where $r \sim U(0, 1)$.

– Selecting points near vertices of the curve, illustrated in Fig. 2(b): For each vertex P_i, generate a new point drawn from a normal distribution centred on P_i with a standard deviation σ (constrained by the bounds of the search space). The step size of the Hilbert curve before scaling is 1, so σ was empirically chosen to be 0.3 to prevent excessive potential overlap with points generated by neighbouring vertices.

Neighbouring vertices on a Hilbert curve are equidistant, so the sequence of vertices has a constant step size. The two randomisation techniques introduce variation into the step size. Figure 3 shows that the randomisation near vertices strategy produces step sizes with a more normal distribution. For all further experiments we have used the randomisation around vertices strategy.

3.2 Search Space Coverage

We now investigate the extent to which Hilbert curve sampling covers the search space compared to competing strategies such as Latin hypercube sampling (LHS). To investigate the search space coverage, the following three

Table 1. Exponential growth in Hilbert curve vertices with dimension and curve order. Values that exceed a value of $10^3 \times d$ are formatted in bold.

		Dimension				
		2	3	5	10	20
Order	1	4	8	32	1024	$\mathbf{1.05 \times 10^6}$
	2	16	64	1 024	$\mathbf{1.05 \times 10^6}$	$\mathbf{1.10 \times 10^{12}}$
	3	64	512	$\mathbf{32\,768}$	$\mathbf{1.07 \times 10^9}$	$\mathbf{1.15 \times 10^{18}}$
	4	256	$\mathbf{4\,096}$	$\mathbf{1.05 \times 10^6}$	$\mathbf{1.10 \times 10^{12}}$	$\mathbf{1.21 \times 10^{24}}$
	5	1 024	$\mathbf{32\,768}$	$\mathbf{3.36 \times 10^7}$	$\mathbf{1.13 \times 10^{15}}$	$\mathbf{1.27 \times 10^{30}}$
	6	$\mathbf{4\,096}$	$\mathbf{262\,144}$	$\mathbf{1.07 \times 10^9}$	$\mathbf{1.15 \times 10^{18}}$	$\mathbf{1.33 \times 10^{36}}$
Budget ($10^3 \times d$)		2 000	3 000	5 000	10 000	20 000

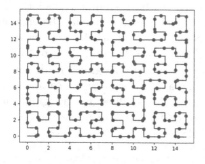

(a) Randomising points along the edges

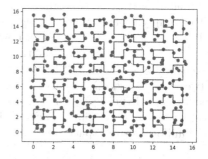

(b) Randomising points near vertices

Fig. 2. Illustration in 2D of adding stochasticity to a Hilbert curve

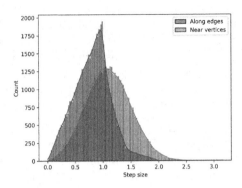

Fig. 3. Effect of randomisation strategy on step size. Sample was generated from an 8th order Hilbert curve in 2D space resulting in a total of 65 535 points. The x-axis represents the distance between two consecutive points.

strategies were used to draw samples from search space $[-5,5]^d$ of sizes $n \in \{100d, 316d, 1000d\}$ for dimensions $d \in \{5, 10, 20, 30\}$:

1. Hilbert curve: randomly select the required number of points from the curve (without replacement).
2. LHS: generate the required sample using a Latin hypercube design [23].
3. Random walk: use a simple random walk [22] with maximum step size of 1 to generate the sample.

For each sample size and dimension combination, a random uniform sample was drawn as a reference set. Thirty independent runs of each sampling strategy were implemented, and for each run, the Hausdorff distance [9] to the reference set was calculated. The Hausdorff distance as a measure of search space coverage was first proposed and investigated by Lang and Engelbrecht [15], and their methodology is followed in this study. Statistical significance tests were performed for each sampling strategy as proposed by Derrac et al. [3] and described by Lang and Engelbrecht [15].

Table 2 shows that the Hilbert curve and LHS have similar Hausdorff distances, implying a similar coverage of the search space, whereas the random walk provides the worst coverage. Lower values were achieved in most cases by the Hilbert curve sampler, confirmed by the statistical significance tests presented in Table 3. The null hypothesis is that there is no significant difference between the Hausdorff distances for each of the sampling strategies. Results show that the null hypothesis can be rejected, given the p-values and a significance level of 0.05. We therefore conclude that the Hilbert curve sampler achieves a more uniform search space coverage than LHS.

Table 2. Average Hausdorff distances between samplers and a uniform random sample for different sample sizes and dimensions d. Best and worst distances are highlighted in cyan and red, respectively.

D	Sample size		Hilbert curve	Latin hypercube	Random walk
5	$100 \times d$	500	**2.0642 (±0.0300)**	2.0866 (±0.0242)	3.9168 (±0.2967)
5	$316 \times d$	1580	**1.6065 (±0.0134)**	1.6218 (±0.0121)	3.9127 (±0.2731)
5	$1000 \times d$	5000	1.2754 (±0.0097)	1.2690 (±0.0044)	3.8658 (±0.2952)
10	$100 \times d$	1000	**4.9716 (±0.0295)**	5.0651 (±0.0300)	7.6117 (±0.3323)
10	$316 \times d$	3160	**4.3345 (±0.0108)**	4.4297 (±0.0143)	7.5796 (±0.3217)
10	$1000 \times d$	10000	**3.8045 (±0.0064)**	3.8781 (±0.0055)	7.5310 (±0.3035)
20	$100 \times d$	2000	**9.6778 (±0.0265)**	9.9578 (±0.0254)	12.5364 (±0.2787)
20	$316 \times d$	6320	**8.9312 (±0.0121)**	9.2413 (±0.0087)	12.5213 (±0.2007)
20	$1000 \times d$	20000	**8.2919 (±0.0082)**	8.5967 (±0.0051)	12.5013 (±0.2640)
30	$100 \times d$	3000	**13.5314 (±0.0263)**	13.8606 (±0.0205)	16.3195 (±0.2825)
30	$316 \times d$	9480	**12.7478 (±0.0177)**	13.1485 (±0.0065)	16.2629 (±0.2186)
30	$1000 \times d$	30000	**12.0615 (±0.0123)**	12.4913 (±0.0057)	16.3055 (±0.2677)

Table 3. Ranks achieved by each of the sampling strategies with the p-value of the significance tests. Best (lowest) values highlighted in bold.

Sampler	Friedman	Friedman-aligned	Quade
Hilbert curve	**1.09**	**289.23**	**1.04**
Latin Hypercube	1.91	431.78	1.96
Random walk	3.00	900.50	3.00
p-value	<1e-5	<1e-5	<1e-5

3.3 Computational Cost

We now compare the cost of the Hilbert curve as a sampler to alternative sampling strategies. Generating the Hilbert curve is not computationally cheap, but we expect that this upfront investment will be justified if the sample is subsequently used to calculate landscape metrics that require ordered samples.

To evaluate the computational cost, a performance counter[2] was used to keep track of the time to a) generate the sample and b) calculate information content metrics for both the Hilbert curve and LHS. Using the `pflacco` library, the information content (`ic`) metrics were calculated for each of the 24 BBOB functions defined as part of the COCO platform [7]. This was done for each dimension and sample size listed in Table 2. Note that for LHS to be used for information content metrics, the sample needs to be ordered. We evaluated both the random ordering and the nearest neighbour ordering strategies. Also note that the

---

[2] Python's `time.perf_counter`.

information content function provided by the `pflacco` library was modified to accept `none` as an ordering argument to use the pre-ordered Hilbert curve samples.

Figure 4(a) shows that the Hilbert curve sampler is slower than LHS to generate a sample of the same size. While the computational cost of the Hilbert curve sampler seems to grow at least quadratically, it remains acceptable even for large sample sizes (just over 1.2 s to generate a sample of 30 000 points). Figure 4(b) shows that for the information content metrics, Hilbert curve sampling strategy is substantially faster (especially at larger sample sizes) than LHS with nearest neighbour ordering.

3.4 Predictive Performance of Hilbert Curve Samples

We have shown that Hilbert curve sampling provides an efficient method for generating spatially correlated sequences of solutions from continuous search spaces, and that these sequences provide good coverage of the search space. We now investigate whether the landscape features extracted from these samples provide good predictive performance for algorithm selection.

Following the approach by Muñoz et al. [25], we use the task of predicting the class of each BBOB function as a proxy for algorithm selection: if the landscape features of function instances can be used to discriminate between problem classes, then they should also be effective predictor variables for algorithm selection. Five target classes corresponding to the BBOB function groupings as outlined in [8] were used, namely (1) separable functions: $\{f_1 \dots f_5\}$, (2) functions with low or moderate conditioning: $\{f_6 \dots f_9\}$, (3) unimodal functions and functions with high conditioning: $\{f_{10} \dots f_{14}\}$, (4) multimodal functions with

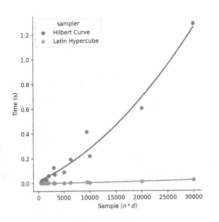

(a) Time required to generate sample only

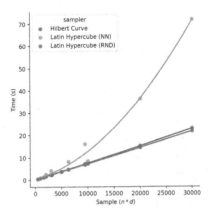

(b) Time to generate sample and calculate information content metrics

Fig. 4. Comparison of time (in seconds) to generate samples and calculate information content metrics for Hilbert curve sampling and LHS using nearest neighbour and random ordering. Trendlines are polynomials of order 2.

adequate global structure: $\{f_{15} \ldots f_{19}\}$, (5) functions with low or moderate conditioning: $\{f_{20} \ldots f_{24}\}$.

The experimental setup is as follows: 24 BBOB functions in dimensions $d \in \{2, 5, 10, 20\}$ are used, with an evaluation budget of $1\,000 \times d$. Each BBOB function has 15 predefined instances, and three random instances per function are held out as testing data[3]. For each combination of instance and dimension, $1\,000d$ samples are drawn from $[-5, 5]^d$, using three sampling strategies: Hilbert curve (HC), Latin hypercube (LH) and random walk (RW). Four feature sets from the pflacco package were selected that did not require further function evaluations, namely: dispersion (disp), ELA y-distribution (ela_distr), ELA meta model (ela_meta), and information content (ic).

Table 4 gives the testing accuracy of a decision tree, k-nearest neighbour, and random forest classifier for the task of predicting the function class from the landscape metrics. The scikit-learn version 1.1.3 default settings were used in all instances. Results show that a random forest model was the most effective at predicting the function class across all sampling strategies. Overall, the RW strategy emerges as the least competitive performer, while the Hilbert curve and Latin hypercube sampling strategies were comparable for all three classifiers.

Table 5 gives the performance for each of the three sampling strategies based on the four feature sets separately. An observable trend is that both the ela_meta and ic features are very salient features and allow the classifiers to accurately discriminate between the function classes, confirming results of Renau et al. [29].

4 Hilbert Curves as an Ordering Tool

In this section, we investigate the use of Hilbert curves as an ordering aid for samples generated from other sampling methodologies. A Hilbert curve of order p maps each point on a 1-dimensional curve $[0, \ldots, 2^{dp}]$ to a point in $[0, \ldots, 2^p - 1]^d$ and vice versa, as illustrated earlier in Fig. 1. The mapping is bijective, and allows for an efficient mapping from a point in the d-dimensional space to a point on the 1-dimensional Hilbert curve.

This d-D to 1-D mapping can be exploited, given that a Latin hypercube sample of n points divides each axis into n intervals. By selecting the appropriate curve order p, i.e. $p = \lceil log_2(n + 1) \rceil$, we can ensure that the resulting Hilbert curve is of sufficient length to accurately map all LHS points to points on the Hilbert curve. The ordering of the points on the Hilbert curve can then be used to provide a spatially correlated ordering of the Latin hypercube sample.

---

[3] This leave-one-instance out approach is used instead of the leave-one-problem out approach due to the small number of classes.

Table 4. Accuracy at predicting the function class using four feature sets based on samples from the three different sampling strategies

	Decision Tree	k-Nearest Neighbour	Random Forest
HC	93.50%($\pm$5.62%)	70.15%($\pm$7.41%)	97.38%($\pm$1.64%)
LH	93.87%($\pm$5.01%)	72.31%($\pm$7.34%)	97.64%($\pm$2.36%)
RW	88.78%($\pm$7.06%)	63.97%($\pm$6.87%)	93.60%($\pm$3.93%)

Table 5. Testing accuracy for predicting the function class for different classifiers and different sampling strategies: Hilbert curve (HC), Latin hypercube (LH), random walk (RW), under different dimensions (D).

Features	D	Decision tree			k-Nearest Neighbour			Random Forest		
		HC	LH	RW	HC	LH	RW	HC	LH	RW
disp	2	66.67%	52.78%	36.11%	70.83%	72.22%	34.72%	72.22%	69.44%	40.28%
	5	61.11%	55.56%	37.50%	61.11%	56.94%	36.11%	54.17%	69.44%	25.00%
	10	55.56%	55.56%	25.00%	52.78%	65.28%	20.83%	66.67%	63.89%	26.39%
	20	58.33%	65.28%	34.72%	65.28%	63.89%	31.94%	63.89%	66.67%	22.22%
ela_distr	2	58.33%	48.61%	34.72%	52.78%	55.56%	33.33%	51.39%	51.39%	31.94%
	5	66.67%	66.67%	33.33%	58.33%	62.50%	30.56%	63.89%	68.06%	40.28%
	10	73.61%	79.17%	40.28%	70.83%	68.06%	45.83%	77.78%	75.00%	36.11%
	20	79.17%	75.00%	41.67%	69.44%	63.89%	43.06%	76.39%	76.39%	43.06%
ela_meta	2	79.17%	79.17%	59.72%	62.50%	66.67%	48.61%	86.11%	81.94%	79.17%
	5	83.33%	86.11%	83.33%	75.00%	79.17%	59.72%	87.50%	93.06%	84.72%
	10	95.83%	90.28%	86.11%	76.39%	87.50%	52.78%	91.67%	94.44%	88.89%
	20	91.67%	88.89%	83.33%	81.94%	87.50%	50.00%	94.44%	91.67%	87.50%
ic	2	90.28%	94.44%	81.94%	90.28%	88.89%	84.72%	90.28%	87.50%	87.50%
	5	90.28%	90.28%	84.72%	84.72%	91.67%	86.11%	93.06%	93.06%	91.67%
	10	87.50%	91.67%	91.67%	86.11%	93.06%	86.11%	91.67%	95.83%	94.44%
	20	88.89%	84.72%	94.44%	73.61%	91.67%	88.89%	86.11%	93.06%	94.44%

Given that the current best practice of using a greedy nearest neighbour strategy to order LHS requires calculating at least half of the pairwise distance matrix between all points in the sample, the Hilbert curve ordering strategy has the potential to be significantly faster. Furthermore, the nearest neighbour strategy tends to start with relatively short step sizes, which increase as the number of unvisited points decrease. Hilbert curve ordering is likely to provide more consistent step sizes.

4.1 Step Size Consistency

Figure 5 shows a visual comparison of the Hilbert curve and nearest neighbour ordering strategies on a $2D$ Sphere function sample. The underlying sample X for both strategies is identical, and was obtained using LHS. In the figures the orderings produced by the Hilbert curve (HC) and the nearest neighbour (NN) approach are depicted as a red line. Colour scale is used to indicate the fitness values of the sampled points. It is evident from Fig. 5 that the maximum step size is smaller for HC than for NN. The step size distributions shown in Figs. 5(c) and 5(d) confirm that the maximum step size of HC is lower, and show a greater skew in the step size distribution for NN.

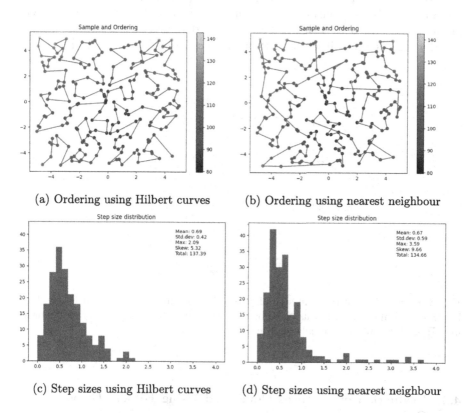

(a) Ordering using Hilbert curves   (b) Ordering using nearest neighbour

(c) Step sizes using Hilbert curves   (d) Step sizes using nearest neighbour

Fig. 5. Visualisation of sample ordering using Hilbert curve (left) and nearest neighbour (right) ordering strategies.

4.2 Computational Cost

To evaluate the computational cost of the Hilbert curve ordering, we compare it to the nearest neighbour and random (RND) ordering strategies applied to a sample generated using LHS. The Latin hypercube samples of sizes $n \in \{100d, 316d, 1000d\}$ were drawn from search space $[-5, 5]^d$ for dimensions $d \in \{5, 10, 20, 30\}$. For each configuration, 24 independent samples were drawn, and timed with a performance counter (Python's `time.perf_counter`).

Figure 6 shows the performance comparison between the various ordering strategies. Hilbert curve sampling as discussed in Sect. 3 is included for completeness, as it does not require additional ordering. From Fig. 6, NN ordering performs the worst, especially for large sample sizes. Furthermore, RND sorting and Hilbert curve *sampling* are equally fast (red and blue curves overlap). Hilbert curve *ordering* produces an intermediate result, slightly slower than random ordering, but significantly faster than the NN ordering. Since the underlying sample is generated by LHS, Hilbert curve ordering retains all the benefits of LHS for ELA features insensitive to order (such as ela_meta, or disp).

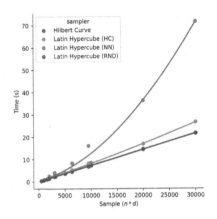

Fig. 6. Comparison of time (in seconds) to order the samples generated by the Latin hypercube sampling strategy using Hilbert curve, nearest neighbour, and random ordering. Trendlines are polynomials of order 2.

4.3 Evaluation of Features Generated Using Hilbert Curve Ordering

We now compare the information content features generated from a Latin hypercube sample using the HC, NN, and RND ordering. Since the feature values are evaluated rather than the computational cost, this experiment was restricted to dimensions $d \in \{2, 5\}$ and sample sizes $n \in \{100d, 1000d\}$. A single sample set X was generated and evaluated on each of the BBOB functions to generate corresponding $y = f(X)$ results. Information content metrics were then calculated on these (X, y) pairs using each of the ordering strategies. This procedure was repeated 30 times.

Figures 7 and 8 present initial partial information (M_0) and maximum information content (H_{max}) landscape features based on the entropy of the sample. We consider BBOB problems in $5D$ (problem indices listed on the x-axis), and calculate feature values based on NN, RND, and HC ordering for $n = 1000d$.

It is evident from Fig. 7 that the HC ordering produces values that lie between RND and NN, but follow a pattern similar to that of NN. It is clear that M_0^{RND} converges to the same value regardless of the underlying function (confirming the results of [25]). M_0^{NN} and M_0^{HC}, on the other hand, produce values that allow for discrimination between functions. Since larger values for M_0 indicate a more rugged landscape (more changes in concavity over the length of the walk), the HC ordering would on average indicate a more rugged landscape than the NN ordering.

According to Fig. 8, H_{max}^{RND} varies over a wider range than M_0^{RND}. However, the H_{max}^{RND} values still tend to cluster around the upper edge of the graph, closer to the maximum value of 1. In comparison, H_{max}^{NN} and H_{max}^{HC} vary over a larger part of the $[0, 1]$ range. Larger values for H_{max} indicate a larger variety of objects in the landscape (pits, peaks, plateaus). Thus, HC ordering again indicates a more rugged landscape than the NN ordering.

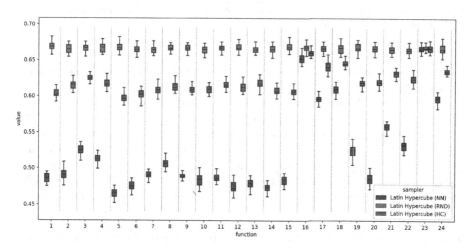

Fig. 7. M_0 in $5D$ as calculated using nearest neighbour (NN), random (RND) and Hilbert curve (HC) ordering methodologies ($n = 1000d$).

To evaluate the relative saliency of the features produced by the various ordering strategies, permutation-based feature importance technique [4] is employed. We train a random forest (RF) classifier using information content features as input ($\{\epsilon_s, \epsilon_{max}, \epsilon_{0.5}, H_{max}$ and $M_0\}$, as defined in [25]), and the function class as output (as defined in [25] and used in Sect. 3). Feature saliency is gauged by randomly permuting each input feature in turn across the dataset, and observing the corresponding reduction in RF accuracy. For this purpose, the features are grouped by the ordering strategy, and a RF is trained on $\frac{2}{3}$ of the data, with the remaining $\frac{1}{3}$ held out for testing.

After training, the base accuracy of the RF classifier is evaluated on the held out data. The accuracy of the RF classifier is then re-evaluated as each feature in the held-out dataset is randomly permuted in turn, and the difference to the

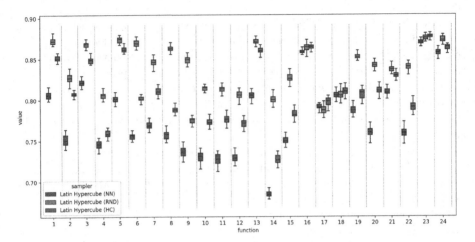

Fig. 8. H_{max} in $5D$ as calculated using nearest neighbour (NN), random (RND) and Hilbert curve (HC) ordering methodologies ($n = 1000d$).

base accuracy is recorded. This procedure is repeated 10 times for each sampling strategy. The results are shown in Table 6.

Table 6 shows that ϵ_s was the most salient feature across all sample orderings, with the RF trained on the HC ordered features relying on ϵ_s more heavily than the RF trained on the NN or RND ordered features. RND sample ordering made the M_0 feature near useless for the classifier, in line with the observations made in Fig. 7. Conversely, RF trained on the NN ordering relied on M_0 more than RF trained on the HC ordered features. Overall, HC ordering yielded marginally better RF performance compared to the NN and RND ordering approaches.

Table 6. Impact of permuting information content features on the accuracy of a random forest (RF) classifier. Classification target was the function groupings.

Sampler	LHS (HC)	LHS (NN)	LHS (RND)
RF base accuracy	98.44% ($\pm$0.39%)	98.25% ($\pm$0.37%)	97.44% ($\pm$0.54%)
ϵ_{max} (ic.eps_max)	-8.17% ($\pm$2.22%)	-8.46% ($\pm$1.98%)	-14.36% ($\pm$2.76%)
$\epsilon_{0.5}$ (ic.eps_ratio)	-20.62% ($\pm$3.28%)	-10.45% ($\pm$1.64%)	-27.82% ($\pm$3.23%)
ϵ_s (ic.eps_s)	-52.40% ($\pm$2.47%)	-47.34% ($\pm$1.69%)	-39.27% ($\pm$2.56%)
H_{max} (ic.h_max)	-8.56% ($\pm$0.87%)	-6.14% ($\pm$0.65%)	-15.08% ($\pm$1.76%)
M_0 (ic.m0)	-12.02% ($\pm$1.02%)	-20.98% ($\pm$1.60%)	-0.09% ($\pm$0.52%)

5   Conclusion

This paper proposed the use of Hilbert space-filling curves in the context of opti-misation problem landscape analysis for the purpose of (1) sampling the search space in a spatially correlated manner that also guarantees uniform coverage, and (2) spatially ordering samples generated using other sampling algorithms such as the Latin hypercube. Experiments were conducted to evaluate the rel-ative computational efficiency of the Hilbert curves, as well as the saliency of the landscape features extracted using Hilbert curve sampling and Hilbert curve ordering. In the context of sampling, Hilbert curves were significantly faster than Latin hypercube sampling for the purpose of generating order-sensitive features such as the information content metrics. Features extracted by the Hilbert curves were informative, and allowed for successful discrimination between problem classes. As an ordering tool, Hilbert curves performed significantly faster than the commonly used nearest neighbour ordering, and also yielded salient land-scape features. Thus, Hilbert curves present a viable computationally efficient alternative to both Latin hypercube sampling, and nearest neighbour ordering of a sample for the purpose of landscape analysis.

References

1. Abel, D.J., Mark, D.M.: A comparative analysis of some two-dimensional order-ings. Int. J. Geogr. Inf. Syst. **4**(1), 21–31 (1990). https://doi.org/10.1080/02693799008941526
2. Beham, A., Wagner, S., Affenzeller, M.: Algorithm selection on generalized quadratic assignment problem landscapes. In: Proceedings of the Genetic and Evolutionary Computation Conference, GECCO 2018, pp. 253–260. Association for Computing Machinery, New York (2018). https://doi.org/10.1145/3205455.3205585
3. Derrac, J., García, S., Molina, D., Herrera, F.: A practical tutorial on the use of nonparametric statistical tests as a methodology for comparing evolutionary and swarm intelligence algorithms. Swarm Evol. Comput. **1**(1), 3–18 (2011). https://doi.org/10.1016/j.swevo.2011.02.002
4. Dwivedi, R., et al.: Explainable ai (xai): core ideas, techniques, and solutions. ACM Comput. Surv. **55**(9) (2023). https://doi.org/10.1145/3561048
5. Falconer, K.: Fractal geometry: mathematical foundations and applications. John Wiley & Sons (2004)
6. Faloutsos, C., Roseman, S.: Fractals for secondary key retrieval. In: Proceed-ings of the Eighth ACM SIGACT-SIGMOD-SIGART Symposium on Principles of Database Systems, PODS 1989. pp. 247–252. Association for Computing Machin-ery, New York (1989). https://doi.org/10.1145/73721.73746
7. Hansen, N., Auger, A., Ros, R., Mersmann, O., Tušar, T., Brockhoff, D.: COCO: A platform for comparing continuous optimizers in a black-box setting. Optimiz. Methods Softw. **36**(1), 114–144 (2020). https://doi.org/10.1080/10556788.2020.1808977
8. Hansen, N., Finck, S., Ros, R., Auger, A.: Real-Parameter Black-Box Optimization Benchmarking 2009: Noiseless Functions Definitions. Research Report RR-6829, INRIA (2009). https://inria.hal.science/inria-00362633

9. Heinonen, J.: Lectures on Analysis on Metric Spaces. Springer Science & Business Media (2001). https://doi.org/10.1007/978-1-4613-0131-8
10. Hilbert, D.: Über die stetige abbildung einer linie auf ein flächenstück. In: Dritter Band: analysis· Grundlagen der Mathematik· Physik Verschiedenes, pp. 1–2. Springer (1935)
11. Jones, T., Forrest, S.: Fitness distance correlation as a measure of problem difficulty for genetic algorithms. In: Proceedings of the Sixth International Conference on Genetic Algorithms, pp. 184–192 (1995)
12. Kerschke, P., Trautmann, H.: Comprehensive feature-based landscape analysis of continuous and constrained optimization problems using the r-package flacco. In: Bauer, N., Ickstadt, K., Lübke, K., Szepannek, G., Trautmann, H., Vichi, M. (eds.) Applications in Statistical Computing. SCDAKO, pp. 93–123. Springer, Cham (2019). https://doi.org/10.1007/978-3-030-25147-5_7
13. Kostovska, A., Jankovic, A., Vermetten, D., Džeroski, S., Eftimov, T., Doerr, C.: Comparing algorithm selection approaches on black-box optimization problems. In: Proceedings of the Companion Conference on Genetic and Evolutionary Computation. ACM (Jul 2023). https://doi.org/10.1145/3583133.3590697
14. Kuk, J., Goncalves, R., Pozo, A.: Combining fitness landscape analysis and adaptive operator selection in multi and many-objective optimization. In: 2019 8th Brazilian Conference on Intelligent Systems (BRACIS). IEEE (Oct 2019). https://doi.org/10.1109/bracis.2019.00094
15. Lang, R.D., Engelbrecht, A.P.: Decision space coverage of random walks. In: 2020 IEEE Congress on Evolutionary Computation (CEC), pp. 1–8. IEEE (2020), https://doi.org/10.1109/CEC48606.2020.9185623
16. Liefooghe, A., Malan, K.M.: Adaptive landscape-aware constraint handling with application to binary knapsack problem. In: Proceedings of the Companion Conference on Genetic and Evolutionary Computation. ACM (Jul 2023). https://doi.org/10.1145/3583133.3596405
17. Malan, K.M.: Landscape-aware constraint handling applied to differential evolution. In: Fagan, D., Martín-Vide, C., O'Neill, M., Vega-Rodríguez, M.A. (eds.) TPNC 2018. LNCS, vol. 11324, pp. 176–187. Springer, Cham (2018). https://doi.org/10.1007/978-3-030-04070-3_14
18. Malan, K.M.: A survey of advances in landscape analysis for optimisation. Algorithms 14(2), 40 (2021). https://doi.org/10.3390/a14020040
19. Malan, K.M., Engelbrecht, A.P.: Quantifying ruggedness of continuous landscapes using entropy. In: 2009 IEEE Congress on Evolutionary Computation, pp. 1440–1447 (2009). https://doi.org/10.1109/CEC.2009.4983112
20. Malan, K.M., Engelbrecht, A.P.: Ruggedness, funnels and gradients in fitness landscapes and the effect on PSO performance. In: 2013 IEEE Congress on Evolutionary Computation. IEEE (Jun 2013). https://doi.org/10.1109/cec.2013.6557671
21. Malan, K.M., Engelbrecht, A.P.: A survey of techniques for characterising fitness landscapes and some possible ways forward. Inf. Sci. 241, 148–163 (2013). https://doi.org/10.1016/j.ins.2013.04.015
22. Malan, K.M., Engelbrecht, A.P.: A progressive random walk algorithm for sampling continuous fitness landscapes. In: 2014 IEEE Congress on Evolutionary Computation (CEC), pp. 2507–2514. IEEE (2014). https://doi.org/10.1109/CEC.2014.6900576
23. McKay, M.D., Beckman, R.J., Conover, W.J.: A comparison of three methods for selecting values of input variables in the analysis of output from a computer code. Technometrics 42(1), 55–61 (2000). https://doi.org/10.1080/00401706.2000.10485979

24. Mersmann, O., Bischl, B., Trautmann, H., Preuss, M., Weihs, C., Rudolph, G.: Exploratory landscape analysis. In: Proceedings of the 13th Annual Conference on Genetic and Evolutionary Computation, pp. 829–836 (2011). https://doi.org/10.1145/2001576.2001690

25. Muñoz, M.A., Kirley, M., Halgamuge, S.K.: Exploratory landscape analysis of continuous space optimization problems using information content. IEEE Trans. Evol. Comput. **19**(1), 74–87 (2015). https://doi.org/10.1109/TEVC.2014.2302006

26. Ochoa, G., Tomassini, M., Vérel, S., Darabos, C.: A Study of NK Landscapes' Basins and Local Optima Networks. In: Proceedings of Genetic and Evolutionary Computation Conference, pp. 555–562 (July 2008)

27. Pitzer, E., Affenzeller, M., Beham, A., Wagner, S.: Comprehensive and automatic fitness landscape analysis using HeuristicLab. In: Moreno-Díaz, R., Pichler, F., Quesada-Arencibia, A. (eds.) EUROCAST 2011. LNCS, vol. 6927, pp. 424–431. Springer, Heidelberg (2012). https://doi.org/10.1007/978-3-642-27549-4_54

28. Renau, Q., Doerr, C., Dreo, J., Doerr, B.: Exploratory landscape analysis is strongly sensitive to the sampling strategy. In: Bäck, T., et al. (eds.) PPSN 2020. LNCS, vol. 12270, pp. 139–153. Springer, Cham (2020). https://doi.org/10.1007/978-3-030-58115-2_10

29. Renau, Q., Dreo, J., Doerr, C., Doerr, B.: Expressiveness and robustness of landscape features. In: Proceedings of the Genetic and Evolutionary Computation Conference Companion, GECCO 2019, pp. 2048–2051. Association for Computing Machinery, New York (2019). https://doi.org/10.1145/3319619.3326913

30. Rivest, R.L.: Partial-match retrieval algorithms. SIAM J. Comput. **5**(1), 19–50 (1976). https://doi.org/10.1137/0205003

31. Sagan, H.: Space-filling curves. Springer Science & Business Media (2012). https://doi.org/10.1007/978-1-4612-0871-6

32. Sallam, K.M., Elsayed, S.M., Sarker, R.A., Essam, D.L.: Landscape-assisted multi-operator differential evolution for solving constrained optimization problems. Expert Syst. Appl. **162**, 113033 (2020). https://doi.org/10.1016/j.eswa.2019.113033

33. Skilling, J.: Programming the Hilbert curve. In: Bayesian Inference and Maximum Entropy Methods in Science and Engineering. American Institute of Physics Conference Series, vol. 707, pp. 381–387 (Apr 2004). https://doi.org/10.1063/1.1751381

34. van Aardt, W.A., Bosman, A.S., Malan, K.M.: Characterising neutrality in neural network error landscapes. In: 2017 IEEE Congress on Evolutionary Computation (CEC), pp. 1374–1381 (2017). https://doi.org/10.1109/CEC.2017.7969464

35. Vassilev, V.K., Fogarty, T.C., Miller, J.F.: Information characteristics and the structure of landscapes. Evol. Comput. **8**(1), 31–60 (2000). https://doi.org/10.1162/106365600568095

36. Vassilev, V.K., Fogarty, T.C., Miller, J.F.: Smoothness, ruggedness and neutrality of fitness landscapes: from theory to application. In: Advances in evolutionary computing, pp. 3–44. Springer (2003). https://doi.org/10.1007/978-3-642-18965-4_1

37. Verel, S., Liefooghe, A., Jourdan, L., Dhaenens, C.: On the structure of multiobjective combinatorial search space: MNK-landscapes with correlated objectives. Eur. J. Oper. Res. **227**(2), 331–342 (2013). https://doi.org/10.1016/j.ejor.2012.12.019

38. Weinberger, E.: Correlated and uncorrelated fitness landscapes and how to tell the difference. Biol. Cybern. **63**(5), 325–336 (1990)

Predicting Algorithm Performance in Constrained Multiobjective Optimization: A Tough Nut to Crack

Andrejaana Andova[1,2](✉) , Jordan N. Cork[1,2] , Aljoša Vodopija[1,2] ,
Tea Tušar[1,2] , and Bogdan Filipič[1,2]

[1] Jožef Stefan Institute, Ljubljana, Slovenia
{andrejaana.andova,jordan.cork,aljosa.vodopija,tea.tusar,
bogdan.filipic}@ijs.si
[2] Jožef Stefan International Postgraduate School, Ljubljana, Slovenia

Abstract. Predicting algorithm performance is crucial for selecting the best performing algorithm for a given optimization problem. While some research on this topic has been done for single-objective optimization, it is still largely unexplored for constrained multiobjective optimization. In this work, we study two methodologies as candidates for predicting algorithm performance on 2D constrained multiobjective optimization problems. The first one consists of using state-of-the-art exploratory landscape analysis (ELA) features, designed specifically for constrained multiobjective optimization, as input to classical machine learning methods, and applying the resulting models to predict the performance classes. As an alternative methodology, we analyze an end-to-end deep neural network trained to predict algorithm performance from a suitable problem representation, without relying on ELA features. The experimental results obtained on benchmark problems with three multiobjective optimizers show that neither of the two methodologies is capable of substantially outperforming a dummy classifier. This suggests that, with the current benchmark problems and ELA features, predicting algorithm performance in constrained multiobjective optimization remains a challenge.

Keywords: Constrained multiobjective optimization · Exploratory landscape analysis · Algorithm performance prediction · Empirical cumulative distribution function · Machine learning · Deep learning

1 Introduction

When attempting to solve an optimization problem, the choice of which optimization algorithm to use is crucial for obtaining satisfying results in a limited time. It is, therefore, necessary to develop a method that identifies which algorithm performs best on a particular optimization problem. The task of selecting a single algorithm that performs best for a given optimization problem is called the algorithm selection task.

© The Author(s), under exclusive license to Springer Nature Switzerland AG 2024
S. Smith et al. (Eds.): EvoApplications 2024, LNCS 14635, pp. 310–325, 2024.
https://doi.org/10.1007/978-3-031-56855-8_19

Solving an algorithm selection task requires a collection of algorithms from which to choose. It also requires a collection of diverse problems, which elicit different performance out of the algorithms. Constrained multiobjective problems (CMOPs) are both lesser in quantity and diversity and greater in complexity than unconstrained and/or single objective problems. Therefore, solving the constrained multiobjective algorithm selection task is an ambitious goal. As a first step towards solving it, we aim to develop a method for predicting algorithm performance on a given CMOP.

In recent years, many researchers have tried to predict algorithm performance [21,28]. They generally do so by extracting exploratory landscape analysis (ELA) features from a population of solutions. These are then used as input to a machine learning classifier, which identifies the optimization algorithm that performs best on the given problem. Many ELA features have been proposed for single-objective optimization, and the package `flacco` [11] contains a broad collection of these. However, ELA features for more complex problems, like CMOPs, are still under development, with only a few related works [2,15,30]. This adds to the difficulty of predicting algorithm performance on these problems.

In a previous work [3], we tried to predict algorithm performance on CMOPs by using the state-of-the-art collection of CMOP ELA features proposed in [2]. These features were used as inputs into classical machine learning regression models. We attempted to predict algorithm performance on three benchmark suites, for 2D, 3D, and 5D CMOPs. The target of our prediction task was the area under an algorithm performance curve (explained in Sect. 2.3). However, the obtained results were not encouraging and, therefore, we are trying to improve upon them. In this work, we have increased the number of CMOPs used in the learning process, changed the prediction target and utilized an end-to-end deep neural network (DNN) methodology that does not use ELA features.

The paper is further organized as follows. In Sect. 2, we introduce the background of our study. In Sect. 3, we explain the applied methodology. In Sect. 4 we present the experimental setup and, in Sect. 5, the obtained results. Finally, in Sect. 6, we provide a conclusion and outline ideas for future work.

2 Background

In this section, we introduce constrained multiobjective optimization, explain ELA for this kind of optimization, present the recently proposed performance indicator specifically developed for CMOPs, and outline deep neural networks.

2.1 Constrained Multiobjective Optimization

A CMOP is formulated as:

$$
\begin{aligned}
\text{minimize} \quad & f_m(\mathbf{x}), \quad m = 1, \ldots, M, \\
\text{subject to} \quad & g_j(\mathbf{x}) \leq 0, \quad j = 1, \ldots, J, \\
& h_k(\mathbf{x}) = 0, \quad k = 1, \ldots, K,
\end{aligned}
\tag{1}
$$

where $\mathbf{x} = (x_1, \ldots, x_D)$ is a D dimensional *solution vector*, $f_m(\mathbf{x})$ are the *objective functions*, and $g_j(\mathbf{x})$ and $h_k(\mathbf{x})$ are the inequality and equality *constraint functions*, respectively. M is the number of objectives, and J and K are the number of inequality and equality constraints, respectively.

A solution $\mathbf{x}$ is *feasible*, if it satisfies all constraints, $g_j(\mathbf{x}) \leq 0$, for $j = 1, \ldots, J$ and $h_k(\mathbf{x}) = 0$, for $k = 1, \ldots, K$. A feasible solution $\mathbf{x}$ is said to *dominate* another feasible solution $\mathbf{y}$ if $f_m(\mathbf{x}) \leq f_m(\mathbf{y})$ for all $1 \leq m \leq M$, and $f_m(\mathbf{x}) < f_m(\mathbf{y})$ for at least one $1 \leq m \leq M$. A feasible solution $\mathbf{x}^*$ is a *Pareto-optimal solution* if there exists no feasible solution $\mathbf{x} \in S$ that dominates $\mathbf{x}^*$. All feasible solutions constitute the *feasible region* F. All nondominated feasible solutions form the *Pareto set* S_o, and the image of the Pareto set in the objective space is the *Pareto front*, $P_o = \{f(\mathbf{x}) \mid \mathbf{x} \in S_o\}$.

2.2 Exploratory Landscape Analysis for Constrained Multiobjective Optimization

ELA is a methodology whereby features of an optimization problem are extracted from a sample of solutions [19]. These features are generally expertly designed statistical relations between solutions. While many ELA feature sets have been designed for single-objective optimization problems, only a few exist for CMOPs.

For CMOPs, state-of-the-art features were collected by Alsouly et al. [2]. They proposed additional features on top of the fast-computing features for CMOPs from the related work. The combined set of features is divided into three groups that describe: the multiobjective landscape, the violation landscape, and a combination of the two – the multiobjective violation landscape.

Features describing the objectives and their internal relations belong to the *multiobjective landscape group*. Global features in this group include the proportion of unconstrained Pareto optimal solutions, the hypervolume of the unconstrained Pareto front, and the correlation between the objective values, among others. Statistics on the distance between random walk neighbors in the objective space make up the random walk features.

Features describing the problem constraints belong to the *violation landscape group*. Global features in this group are devoted to global constraint violation statistics, while the random walk features consist of constraint violation statistics between random walk neighbors.

Features describing the relations between the objective and the constraints belong to the *multiobjective violation landscape group*. Global features include the proportion of feasible solutions, the proportion of Pareto optimal solutions, the hypervolume, statistics on the correlations between objectives and constraints, and others. Statistics on the dominance relations between random walk neighbors make up the random walk features.

2.3 Empirical Cumulative Distribution Functions

In constrained multiobjective optimization, there is a drawback to using the hypervolume of feasible solutions as the quality indicator, because it does not record algorithm performance until feasible solutions are reached. However, recently, [29] introduced a new quality indicator for constrained multiobjective optimization, I_{CMOP}, to address the gap in this area. The new indicator generalizes the hypervolume-based quality indicator I_{HV+} from [10]. Notably, both I_{HV+} and I_{CMOP} assume that low quality indicator values indicate better sets of solutions and vice versa. I_{CMOP} can be defined as follows:

1. When all solutions in the set are infeasible, the I_{CMOP} quality indicator takes on the smallest constraint violation of all solutions in the set, plus a threshold τ^*.
2. When the set contains at least one feasible solution, the quality indicator equals the value of I_{HV+} bounded above by the threshold τ^*, i.e., it equals $\min\{I_{HV+}, \tau^*\}$.

The threshold value τ^* ensures that an infeasible solution will always be deemed worse than a feasible one.

Also, to be able to compare different CMOPs, one first needs to normalize the I_{HV+} value and the constraint violation value, based on a sample of 100 solutions. The details of how this is done can be found in [29].

For algorithm performance measurement during the algorithm run, we track the number of function evaluations (*runtimes*) needed to reach a particular quality indicator value (*target*). This is carried out for a set of targets and the runtimes are visualized using the Empirical Cumulative Distribution Function (ECDF) [10]. The ECDF shows the proportion of targets achieved by the algorithm at a certain runtime and increases as the algorithm achieves further targets. The maximum value achievable by an algorithm is 1, meaning it reached all targets. One way to express algorithm performance in a single number is by computing the area under the curve of the ECDF – larger values correspond to a better/faster algorithm performance.

2.4   Deep Neural Networks

Deep Neural Networks (DNNs) are one of the most widely used prediction models at the moment. For more details on how they work, refer to [20]. Here, we briefly introduce the three DNN architectures used in our work. They are as follows:

– A *feedforward neural network* (FNN) is a standard deep neural network, consisting of layers whose neurons are fully connected to the neurons from the neighboring layers.
– *Convolutional neural networks* are DNNs consisting of convolutional layers followed by activation layers and, sometimes, pooling layers. They are most often used in computer vision, as they are good at describing the local properties of the images, using filters that can be of different sizes.
– An *autoencoder* is a DNN architecture that consists of an encoder and a decoder part. These parts are usually symmetrical, therefore, the input and the output of an autoencoder neural network have the same shape. The goal of this neural network architecture is to compress the data. Thus, the encoder

compresses the data, and the decoder decompresses it. Essentially, the autoencoder can also be an FNN or a convolutional neural network as long as it performs data compression.

3   Methodology

In this section, we present the methods applied in this study. First, we explain how the ECDF of the I_{CMOP} indicator was used to define three different classification tasks. We then describe various methods for solving these tasks – machine learning methods that predict algorithm performance based on ELA features and the newly proposed end-to-end DNN, which circumvents the ELA features by using the problem landscape samples directly.

3.1   Classification Tasks

The ECDF of the performance indicator, described in Sect. 2.3, shows the number of targets achieved at each evaluation step. As explained in [29], to compare targets between different CMOPs, we normalize the targets using a sample of 100 solutions, and we set $\tau^* = 1$. Also, the authors state that a good set of target precision values corresponds to $\tau^\epsilon = \tau^{\text{ref}} + \epsilon$, where $\epsilon \in \{10^p | p \in \{-5, -4.9, \ldots, 0\}\} \cup \{1 + 10^p | p \in \{-5, -4.9, \ldots, 0\}\}$, and τ^{ref} is the hypervolume of the true Pareto front, or an approximation of it. We used the same for our target precision values.

In a previous work [3], we were predicting algorithm performance using the area under the curve of the ECDF. This turned out to be a very difficult regression task. Therefore, to alleviate it, this work makes two changes to the methodology: (1) instead of the area under the curve, we predict the number of evaluations needed to reach three chosen target proportions, and (2), we predict ranges of values instead of exact numbers, transforming a regression task into a classification one.

More specifically, the target proportions of interest are:

- The number of evaluations needed until a feasible solution is obtained, which due to the choice of targets, corresponds to satisfying 50% of the targets.
- The number of evaluations needed to satisfy 70% of the targets.
- The number of evaluations needed to satisfy 90% of the targets.

Predicting the exact number of evaluations needed to satisfy a given percentage of targets, is difficult. Additional challenges arise from the fact that an algorithm may never reach the most difficult targets on some of the problems, which then requires special handling of such cases. Because of this, we group the number of evaluations into classes and treat their prediction as a classification task, which is expected to be easier to solve.

The number of evaluations of interest depends on the experimental setup. In our case, we will be performing at most 24 000 evaluations and use algorithms with a population size of 200. Therefore, we form the following classes:

- Class 0: The goal is achieved between 1 and 200 evaluations (in the initial generation),
- Class 1: The goal is achieved between 201 and 2 000 evaluations,
- Class 2: The goal is achieved between 2 001 and 8 000 evaluations,
- Class 3: The goal is achieved between 8 001 and 24 000 evaluations,
- Class 4: The goal is never achieved.

3.2   Classical Machine Learning

For the machine learning part, we use the ELA features outlined in Sect. 2.2 as input to three classical machine learning algorithms – Decision Trees [16], Random Forest Classification [5], and C-Support Vector Classification (SVC) [23]. We also include a dummy model in the comparison, which predicts the most frequent class in the training data. We utilize the scikit-learn implementations of these methods with default parameter settings [22].

3.3   DNN

Inspired by developments in computer vision, we decided to test whether methodologies from that field could be used for algorithm performance prediction on CMOPs. Some experiments have already been done in single-objective optimization [24,26], but they did not show promising results compared to the results obtained by the well-developed ELA features for single-objective optimization.

For a proof of concept, we limit the dimensionality of the search and objective spaces to 2D. In this way, no additional manipulation, such as dimensionality reduction, is required. More specifically, in our approach, we treat the search space as an image, discretized into 32 × 32 pixels. Each pixel contains the red, green, and blue color components, representing the two objectives and the overall constraint violation, respectively.

Data Generation. To generate images of the search spaces, we use the following sampling technique. First, we divide the 2D search space into "pixels", by splitting each dimension of the search space into 32 equally sized intervals. Then, for each pixel, we randomly generate a solution within it, and use its objective values and the overall constraint violation value to assign the color to the pixel. A visual representation of a sample generated using this technique is shown in Fig. 1.

DNN Architecture. The architecture of the DNN is composed of a convolutional neural network autoencoder and an FNN. The encoder part of the autoencoder is used as input to the FNN, whose target is the prediction class defined in Sect. 3.1. The DNN architecture is shown in Fig. 2.

Each flat rectangle in the figure represents one layer of the DNN architecture. It contains information about the keras library [1] layer class that we used on

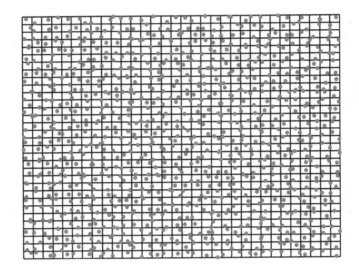

Fig. 1. An example sample of size 32 × 32 for the DNN.

the left side, and the shape of its input/output on the right side. For example, the first layer in the DNN is an InputLayer, and it takes as input images of size 32 × 32 with 3 channels.

The top part of the figure presents the encoder, which consists of three pairs of convolutional and max-pooling layers. The bottom part is divided into the decoder (on the left) and the FNN (on the right). The decoder is symmetrical to the encoder, whereas the FNN contains Dense and Dropout layers. The last layer in the FNN has an output of 5 neurons, each one assigned to one of the prediction classes presented in Sect. 3.1.

The idea behind this architecture is that, by providing the autoencoder with the same image as input and output, we force it to encode the input image so that the least amount of information is lost in the training process. The encoded part can be seen as landscape features that the autoencoder automatically extracts from the input data.

To cause the DNN to encode the properties that are useful for predicting algorithm performance, we use the encoded part as input to an FNN. Both parts of the DNN are trained simultaneously, with a combined loss function (mean absolute error for the decoder, and categorical cross-entropy for the FNN).

Data Preprocessing. The objectives and overall constraint violation have different value ranges across different problems. For this reason, as a preprocessing step, we normalize each of these functions. The min-max normalization procedure applies the normalization over all samples of a given problem, using the minimum and maximum value of the given objective. Furthermore, we normalize the constraints by assigning a 0 value to the feasible solutions, and a 1 value

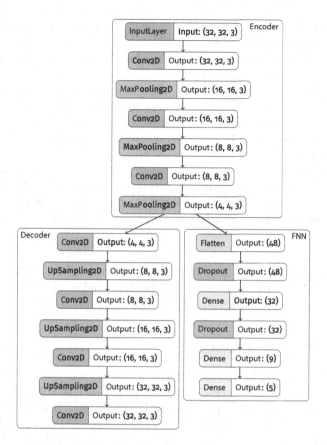

Fig. 2. The applied DNN architecture consisting of the encoder, decoder and FNN.

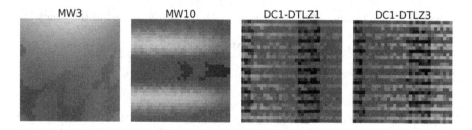

Fig. 3. Four example CMOP inputs, as images, for use with the DNN. Red represents the value of the first objective function, green the value of the second objective function, and blue the constraint violation. Prior to this encoding, the objective values are normalized using the minimum and maximum objective values of the problem samples. (Color figure online)

to the infeasible solutions. Example visualizations of several input images from different CMOPs are presented in Fig. 3.

Note that there are many ways to normalize the functions. For example, one other possibility to normalize the overall constraint violation is to use its the minimum and maximum values. However, after some preliminary experimentation with different normalization techniques, we found that the obtained algorithm performance prediction results were comparable. Thus, in this paper, we only present the results derived from the normalization techniques described in the paragraph above.

In constrained multiobjective optimization, the order of the objectives should not be important. Thus, we generate two images for each input sample – one where the first objective is assigned the red color and the second objective green, and another image where the ordering is reversed. The blue color always encodes the constraint violation.

DNN Settings. We used the ReLu activation function for each hidden layer in the DNN. We set the batch size to 1 000, the number of epochs to 100, and we used the Adam optimizer [13] with a learning rate of 0.0001.

4   Experimental Setup

Our work is focused on bi-objective CMOPs with 2D search spaces. We used six benchmark suites in the experiments: MW [17], C-DTLZ [12], CTP [7], DAS-CMOP [9], and DC-DTLZ [14], as well as three individual benchmark problems: BNH [4], TNK [27], and SRN [25]. The total number of CMOPs with two variables and two objectives from these suites is 36 (see Table 1 for a break-down over problem suites).

Table 1. The number of bi-objective 2D CMOPs per suite used in this study.

MW	C-DTLZ	CTP	DAS-CMOP	DC-DTLZ	BNH	TNK	SRN
8	5	8	6	6	1	1	1

For the purpose of predicting algorithm performance, three multiobjective optimization algorithms were tested, each with a different constraint handling technique. These algorithms were NSGA-III [12], MOEA/D-IEpsilon [8], and C-TAEA [14]. To handle the variation of the results due to the stochastic nature of the algorithms, 31 runs of each algorithm were conducted on each problem. With this approach, algorithm performance can be estimated more accurately. To extract the target classes, we used the mean of the ECDF values over all 31 algorithm runs. Additionally, we applied the same population size and number of generations to all algorithms, allowing for a fair comparison of results. The population size was set to 200, and the number of generations to 120. To generate reference vectors for NSGA-III and MOEA/D-IEpsilon, we used the Das-Denis approach [6]. The number of reference vectors was 200 for each algorithm.

The ELA features were calculated stochastically, whereby a different sample of solutions was selected each time the feature calculation is begun. This was dealt with by creating 100 samples using Latin hypercube sampling, which resulted in 100 sets of features (i.e., learning instances) for each problem. Similarly, we created 100 samples per problem for the DNN method using the sampling described in Sect. 3.3.

For easier reproducibility of the stochastic learning models, we report that the random number generator was seeded with the value of ten to obtain the results in the following section. Moreover, experiments with alternative seeds resulted in comparable results.

To evaluate the performance of each classifier, we used the leave-one-problem-out evaluation methodology. In this approach, no information about the target problem is available in the training data. Thus, all instances of a problem are used as test data, and the instances from the rest of the problems as training data. This process is repeated for each problem and the average mean absolute error is used as an evaluation metric.

5   Results

The classification accuracy for the desired target percentages for all learning methods is presented in Table 2. From the results we can see that none of the learning models drastically outperforms the dummy classifier. The only exception is the Random forest model. This performs better than the dummy classifier in most cases, except for MOEA/D and C-TAEA when predicting the evaluation class with at least 90% of the targets achieved.

To analyze more thoroughly the predictions by the Random forest model, we provide its confusion matrices for all three classification tasks in Fig. 4. In addition, Fig. 5 shows problem samples in the ELA feature space, reduced to 2D using the t-distributed stochastic neighbor embedding (t-SNE) method [18].

In Table 2, we can see that the classification accuracy is the same across all optimization algorithms when tackling the first classification task, that being to achieve 50% of the targets, i.e., to reach the border between the infeasible and feasible regions. An explanation for this can be derived from the confusion matrices in Fig. 4. These show that most of the optimization algorithms find a feasible solution in the initial population. They are, therefore, labeled with class 0. Otherwise, they achieve a feasible solution in at most 2 000 evaluations.

As shown in Fig. 5, with t-SNE dimension reduction, the instances from the same CMOP form clusters. This means the ELA features from the same problem do not provide the diversity required by the machine learning models. Consequently, during prediction, the learning models usually have either a 100% or 0% accuracy for a given CMOP. This is manifested in the nearly fully rounded results present in the confusion matrices in Fig. 4. A similar behavior can be observed for the DNN method, although this method does not rely on ELA features.

Table 2. Classification accuracy of the learning models predicting the algorithm performance classes.

Targets	Classifier	NSGA-III	MOEA/D	C-TAEA
50%	Dummy	0.916	0.916	0.916
	Decision tree	0.916	0.916	0.916
	Random Forest	0.944	0.944	0.944
	SVC	0.888	0.888	0.888
	DNN	0.916	0.916	0.916
70%	Dummy	0.416	0.416	0.583
	Decision tree	0.446	0.376	0.658
	Random Forest	0.576	0.549	0.674
	SVC	0.406	0.406	0.588
	DNN	0.381	0.304	0.583
90%	Dummy	0.638	0.666	0.722
	Decision tree	0.581	0.668	0.687
	Random Forest	0.677	0.638	0.703
	SVC	0.623	0.650	0.727
	DNN	0.638	0.666	0.722

The DNN, proposed as a novelty in this work, unfortunately never outperforms the dummy classifier and sometimes performs even worse than it, although the loss was observed to decrease during training. Worse performance is, for example, seen when predicting the number of evaluations needed by the NSGA-III and MOEA/D algorithms to achieve 70% of the targets. A reason for the poor DNN performance could be that we used only 35 CMOPs for training. Although we generated 100 samples for each problem, this may still not provide enough diversity and the DNN is not able to learn the patterns of the search space. Namely, it is known that DNN's need huge amounts of data to learn adequately. The classical machine learning models, on the other hand, are designed to be able to handle small amounts of data, but, as stated before, their performance was not found to be promising either.

A reason for poor performance of the classical machine learning models on CMOPs could be that, just like the DNN, they also need more data (although probably less so than the DNN). The small number of CMOPs used for training is certainly a difficulty, but the similarity of some properties across different CMOPs is also a potential reason for the low prediction performance and (likely)

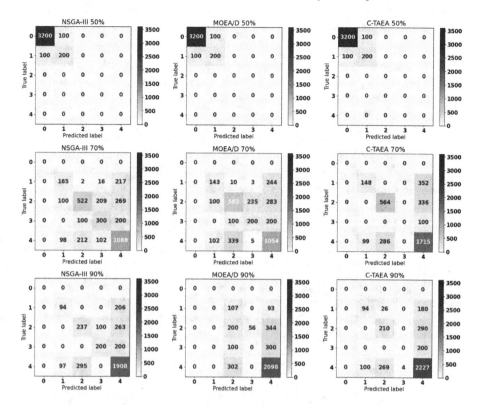

Fig. 4. Confusion matrices of the random forest models for the three desired target percentages. Each confusion matrix refers to the algorithm performance classes explained in Sect. 3.1 in more detail.

overfitting of the data. For example, as shown in Fig. 3, DC1-DTLZ1 and DC1-DTLZ3 have very similar landscapes, and, given that the order of objectives in CMOPs is insignificant, the red and the green sectors may be swapped.

Another reason for the poor performance of feature-based performance prediction might be the recency of research into ELA features for CMOPs. Possibly not all informative characteristics of CMOPs are included in the feature set, as of yet.

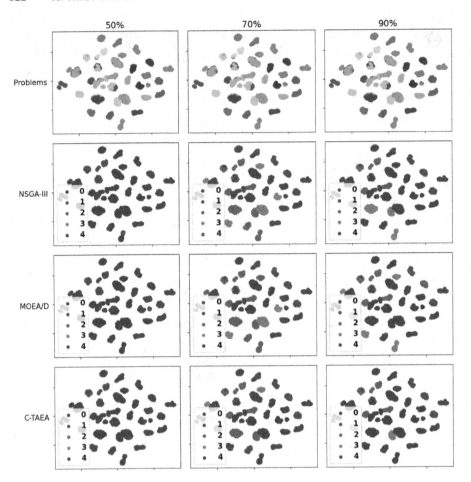

Fig. 5. Visualizations of ELA features for the three desired target percentages, reduced in dimensionality using the t-SNE method. The colors in the first row of the plots represent the problems included in the experiment. In the remaining rows, the colors identify the classes representing the number of evaluations needed to achieve a percentage of targets. (Color figure online)

6   Conclusion

In this work, we tried to improve upon our previous attempt at algorithm performance prediction for three widely used multiobjective optimization algorithms, NSGA-III, MOEA/D-IEpsilon, and C-TAEA, on 2D, 3D, and 5D CMOPs. Previously, we worked on predicting the area under the curve of the ECDF for the I_{CMOP} quality indicator proposed in [29]. We used classical machine learning regression models, whose inputs were the ELA features proposed in [2]. Unfortunately, the obtained results were not encouraging. Consequently, in this work, we focused on 2D CMOPs. We used a total of nine benchmark suites and problems, which resulted in 36 CMOPs. This is significantly larger than in the previous work where only 13 were used. Furthermore, we changed the prediction task – in this work, we were predicting the number of evaluations needed to achieve 50%, 70%, and 90% of the ECDF targets. Moreover, because predicting the number of evaluations is a hard task, we discretized the number of evaluations needed into five classes.

The results from the previous work left questions as to whether the prediction performance was poor because of the small number of CMOPs used for training, or the underdevelopment of the CMOP ELA feature set. To eliminate the second issue, we proposed an end-to-end DNN, that does not include ELA features to predict algorithm performance. As far as we are aware, this is the first time an end-to-end DNN has been used to predict algorithm performance on CMOPs.

Unfortunately, the newly proposed method did not outperform the dummy prediction model. Nonetheless, the reason for this might be that using merely 36 CMOPs is not enough for training a DNN. Thus, this left us with the dilemma of poor algorithm performance prediction – are more CMOPs required to predict algorithm performance, or better ELA features? Moreover, the tested evolutionary algorithms performed comparably on the benchmark problems. This calls for involving a larger set of algorithms that would potentially show different performance.

In the future, we plan to extend our research on end-to-end DNNs for algorithm performance prediction by applying publicly available pretrained DNNs. The idea is to enhance the performance of the proposed architecture. This is a standard practice in deep learning when dealing with small datasets and thus, although none of the pretrained models was trained on problem landscapes, their learned patterns might still help with our prediction task.

Another way forward is to utilize a larger CMOP benchmark suite. This can be constructed by combining the objectives and constraints of constrained single-objective problems from various benchmark suites. This way, we could include a much larger number of CMOPs in the data, possibly helping both the classical machine learning methods and the deep learning methods better predict algorithm performance. A drawback to this approach is that running the algorithms 31 times for each problem combination would be a time-consuming task. It is possible, however, that already a small proportion of the problem combinations would contribute diversity to the extended benchmark suite.

Ideally, in future work, both the ideas stated above would be combined. Tests on an extended CMOP benchmark suite are sure to answer whether more CMOPs are needed to better predict algorithm performance, while the inclusion of knowledge from pretrained DNNs is likely to provide insights into the possibility of improving the current ELA features for CMOPs.

Acknowledgements. The authors acknowledge the financial support from the Slovenian Research and Innovation Agency (young researcher program, research core funding No. P2-0209, and project No. N2-0254 "Constrained Multiobjective Optimization Based on Problem Landscape Analysis"). The publication is also based upon work from COST Action CA22137 "Randomised Optimisation Algorithms Research

Network" (ROAR-NET), supported by European Cooperation in Science and Technology (COST).

References

1. Keras. https://github.com/fchollet/keras (Accessed 27 September 2023)
2. Alsouly, H., Kirley, M., Muñoz, M.A.: An instance space analysis of constrained multi-objective optimization problems. IEEE Trans. Evol. Comput. **27**(5), 1427–1439 (2023). https://doi.org/10.1109/TEVC.2022.3208595
3. Andova, A., Vodopija, A., Cork, J., Tušar, T., Filipič, B.: An attempt at predicting algorithm performance on constrained multiobjective optimization problems. In: Slovenian Conference on Artificial Intelligence: Proceedings of the 26th International Multiconference Information Society, IS 2023 (2023)
4. Binh, T.T., Korn, U.: Mobes: a multiobjective evolution strategy for constrained optimization problems. In: Proceedings of the 3rd International Mendel Conference on Genetic Algorithms, MENDEL 1997, pp. 176–182 (1997)
5. Breiman, L.: Random forests. Mach. Learn. **45**, 5–32 (2001). https://doi.org/10.1023/A:1010933404324
6. Das, I., Dennis, J.E.: Normal-boundary intersection: a new method for generating the Pareto surface in nonlinear multicriteria optimization problems. SIAM J. Optim. **8**(3), 631–657 (1998). https://doi.org/10.1137/S1052623496307510
7. Deb, K., Pratap, A., Meyarivan, T.: Constrained test problems for multi-objective evolutionary optimization. In: Zitzler, E., Thiele, L., Deb, K., Coello Coello, C.A., Corne, D. (eds.) EMO 2001. LNCS, vol. 1993, pp. 284–298. Springer, Heidelberg (2001). https://doi.org/10.1007/3-540-44719-9_20
8. Fan, Z., et al.: An improved epsilon constraint-handling method in MOEA/D for CMOPs with large infeasible regions. Soft. Comput. **23**, 12491–12510 (2019). https://doi.org/10.1007/s00500-019-03794-x
9. Fan, Z., et al.: Difficulty adjustable and scalable constrained multiobjective test problem toolkit. Evol. Comput. **28**(3), 339–378 (2020). https://doi.org/10.1162/evco_a_00259
10. Hansen, N., Auger, A., Brockhoff, D., Tušar, T.: Anytime performance assessment in blackbox optimization benchmarking. IEEE Trans. Evol. Comput. **26**(6), 1293–1305 (2022). https://doi.org/10.1109/TEVC.2022.3210897
11. Hanster, C., Kerschke, P.: flaccogui: exploratory landscape analysis for everyone. In: Proceedings of the Genetic and Evolutionary Computation Conference Companion, GECCO 2017, pp. 1215–1222. ACM (2017). https://doi.org/10.1145/3067695.3082477
12. Jain, H., Deb, K.: An evolutionary many-objective optimization algorithm using reference-point based nondominated sorting approach, Part II: Handling constraints and extending to an adaptive approach. IEEE Trans. Evol. Comput. **18**(4), 602–622 (2013). https://doi.org/10.1109/TEVC.2013.2281534
13. Kingma, D.P., Ba, J.: Adam: a method for stochastic optimization (2014). https://doi.org/10.48550/arXiv.1412.6980 arXiv preprint arXiv:1412.6980
14. Li, K., Chen, R., Fu, G., Yao, X.: Two-archive evolutionary algorithm for constrained multiobjective optimization. IEEE Trans. Evol. Comput. **23**(2), 303–315 (2018). https://doi.org/10.1109/TEVC.2018.2855411
15. Liefooghe, A., Daolio, F., Verel, S., Derbel, B., Aguirre, H., Tanaka, K.: Landscape-aware performance prediction for evolutionary multiobjective optimization. IEEE

Trans. Evol. Comput. **24**(6), 1063–1077 (2019). https://doi.org/10.1109/TEVC. 2019.2940828

16. Loh, W.Y.: Classification and regression trees. Wiley Interdisciplinary Rev. Data Mining Knowl. Dis. **1**(1), 14–23 (2011). https://doi.org/10.1002/widm.8

17. Ma, Z., Wang, Y.: Evolutionary constrained multiobjective optimization: test suite construction and performance comparisons. IEEE Trans. Evol. Comput. **23**(6), 972–986 (2019). https://doi.org/10.1109/TEVC.2019.2896967

18. Van der Maaten, L., Hinton, G.: Visualizing data using t-SNE. J. Mach. Learn. Res. **9**(11), 2579–2605 (2008). https://jmlr.org/papers/v9/vandermaaten08a.html

19. Mersmann, O., Bischl, B., Trautmann, H., Preuss, M., Weihs, C., Rudolph, G.: Exploratory landscape analysis. In: Proceedings of the Genetic and Evolutionary Computation Conference, GECCO 2011, pp. 829–836. ACM (2011). https://doi. org/10.1145/2001576.2001690

20. Nielsen, M.A.: Neural Networks and Deep Learning. Determination Press (2015)

21. Nikolikj, A., Doerr, C., Eftimov, T.: Rf+clust for leave-one-problem-out performance prediction. In: Applications of Evolutionary Computation: 26th International Conference, pp. 285–301. Springer (2023). https://doi.org/10.1007/978-3-031-30229-9_19

22. Pedregosa, F., et al.: Scikit-learn: machine learning in Python. J. Mach. Learn. Res. **12**, 2825–2830 (2011). https://www.jmlr.org/papers/v12/pedregosa11a.html

23. Platt, J.C.: Probabilities for SV machines. In: Smola, A.J., Bartlett, P.L., Schölkopf, B., Schuurmans, D. (eds.) Advances in Large Margin Classifiers, pp. 61–73. MIT Press (2000)

24. Prager, R.P., Seiler, M.V., Trautmann, H., Kerschke, P.: Automated algorithm selection in single-objective continuous optimization: a comparative study of deep learning and landscape analysis methods. In: International Conference on Parallel Problem Solving from Nature, PPSN 2022. pp. 3–17. Springer (2022). https://doi. org/10.1007/978-3-031-14714-2_1

25. Srinivas, N., Deb, K.: Multiobjective optimization using nondominated sorting in genetic algorithms. Evolutionary Comput. **2**(3), 221–248 (1994). https://doi.org/ 10.1162/evco.1994.2.3.221

26. van Stein, B., Long, F.X., Frenzel, M., Krause, P., Gitterle, M., Bäck, T.: Doe2vec: Deep-learning based features for exploratory landscape analysis. arXiv preprint arXiv:2304.01219 (2023). https://doi.org/10.48550/arXiv.2304.01219

27. Tanaka, M., Watanabe, H., Furukawa, Y., Tanino, T.: GA-based decision support system for multicriteria optimization. In: 1995 IEEE International Conference on Systems, Man and Cybernetics. Intelligent Systems for the 21st Century, vol. 2, pp. 1556–1561 (1995). https://doi.org/10.1109/ICSMC.1995.537993

28. Vermetten, D., Wang, H., Bäck, T., Doerr, C.: Towards dynamic algorithm selection for numerical black-box optimization: Investigating bbob as a use case. In: Proceedings of the Genetic and Evolutionary Computation Conference, GECCO 2020. pp. 654–662. ACM (2020). https://doi.org/10.1145/3377930.3390189

29. Vodopija, A., Tušar, T., Filipič, B.: Characterization of constrained continuous multiobjective optimization problems: A performance space perspective. arXiv preprint arXiv:2302.02170 (2023). https://doi.org/10.48550/arXiv.2302.02170

30. Vodopija, A., Tušar, T., Filipič, B.: Characterization of constrained continuous multiobjective optimization problems: a feature space perspective. Inf. Sci. **607**, 244–262 (2022). https://doi.org/10.1016/j.ins.2022.05.106

On the Latent Structure
of the bbob-biobj Test Suite

Pavel Krömer[1]([✉])[iD], Vojtěch Uher[1][iD], Tea Tušar[2,3][iD], and Bogdan Filipič[2,3][iD]

[1] VSB - Technical University of Ostrava, Ostrava, Czech Republic
{pavel.kromer,vojtech.uher}@vsb.cz
[2] Jožef Stefan Institute, Ljubljana, Slovenia
{tea.tusar,bogdan.filipic}@ijs.si
[3] Jožef Stefan International Postgraduate School, Ljubljana, Slovenia

Abstract. Landscape analysis is a popular method for the characterization of black-box optimization problems. It consists of a sequence of operations that, from a limited sample of solutions, approximate and describe the hypersurfaces formed by characteristic problem properties. The hypersurfaces, called problem landscapes, are described by sets of carefully crafted features that ought to capture their characteristic properties. In this way, arbitrary optimization problems with potentially very different technical parameters, such as search space dimensionality, are projected into specific feature spaces where they can be further studied. The representation of a problem in a feature space can be used, for example, to find similar problems and identify metaheuristic optimization algorithms that have the best track record on the same type of tasks. Because of that, the quality and properties of problem representation in the feature spaces gain importance. In this work, we study the representation properties of the popular bbob-biobj test suite in the space of bi-objective features, analyze the structure naturally emerging in the feature space, and analyze the high-level properties of the projection. The obtained results clearly demonstrate the discrepancies between the latent structure of the test suite and its expert perception.

Keywords: Multiobjective optimization · Problem landscape · Exploratory landscape analysis · Clustering · Data visualization

1 Introduction

Landscape analysis (LA) is a methodology designed to enable an insight into and understanding of complex optimization problems. It aims at problem characterization and the discovery of the relationship between problems and different (metaheuristic) optimization algorithms. LA is a complex process that consists of a series of steps including solution sampling, evaluation of characteristic measures, computation of landscape features, and analysis of problem properties in the feature spaces. The problem characterization in the feature space can be used for different purposes including algorithm performance prediction and behavior explanation [21], automated algorithm selection [13,20], parameter tuning [16], problem classification, and others. Such insights are especially useful when metaheuristic optimization algorithms are going to be applied to (black-box) optimization problems with an unknown nature. Individual metaheuristic algorithms represent different search strategies and their use results in the pursuit of distinct search trajectories through the solution spaces of the solved problems. This leads to different performance of different algorithms on different problems, i.e., inside differently structured search spaces [15]. Overall, LA is an attempt at a computationally feasible characterization of optimization problems through their mapping to well-structured *features spaces* that can be more easily analyzed.

The feature spaces can be used by various downstream algorithms to characterize the problems, find out their mutual relationships, similarities, dissimilarities, and, for example, assign previously unknown problems to existing problem categories. Such analyses can contribute to an efficient application of modern nature-inspired metaheuristics to a growing number of real-world problems.

In this work, LA is used to analyze a test suite from a popular benchmarking platform, COmparing Continuous Optimizers (COCO) [4]. Specifically, we take a detailed look at a feature-based representation of COCO's multiobjective test suite, bbob-biobj. The suite is a set of 55 noiseless, scalable bi-objective test problems (functions) that can be used to evaluate metaheuristic optimization algorithms [2]. It was selected as it is an established set of well-understood, carefully crafted, and expertly annotated benchmarking problems for multiobjective optimization. Each test problem is processed by the LA process and projected to a feature space determined by a set of features designed specifically for continuous multiobjective optimization problems [10]. Then, the structure of the feature space is assessed by cluster analysis and its latent structure is studied. The discovered structure is then matched with three different expert-defined classifications of the test problems. In this way, this work provides useful insight into the differences between the latent and the expert-defined structure of the test suite.

The rest of this paper is organized in the following way. Section 2 gives a short definition of landscape analysis and provides an overview of landscape features often used for the description of single- and multiobjective problems. The computational experiments and obtained results are detailed in Sect. 3. Finally, major conclusions are drawn and future work is outlined in Sect. 5.

2 Landscape Analysis

A problem landscape can be defined with the help of the d-dimensional search (solution) space, $\mathbb{X}$, a neighborhood operator, $n(x)$, that defines for each solution, $x \in \mathbb{X}$, a set of neighboring solutions, and a characteristic function, $\mathcal{C}(x) : \mathbb{X} \to \mathbb{R}$, that maps each solution to a certain quality observed in the context of the solved problem [19]. The tuple, $(\mathbb{X}, n(x), \mathcal{C})$, defines a hypersurface associated with the values of the characteristic function for the investigated problem and is in line with a popular analogy with real-world landscapes referred to as a *landscape* [17]. The information represented by the landscape depends on which problem property is expressed by the characteristic function. It can be, for example, the fitness of the solution (fitness landscape), the level of constraint violation (violation landscape) [12], and so on, leading to potentially several distinct hypersurfaces associated with a single investigated problem. The goal of LA is the characterization of a problem on the basis of the information from $(\mathbb{X}, n(x), \mathcal{C})$. It enables a comprehensive investigation of a problem's search, objective, and other characteristic spaces. This scrutiny relies on the extraction of characteristic features derived from a specifically chosen set of sampling points distributed across the search space.

The outcome of the analysis is affected by each of its stages. *Solution sampling* is a process that selects a series of samples (sampling points, problem solutions) from the problem's search space. The particular sampling strategy it follows decides which patterns of solutions take part in problem characterization and which do not, how dense and how regular is the coverage of the search space, and so on. Popular sampling strategies include pseudorandom (uniform), quasirandom [18], and Latin Hypercube sampling-based sampling [14]. Next, the selection of the *characteristic (performance) measures* evaluated for the sampled solutions defines which solution properties are considered for the problem characterization exercise and which are not. The most used characteristic measures include the values of fitness function(s) [15] and the level of constraint violation [12]. Finally, the choice of the *landscape features* defines how the hyperplanes associated with the values of each metric for the sampled solutions are going to be approximated (summarized) and, in the end, the problems represented in a feature space.

2.1 Landscape Features

The choice of features to approximate problem landscapes depends on a number of factors [8]. Groups of features reflecting different problem landscape properties such as fitness-distance correlation, landscape ruggedness, and information content have been introduced [7]. Later, Mersmann et al. [15] proposed a battery of 6 groups of features that can describe single-objective black-box optimization problems from different perspectives including global and local structure, modality, convexity, curvature, presence of plateaus, separability, and others. The 83 features defined in [15] are often referred to as classical exploratory landscape analysis (ELA) features. Kerschke and Trautmann assembled a battery of 300 features for the characterization of continuous optimization problems with constraints [6]. The features were divided into 17 feature sets and included the classical ELA features proposed by Mersmann et al. [15], cell mapping and generalized cell mapping features, barrier tree features, and many other types of numerical characteristics of problem landscapes. The characterization of multi-objective problems by landscape analysis motivated the development of special features for the description of (continuous) multi-objective landscapes [10]. The features were designed so that they inherently considered multiple fitness values (i.e., multiple fitness landscapes) and reflected, for example, the multimodality, evolvability, and ruggedness of the landscapes.

3 Methodology

In this work, a popular suite of bi-objective benchmarking problems, bbob-biobj, is modeled by landscape analysis, and the properties of its representation in the feature space are analyzed. In particular, we adopt three expert-defined classifications of the test problems and assess how they correspond with the latent structure of the test suite in the feature space obtained by cluster analysis.

3.1 Expert Classification

The bbob-biobj test suite [2] consists of 55 bi-objective benchmark problems termed F_1 to F_{55}. They were constructed by combining all possible pairs of 10 single-objective problems from the bbob suite [5] and were divided into 15 groups reflecting the properties of their single-objective functions. We use the 15 groups, each containing either three or four problems, as the first expert classification. Additionally, 2-D problem instances were manually inspected in [2] and a number of their properties was collected into a table. We provide two other expert classifications depending on these observations. One constructs 7 classes based on the number of distinct Pareto set parts (1, 2, 3, 4, 5, 5–9 or 10+, as in [2]), while the other splits the problems into 2 classes depending on the convexity or non-convexity of the Pareto front. These three expert classifications are shown in the first three images of Fig. 1.

3.2 Cluster Analysis

Cluster analysis is a family of methods for unsupervised data analysis that enable the discovery of structure in data and the classification of objects into meaningful groups (clusters). A (non-overlapping) clustering of a data set $\mathbb{D} = \{d_1, d_2, \ldots, d_n\}$ into k clusters is a set of k disjoint partitions, $\mathcal{S} = \{S_1, S_2, \ldots, S_k\}$, subject to $S_i \subset \mathbb{D}$ and $S_i \neq \emptyset$ for each $S_i \in \mathcal{S}$, $S_i \cap S_j = \emptyset$ for each $S_i, S_j \in \mathcal{S}$ and $i \neq j$, and $\bigcup_{i=1}^{k} \{S_i\} = \mathbb{D}$. Because many different types of clusters may be found in the data, cluster analysis includes various clustering algorithms that can assign objects to clusters. Individual methods are often based on different assumptions about the data and the nature of the clusters and, therefore, can lead to different assignments of objects to clusters. To assess the validity of clustering on a data set, so-called cluster validity metrics can be used.

In this work, we use cluster analysis to analyze the structure of the representation of a bi-objective test suite in a feature space created by a particular landscape analysis pipeline. We analyze the latent structure of the representation by grouping the feature vectors into different numbers of clusters by the K-Means algorithm. It was selected as a popular first-choice clustering method

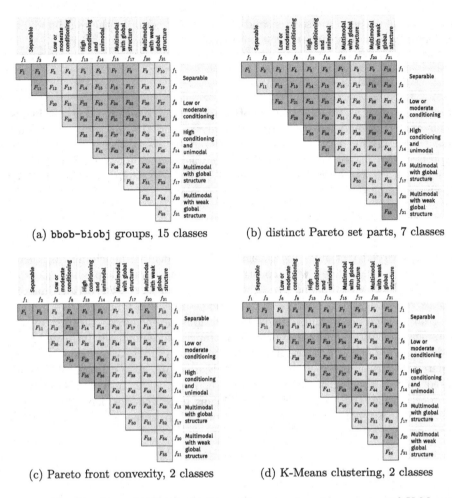

(a) bbob-biobj groups, 15 classes

(b) distinct Pareto set parts, 7 classes

(c) Pareto front convexity, 2 classes

(d) K-Means clustering, 2 classes

Fig. 1. Classifications of 55 bbob-biobj problems as done by experts and K-Means. The colors represent different classes: (a) 15 bbob-biobj groups, (b) 7 classes based on the number of distinct Pareto set parts, (c) 2 classes based on the convexity of the Pareto front, and (d) 2 classes found in this work by K-Means for Sobol sampling, dimension $d = 3$ and sample set size $n = 2^{16}$. Additionally, gray rectangles on the top and right of each image show the five groups of single-objective functions.

to investigate the feasibility of feature space analysis through clustering. Then, each K-Means clustering is assessed by two cluster validity measures, the Silhouette index and the Variance ratio, to obtain information about the real number and quality of clusters emerging in the data. A brief description of the used methods follows.

K-Means. K-Means is a widespread clustering algorithm that can split n data points into k clusters ($k \leq n$) represented by the centers of mass (centroids). The computation of K-Means clustering is NP-hard, but there is an efficient iterative heuristic algorithm that can be used to approximate the clusters. Its steps can be summarized as follows:

1. Initialize k centroids, $\mu_1, \mu_2, \ldots, \mu_k$, either randomly or using a specific initialization strategy.
2. Assign each data point, x_i, to the cluster with the nearest center, $s_i = \arg\min_j \|d_i - \mu_j\|^2$.
3. Update the centroids, $\mu_j = \frac{1}{|S_j|} \sum_{x_i \in S_j} d_i$.
4. Repeat steps 2 and 3 until the assignment of points to clusters stabilizes.

Silhouette Index. The Silhouette index (SIL) is a clustering validity measure to assess the validity of an arbitrary clustering of a data set [1]. Essentially, it is defined as the mean ratio of the difference between the average intra-cluster and inter-cluster distances for all points in the data set. The silhouette value (width) of a single point, x_i, is defined as

$$\text{SIL}(d_i) = \frac{b(d_i) - a(d_i)}{\max\{a(d_i), b(d_i)\}}, \tag{1}$$

where $a(x_i)$ is the average distance between d_i and the points in the same cluster, S_i, and $b(d_i)$ is the average distance between d_i and the points in other clusters. The overall Silhouette index for the clustering is the average of the silhouette values for all points, $\text{SIL} = \frac{1}{n} \sum_{i=1}^{n} \text{SIL}(d_i)$, where n is the total number of data points. The best possible value of the Silhouette index is 1 and the worst is -1. Values close to 0 indicate overlapping clusters. Negative values generally suggest that a sample has been assigned to the wrong cluster because there is another cluster with more similar (less distant) points [1].

Variance Ratio. The Variance ratio (Calinski-Harabasz Index, CHI) evaluates how similar a point, x_i, is to other points in the same cluster (cohesion) and how dissimilar it is from other clusters (separation). Cohesion is based on the distances from the data points in a cluster to its centroid and separation is estimated on the basis of the distance of the cluster centroids from the center of mass of the entire data set:

$$\text{CHI} = \frac{\sum_{l=1}^{k} (|S_l| \cdot \|\mu_l - \mu\|^2)}{k - 1} \cdot \frac{n - k}{\sum_{l=1}^{k} \sum_{i=1}^{n_l} \|d_i - \mu_l\|^2}, \tag{2}$$

where S_l is the l-th cluster, μ_l is its centroid, and μ is the center of mass of the entire data set. A clustering with a higher CHI consists of clusters that are more dense and better separated than in a clustering with a lower CHI value.

3.3   Experiment Design

In order to study the feature-based representation of `bbob-biobj` problems under a wide variety of conditions, the landscape analysis was done for each problem with three problem dimensions ($d \in \{3, 5, 10\}$), four different sampling strategies (Uniform random, Sobol and Halton sequence-based, and optimized Latin Hypercube Sampling (LHSO)-based), and six sample sizes ($n \in \{2^5, 2^8, 2^{10}, 2^{12}, 2^{14}, 2^{16}\}$). This led to 72 different experimental configurations (combinations of problem dimension, sampling strategy, and sample size). Although the test suite contains multiple instances of each problem, only the first one (i.e., the one consisting of instance 02 of the first and instance 04 of the second underlying function) was used. To obtain robust approximations of the problem landscapes, each sampling was randomized 31 times, and as a result, each experimental configuration was represented by 31 bi-objective fitness landscapes. The landscapes were described by the multiobjective features from [10]. The features were engineered specifically for continuous multi-objective landscapes and express their global properties, multimodality, evolvability, and ruggedness. In the end, 31 different representations of each problem in each configuration were obtained in the feature space.

In each representation, we analyzed the emerging structure by the K-Means algorithm with a different number of clusters, $k \in \{2, 3, \ldots, 15\}$, and assessed the quality of the clustering by SIL and CHI cluster validity measures. In addition, we compared the clusters emerging in the feature space with problem classes defined by experts.

4   Results

The results of the computational experiments are presented and discussed in this section. It first describes how the optimum number of clusters was selected and then provides a comprehensive analysis and interpretation of the discovered latent clusters.

4.1   Number of Clusters

The results of unsupervised clustering are visually summarized in Fig. 2. The figure consists of 72 plots with average SIL (blue) or CHI (red) values for K-Means clusters with $k \in \{2, 3, \ldots, 15\}$ for each problem configuration. The mosaic clearly demonstrates that the highest average values of SIL and CHI, indicating clusters of best quality, were obtained for the lowest value of k, $k = 2$, no matter what problem dimension, sampling strategy, and sample size were used in the LA.

A more detailed illustration of the values of SIL and CHI for a selected sampling strategy and problem dimension is provided in Fig. 3. The plots also illustrate the quality of the three expert-defined classes by boxplots drawn at $x = 2$, $x = 7$, and $x = 15$ (their corresponding number of classes). Besides confirming that the problem representations in the feature space are better aligned

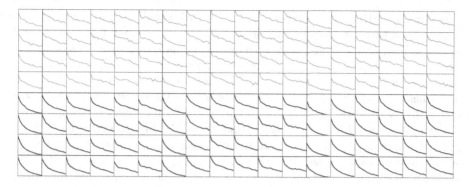

Fig. 2. A summary of the trends of SIL (blue) and CHI (red) values for $k \in \{2, 3, \ldots, 15\}$ for all experimental configurations (higher is better). (Colour figure online)

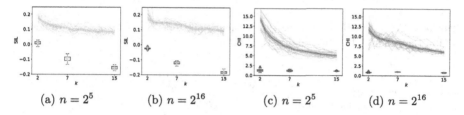

(a) $n = 2^5$ (b) $n = 2^{16}$ (c) $n = 2^5$ (d) $n = 2^{16}$

Fig. 3. The values of SIL (blue) and CHI (red) for K-Means clusters with $k \in \{2, 3, \ldots, 15\}$ and 2, 7, and 15 expert classes (boxplots at $x = 2$, $x = 7$, and $x = 15$) for two sample sizes, $n \in \{5, 16\}$, Sobol sequence-based sampling and problem dimension $d = 3$. The thin lines are the validity indices for clusters found in randomized samples. The solid thick line shows the mean values of SIL/CHI and the bands around it represent the 95% confidence intervals around the mean. (Colour figure online)

with a smaller number of clusters (with the best cluster validity achieved for $k = 2$), the plots also show that the expert-defined clusters are not mapped to meaningful structures in the feature space. This is demonstrated by the significantly lower values of SIL and CHI for the expert-defined clusterings (see the boxplots). Finally, we note that although the example shows the results for only a few experimental configurations, the results for other combinations of problem dimensions, sampling strategies, and sample sizes demonstrate the same trends.

4.2   Analysis of Latent Clusters

The rest of the experiments focused on a closer investigation of the latent clusters with $k = 2$ (shown in Fig. 1d), the feature space representation of the expert-defined clusters, and their comparison. The validity of the latent clusters, discovered by the K-Means algorithm for $k = 2$, is compared for all experimental configurations in Tables 1 and 2. The tables illustrate several trends. First, it can be seen that although the validity of the clusters for $k = 2$ is the highest, they are not perfect. For example, the average SIL values around 0.2 in Table 1 suggest that there is an emerging structure, but it does not fully correspond to two well-separated dense clusters, as would be indicated by SIL close to 1.0. Next, the tables show the clear effects of different sample sizes and sampling strategies on the quality of clusters in the feature space. Generally, they show that when the sample size is equal to 2^{10} or higher, the quality of the clusters is better than when it is smaller, underlining the need for sufficiently large problem samples. The tables also illustrate that the used sampling strategies have an effect on the structure of the feature space. Although the results clearly depend on the experimental configuration, it can be seen that, in the majority of cases, more advanced sampling strategies lead to better-structured feature spaces than those obtained with Uniform random sampling. Finally, Table 2 highlights that the structure of the feature-space representation of problems with higher dimensions is affected by small sample sizes much more than the structure of problems with a lower dimension.

Table 1. Average SIL values for the K-Means clusters with $k = 2$. The values fall within the range $[-1, 1]$, the higher the better.

Probl. dim.	Sampling strategy	Sample size (n)					
		2^5	2^8	2^{10}	2^{12}	2^{14}	2^{16}
$d = 3$	Uniform	0.16858	0.17781	0.17848	0.17072	0.17029	0.19833
	Sobol'	0.17559	0.18223	0.18321	0.18500	0.17099	0.20663
	LHSO	0.17640	0.19088	0.19165	0.18685	0.19287	0.19753
	Halton	0.17659	0.17968	0.18101	0.19030	0.16831	0.17988
$d = 5$	Uniform	0.16745	0.20397	0.20444	0.20578	0.19567	0.19842
	Sobol'	0.17410	0.19161	0.21940	0.21072	0.22450	0.19939
	LHSO	0.16821	0.19430	0.21233	0.20997	0.22515	0.20409
	Halton	0.17013	0.19971	0.21988	0.21547	0.22791	0.20975
$d = 10$	Uniform	0.15181	0.16620	0.17792	0.17534	0.18179	0.17839
	Sobol'	0.14589	0.16351	0.16968	0.18216	0.18229	0.17991
	LHSO	0.14689	0.16627	0.16854	0.17796	0.18022	0.18451
	Halton	0.15763	0.16152	0.17479	0.17816	0.18483	0.19565

For a better understanding and comparison, the expert classification-based and the latent clusters from the features space have been visualized by the t-distributed stochastic neighbor embedding (t-SNE) algorithm [11] with default parameters as used in the scikit-learn Python library. An example of the visualization for a single experimental configuration for problems with dimension $d = 3$, Uniform random sampling, and sample size $n = 2^{16}$ is shown in Fig. 4. The figure clearly illustrates the properties of the clusters suggested by the cluster validity measures discussed in the previous paragraphs. It can be seen that the expert-defined clusters (Fig. 4a–Fig. 4c) are spread across the feature space and do not match with its emerging structure well. On the other hand, the latent clusters (Fig. 4d) are much more in line with the organization of the feature space.

Table 2. Average CHI values for the K-Means clusters with $k = 2$. CHI ≥ 0, the higher the better.

Probl. dim.	Sampling strategy	Sample size (n)					
		2^5	2^8	2^{10}	2^{12}	2^{14}	2^{16}
$d = 3$	Uniform	13.1341	13.7153	13.4966	13.1072	13.0389	12.0623
	Sobol'	14.0512	14.4354	14.2152	13.0454	12.6862	11.6280
	LHSO	13.8949	14.2946	13.9047	13.2398	12.8289	11.9239
	Halton	13.8517	13.2104	14.1193	13.5609	12.0527	12.7317
$d = 5$	Uniform	13.2541	14.3907	15.4810	14.5754	14.3795	13.6464
	Sobol'	12.9498	15.5466	14.9389	14.9061	14.4485	14.8642
	LHSO	13.2910	15.1948	15.2130	15.0388	14.2593	14.0611
	Halton	13.0484	15.8946	15.4437	15.3982	14.4851	14.5858
$d = 10$	Uniform	11.5130	12.6852	14.4477	14.2677	14.7421	14.4298
	Sobol'	10.9866	13.2073	13.6819	14.3397	14.9993	14.9720
	LHSO	11.4390	13.4044	13.6519	14.5277	14.3506	13.7110
	Halton	11.3276	12.9814	13.9742	14.3852	14.5919	13.8328

As the next step, we studied the stability of the latent clusters and the effect of sampling randomization on the assignment of problems to the emerging classes. We observed for each problem to which of the two K-Means clusters it was assigned across the 31 randomized samplings. The results of this analysis for selected problem configurations are visually illustrated in Fig. 5. The plots in the figure illustrate for $d = 3$, sample size 2^{16}, and two different sampling strategies the percentage of times each problem has been assigned to the first (orange) or the second cluster (blue). Every problem is represented by a single chart. In the top plots, the order of problems is F_1 to F_{55}, in the bottom plots, the problems are sorted according to the stability of their assignment to clusters. The plots clearly illustrate that most problems are in the majority of cases assigned to the same cluster, i.e., their chance of being assigned to one cluster is much higher than the chance of being assigned to the other one. Some problems are assigned to the same cluster in 100% of cases, which suggests good stability of the latent structures in the feature space and a robust representation of the problems by the LA process. Nevertheless, it can be also seen that there are some problems that oscillate between clusters. However, this is not surprising given the heavy use of stochastic principles in the LA pipeline and the clustering process, as well.

4.3   Interpretation of Discovered Classes

Finally, we wish to see whether the results of K-Means clustering can be meaningfully interpreted. To this end, we join the K-Means classifications into two classes for all sampling methods, problem dimensions, and sample sizes used in this study. Since each separate classification labels the two classes arbitrarily, we match the class labels among all classifications by centering them around problems that are very often classified together. For each of the problems, we

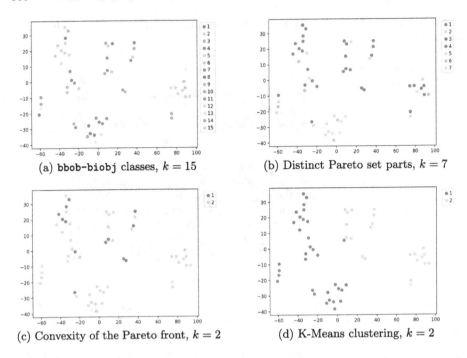

Fig. 4. t-SNE visualization of feature vectors of 55 `bbob-biobj` test problems computed for Uniform random sampling (non-rotated), dimension $d = 3$ and sample set size $n = 2^{16}$. The colors match those from Fig. 1 and represent different classes assigned by experts: (a) 15 `bbob-biobj` classes, (b) 7 classes based on the number of distinct Pareto set parts, (c) the convexity of the Pareto front, or by K-Means clustering: (d) 2 latent classes.

are then able to express how often the problem was classified in the first class with a number between 0 and 1.

We show these results in Fig. 6 with the frequency of classification into the first class on the y-axis. Each problem is depicted using two plots of its 2-D landscape. Both are based on a discretization of the search space with a grid of points (a 501×501 grid was used for all plots in this paper). The top plot (shown also on the left in Fig. 7) visualizes the correlation between objectives. For each grid point, the correlation between objectives is computed as the Pearson correlation coefficient of the objective values in 100 equidistant points on the circle with radius 10^{-6} centered in that grid point. Blue hues denote positive correlations between the objectives and red hues denote negative ones (note that the objectives are anti-correlated on the Pareto set). The bottom plot (also on the right in Fig. 7) depicts the dominance rank ratio for each grid point, which assigns dark hues to points 'close' to the Pareto set (with only a few points that dominate them) and lighter hues to those 'far' from it [2,3]. The Pareto set in these plots is visualized in yellow.

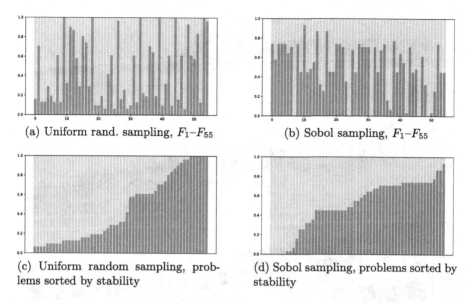

(a) Uniform rand. sampling, F_1–F_{55}

(b) Sobol sampling, F_1–F_{55}

(c) Uniform random sampling, problems sorted by stability

(d) Sobol sampling, problems sorted by stability

Fig. 5. Stacked bar charts of 55 test problems clustered by K-Means into two clusters. The bars represent the normalized portion of each of the two classes into which a problem was assigned during 31 rotations. The figures show Uniform and Sobol samplings for dimension $d = 3$, sample set size $n = 2^{16}$, and $k = 2$ clusters. Figures (a) and (b) illustrate the original order of functions, and (c) and (d) the same problems sorted by one class.

Figure 6 makes it possible to inspect whether there is some visual similarity between problems often classified in the same class by K-Means. We can certainly see this for problems often classified into the first class (top row of problems in Fig. 6), as their landscapes look very much alike. This is not surprising given that the first ten problems all contain the same single-objective function f_{21} with multiple peaks as one of the objectives, while the next seven problems all contain the Schaffer's F7 function f_{17}, which gives the landscapes an appearance of 'waves'. However, the rest of the classification is harder to understand. The bottom problems in Fig. 6 correspond to those most often classified into the second class and while we can find small groups of similar ones, a general trend cannot be found. This demonstrates that the interpretation of the latent relationships between problem representations is not straightforward and its understanding in terms of problem properties is a complex exercise requiring detailed knowledge of the investigated problems.

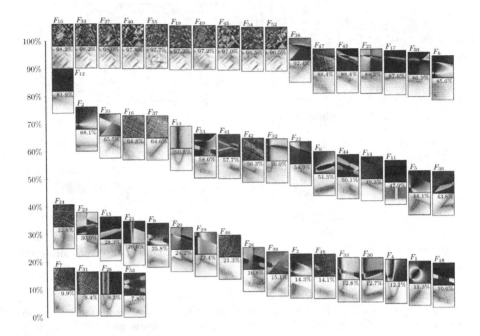

Fig. 6. Two-class classification of the 55 `bbob-biobj` problems by K-Means visualized according to how often each of the problems was classified in the first cluster (y axis). The problems are sorted by this value and stacked along the x axis to enhance readability. For each problem we show two visualizations of its 2-D landscape: the Pearson correlation coefficient on top and the dominance rank ratio on the bottom (see Fig. 7 and text for more information).

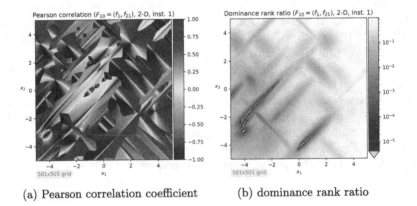

(a) Pearson correlation coefficient   (b) dominance rank ratio

Fig. 7. Problem landscape visualizations for the `bbob-biobj` problem F_{10} of dimension 2, instance 1: (a) the Pearson correlation coefficient and (b) the dominance rank ratio. (Color figure online)

5   Conclusions

This work investigates a popular suite of bi-objective benchmark functions, bbob-biobj, from the landscape analysis point of view. It studies the projections of the test problems to a feature space outlined by a set of features for multi-objective optimization problems. The feature vectors, representing the test problems in the feature space, are thoroughly analyzed by cluster analysis to find out their mutual relationships and potential latent structures that can emerge in this space. The analysis is conducted for a wide range of problem and landscape analysis configurations to obtain robust information about the representation of the test suite in this feature space. The results of the unsupervised analysis suggest that the problems are organized into no more than two classes that are not perfectly separated but still can very well represent some problems that are in the majority of cases assigned to one of these groups. The properties of the problems in the discovered groups of problems were also broadly studied.

The latent classes were further compared with the feature-space representation of three expert classifications of the problems in the test suite from the literature. All expert classifications were based on different fundamental properties of the problems such as the nature of the underlying single-objective test functions or the character of the Pareto-front of the bi-objective problem. The experiments clearly demonstrated that the expert classification is not matched with corresponding structures in the feature space and therefore should be used with caution when dealing with, for example, applications such as landscape analysis-based problem classification. The understanding of problem representation in feature space can be even more important for real-world problems, where the apriori information is based on the knowledge of the underlying domain and can be only loosely connected to the properties of the problem formulation.

Future work on this topic will include an extended analysis of the test suite in spaces defined by other types of landscape features, a thorough evaluation of the potential emerging structure by other types of unsupervised methods (e.g., distance-based clustering), and further investigation into the nature of the discovered latent groups of problems. Additionally, the same clustering-based investigation will be conducted for other suites of multiobjective problems such as those from the CEC 2020 Multimodal Multiobjective Optimization test suite [9].

Acknowledgment. This work is part of the project "Constrained Multiobjective Optimization Based on Problem Landscape Analysis" co-funded by the Czech Science Foundation (grant no. GF22-34873K) and the Slovenian Research and Innovation Agency (project no. N2-0254). Furthermore, the Czech authors acknowledge support from the Student Grant System, grants no. SP2024/006 and SP2024/007, VSB – Technical University of Ostrava, and the Slovenian authors acknowledge additional financial support from the Slovenian Research and Innovation Agency (research core funding no. P2-0209). The publication is also based upon work from COST Action CA22137 "Randomised Optimisation Algorithms Research Network" (ROAR-NET), supported by European Cooperation in Science and Technology (COST).

References

1. Bandyopadhyay, S., Saha, S.: Unsupervised Classification: Similarity Measures, Classical and Metaheuristic Approaches, and Applications. Springer (2012). https://doi.org/10.1007/978-3-642-32451-2
2. Brockhoff, D., Auger, A., Hansen, N., Tušar, T.: Using well-understood single-objective functions in multiobjective black-box optimization test suites. Evol. Comput. **30**(2), 165–193 (2022). https://doi.org/10.1162/evco_a_00298
3. Fonseca, C.M.: Multiobjective genetic algorithms with application to control engineering problems. Ph.D. thesis, University of Sheffield (1995)
4. Hansen, N., Auger, A., Ros, R., Mersmann, O., Tušar, T., Brockhoff, D.: COCO: a platform for comparing continuous optimizers in a black-box setting. Optimiz. Methods Softw. **36**(1), 114–144 (2021). https://doi.org/10.1080/10556788.2020.1808977
5. Hansen, N., Finck, S., Ros, R., Auger, A.: Real-parameter black-box optimization benchmarking 2009: Noiseless functions definitions. Research Report RR-6829, INRIA (2009). https://hal.inria.fr/inria-00362633
6. Kerschke, P., Trautmann, H.: The R-Package FLACCO for exploratory landscape analysis with applications to multi-objective optimization problems. In: 2016 IEEE Congress on Evolutionary Computation (CEC), pp. 5262–5269. IEEE (2016). https://doi.org/10.1109/CEC.2016.7748359
7. Kerschke, P., Trautmann, H.: Automated algorithm selection on continuous black-box problems by combining exploratory landscape analysis and machine learning. Evol. Comput. **27**(1), 99–127 (2019). https://doi.org/10.1162/evco_a_00236
8. Lang, R.D., Engelbrecht, A.P.: An exploratory landscape analysis-based benchmark suite. Algorithms **14**(3), 78 (2021). https://doi.org/10.3390/a14030078
9. Liang, J.J., Suganthan, P.N., Qu, B.Y., Gong, D.W., Yue, C.T.: Problem definitions and evaluation criteria for the CEC 2020 special session on multimodal multiobjective optimization. Technical Report 201912, Computational Intelligence Laboratory, Zhengzhou University, Zhengzhou, China and Nanyang Technological University, Singapore (2019)
10. Liefooghe, A., Verel, S., Lacroix, B., Zǎvoianu, A.C., McCall, J.: Landscape features and automated algorithm selection for multi-objective interpolated continuous optimisation problems. In: Proceedings of the Genetic and Evolutionary Computation Conference (GECCO), pp. 421–429. ACM (2021). https://doi.org/10.1145/3449639.3459353
11. Van der Maaten, L., Hinton, G.: Visualizing data using t-SNE. J. Mach. Learn. Res. **9**(11), 2579–2605 (2008). https://jmlr.org/papers/v9/vandermaaten08a.html
12. Malan, K.M., Moser, I.: Constraint handling guided by landscape analysis in combinatorial and continuous search spaces. Evol. Comput. **27**(2), 267–289 (2019). https://doi.org/10.1162/evco_a_00222
13. Malan, K.M.: A survey of advances in landscape analysis for optimisation. Algorithms **14**(2), 40 (2021). https://doi.org/10.3390/a14020040
14. McKay, M.D., Beckman, R.J., Conover, W.J.: A comparison of three methods for selecting values of input variables in the analysis of output from a computer code. Technometrics **42**(1), 55–61 (2000). https://doi.org/10.2307/1271432
15. Mersmann, O., Bischl, B., Trautmann, H., Preuss, M., Weihs, C., Rudolph, G.: Exploratory landscape analysis. In: Proceedings of the 13th Annual Genetic and Evolutionary Computation Conference (GECCO), pp. 829–836. ACM (2011). https://doi.org/10.1145/2001576.2001690

16. Pikalov, M., Mironovich, V.: Automated parameter choice with exploratory landscape analysis and machine learning. In: Proceedings of the Genetic and Evolutionary Computation Conference (GECCO) Companion, pp. 1982–1985. ACM (2021). https://doi.org/10.1145/3449726.3463213
17. Pitzer, E., Affenzeller, M.: A comprehensive survey on fitness landscape analysis. In: Fodor, J.C., Klempous, R., Suárez Araujo, C.P. (eds.) Recent Advances in Intelligent Engineering Systems, Studies in Computational Intelligence, vol. 378, pp. 161–191. Springer (2012). https://doi.org/10.1007/978-3-642-23229-9_8
18. Renau, Q., Doerr, C., Dreo, J., Doerr, B.: Exploratory landscape analysis is strongly sensitive to the sampling strategy. In: Bäck, T., et al. (eds.) Parallel Problem Solving from Nature - PPSN XVI, pp. 139–153. Springer (2020). https://doi.org/10.1007/978-3-030-58115-2_10
19. Richter, H.: Fitness landscapes: From evolutionary biology to evolutionary computation. In: Richter, H., Engelbrecht, A. (eds.) Recent Advances in the Theory and Application of Fitness Landscapes, pp. 3–31. Springer (2014). https://doi.org/10.1007/978-3-642-41888-4_1
20. Tanabe, R.: Benchmarking feature-based algorithm selection systems for blackbox numerical optimization. IEEE Trans. Evolutionary Comput. 1321–1335 (2022). https://doi.org/10.1109/TEVC.2022.3169770
21. Trajanov, R., Dimeski, S., Popovski, M., Korošec, P., Eftimov, T.: Explainable landscape analysis in automated algorithm performance prediction. In: Jiménez Laredo, J.L., Hidalgo, J.I., Babaagba, K.O. (eds.) Applications of Evolutionary Computation, EvoApplications 2022, pp. 207–222. Springer (2022). https://doi.org/10.1007/978-3-031-02462-7_14

Soft Computing Applied to Games

Strategies for Evolving Diverse and Effective Behaviours in Pursuit Domains

Tyler Cowan and Brian J. Ross[✉]

Department of Computer Science, Brock University, 1812 Sir Isaac Brock Way,
St. Catharines, ON L2S 3A1, Canada
bross@brocku.ca
http://www.cosc.brocku.ca/~bross/

Abstract. Gamer engagement with computer opponents is an important aspect of computer games. Players will be bored if computer opponents are predictable, and the game will be monotonous. Computer opponents that are both challenging and exhibit interesting and novel behaviours are ideal. This research explores different strategies that encourage diverse emergent behaviours for evolved intelligent agents, while maintaining good performance with the task at hand. We consider the pursuit domain, which consists of a single predator agent and twenty prey agents. The predator's controller is evolved through genetic programming, while the preys' controllers are hand-crafted. The fitness of a solution is calculated as the number of prey captured. Inspired by Lehman and Stanley's novelty search strategy, the fitness is combined with a diversity score, determined by combining four rudimentary behaviour measurements. We combine these basic scores using the many objective optimization strategy known as "sum of ranks", which is proven to effectively balance a high number of conflicting objectives in optimization problems. We also examine different population diversity strategies, as well as different weighting schemes for combining fitness and diversity scores. After producing sets of solutions for the above experiments, we manually tabulate higher-level emergent behaviour observed in the evolved predators. The use of K-nearest neighbours (K=32) with population archive, combined with a fitness:diversity weighting of 50:50, gave the best results, as it effectively balanced good fitness performance and diverse emergent behaviour.

1 Introduction

An evolutionary algorithm (EA) is typically designed with specific goals to solve, for example, minimizing a price or maximizing a score. The nature of EA search is that multiple instances or runs of searches often converge to similar solutions. If the fitness of such generic solutions is acceptable and of primary interest, this is not a problem. In some applications, however, the goal is not simply to evolve a fit solution. For example, consider the task of creating an intelligent agent opponent (NPC, or non-playable character) for a video game. A computer opponent that always behaves in a predictable manner is undesirable, since human players will

© The Author(s), under exclusive license to Springer Nature Switzerland AG 2024
S. Smith et al. (Eds.): EvoApplications 2024, LNCS 14635, pp. 345–360, 2024.
https://doi.org/10.1007/978-3-031-56855-8_21

quickly learn to adapt to it, which makes the game monotonous, unchallenging, and not fun to play. But as is the aforementioned nature of EA search, multiple runs to evolve an intelligent agent controller often converge to a similar "generic" agent, which becomes a boring NPC competitor for a human.

There exist many works that attempt to remedy this problem. Liberatore *et al.* explore evolution occurring in real-time [15]. While their work showed promise, compromises had to be made in order to ensure that the real-time evolution did not negatively impact the player's device too drastically. Dockhorn and Kruse explore evolution occurring for multiple NPCs separately [7]. Their work was successful at adding diversity, but required multiple agents, where each agent would exhibit the issue of monotony on its own. Alhejali and Lucas explore the use of evolution occurring throughout multiple environments [1]. Agents could successfully adapt to the new environments, but would still act with monotony among the same environments. Rohlfshagen and Lucas introduce a competition to encourage improvements in general [21]. Competitions like this are great for generating interest in regards to the presented issue, but participants are scored on how *effective* their agents are, not how *fun* their agents are.

There is a practical need for computer automation of intelligent agent design, for example, by using EAs to evolve agents. Hand-coding intelligent agents comes with two significant caveats: (1) developers must spend time and resources designing the behaviours of the NPCs; and (2) players will eventually learn and grow tired of the same behaviours being exhibited by the NPCs. As noted by Bullen and Katchabaw, it is important to produce *enjoyable* and *satisfying* experiences for players [3]. Therefore, we should consider *behavioural diversity*.

1.1   Diversity Search

The search for diversity is a relatively young concept, which has been gaining traction in recent years. Some early approaches to the concept are explored by Lehman and Stanley [11–13], and Mouret and Doncieux [18], who have shown diversity search techniques outperforming their fitness-focused counterparts. These techniques are demonstrated to solve problems like the deceptive maze problem, which is inherently difficult for fitness-based optimization. Fitness relies on static measurements, such as the distance between the agent and the end-goal, and so thorough exploration is often lost to premature convergence. The diversity-based evaluation strategies prove that one can solve this problem by encouraging diverse behaviours among the population.

Diversity search has been applied to the pursuit domain problem. Pozzuoli and Ross improve agent diversity using the MAP-Elites algorithm [20]. Using an age-layered evolutionary algorithm, it is shown that diverse behaviours can be encouraged during evolution. Joseph uses deep learning to evolve diverse high-level emergent predator behaviours [9]. After training a CNN on trace images of predator movements seen during simulations, the CNN is able to inform the fitness evaluation of the degree that these trained behaviours are seen during evolution, and thus allow exploration of new, unique behaviours.

An important note to make when considering the search for diversity is the shift of the overall objective. Fitness-based evaluation relies on a static objective, while diversity-based evaluation is more open-ended [14]. This open-endedness results in a large search space that might contain vast regions of undesirable behaviours. Gomes *et al.* explain that this problem is *"overcome by combining exploratory pressure of novelty search with the exploitative value of fitness-based evolution"* [8]. Mouret successfully explores the use of multi-objectivization to do so [19].

1.2 Contributions

This paper presents a number of contributions to the problem of evolving intelligent agents that are both fit (effective) and diverse (interesting). We examine the pursuit domain problem (*aka* predator vs prey), in which the predator controller is evolved with genetic programming [10]. We are inspired by Lehman and Stanley's novelty search [12], in which both fitness (number of prey captured) and novelty (diverse behaviour) contribute to the optimization function. Combining fitness and diversity in this manner is called *quality-diversity search* [4]. We use the many-objective strategy of *sum of ranks* (or *average rank*) to combine four rudimentary behaviour measurements together, in order to calculate a diversity distance measurement [2,5]. We compare a number of distance calculation strategies, as well as weighting schemes for combining fitness and diversity scores. Finally, we manually observe the emergent behaviour of evolved solutions, to determine whether the diversity measurement combined with fitness indeed results in identifiable novel or "interesting" behaviour. Our primary goal is to encourage behavioural diversity among solutions, while simultaneously maintaining high fitness.

More details of this research are in [6].

1.3 Organization of Paper

Section 2 describes the design of the systems used: the simulation environment, agent controllers, genetic programming language, multi-objective analysis, and evaluation strategies. Section 3 describes the experiment design, outlines the most important experiments conducted throughout this research, and discusses their results. Section 4 summarizes the paper and suggests future research directions.

2 System Design

2.1 Simulation Environment

For efficiency reasons, many works in predator-prey problems limit agent movements to the four discrete directions (up, down, left, right). However, this can constrain variations in behaviour. Our simulation environment is continuous, in which floating point values are used for the coordinates, speeds and rotations. This means the agents are free to move in any direction by any amount (unless restricted by other means). This should allow agents to potentially exhibit a broad range of behaviours.

Every simulation runs for 5000 time units (ticks). During a tick, each agent takes its turn, starting with the predator. Once all the turns are completed, the next simulation tick begins. The predator must catch as many prey as possible before the simulation ends. If a predator and a prey are ever within the same cell, the predator immediately catches the prey and is awarded a fitness point.

2.2   Agent Controllers

Each agent has a two-dimensional location vector (LOC) representing their coordinates within the environment, a two-dimensional rotation vector (ROT) representing the direction they are facing, and a speed value ($SPEED$) representing the maximum distance they can move in a single turn. During the predator's turn, the predator will collect information on the environment and feed it into its GP-evolved tree. The value returned after executing the tree will replace their ROT for this turn. The predator then rotates to match the direction of ROT and moves in that direction. The distance moved is equal to the magnitude of ROT, clamped between 0.0 and $SPEED$.

The prey behave in a similar manner, replacing the GP-evolved tree with their own semi-constant rotation speed value ($ROTSPEED$). During their turn, the prey will rotate their ROT by $ROTSPEED$ and move in that direction 0.7 units. Every five ticks, the prey will modify $ROTSPEED$ by a small random amount. The result is an agent that appears to move in a natural but random motion which can be likened to the movement of a bee searching for a flower.

As shown in Fig. 1, the predator has a field of view of 90° with a distance of 20.0 units, and a sensing radius of 5.0 units If a prey enters either of these zones, the predator will become aware of its location for as long as the prey remains within either of them. When a prey leaves the predator's field of view, the predator will remain aware of the prey's location for another five ticks. When a prey leaves the predator's sensing radius without ever being in the field of view, the predator will lose awareness of the prey immediately.

2.3   Genetic Programming

The Java-based ECJ v.27 [16] is used as the GP platform. The GP parameters used are in Table 1, and most are common in the literature [10].

The GP language is strongly typed [17], and uses three types: conditions (C), floats (F), and 2D vectors (V). The functions and terminals for these types are in Tables 2, 3, and 4. This rich set of language primitives is provided to facilitate the emergence of unexpected and complex behaviours. Most of the language is self-explanatory from the table; see [anon.] for more details. While the language is strongly-typed, functions exist to convert between types: a condition can be

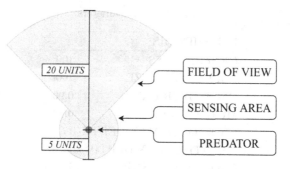

Fig. 1. Predator Agent Perception. *A diagram displaying the sight and sensing areas of a predator.*

made from a float using *ToCondition*; and a vector can be made from two floats using *MakeVector*. The *Get* and *Set* functions provide basic memory, which can share values between ticks in the simulator.

2.4 Computing Diversity

The *fitness* of an individual represents its effectiveness as a predator, which is measured by its prey capturing skill. The *diversity* of an individual is the uniqueness (novelty) of its behaviour during the simulation. The final score used in GP evolution is a (weighted) combination of the individual's fitness and diversity scores, with the goal to maximize both.

The main algorithm steps in computing diversity are as follows:

1. **Calculate individual behaviour vectors**: For every individual in the population, collect its four rudimentary behaviour scores and save them into a raw behaviour vector.
2. **Create a baseline behaviour vector for the population comparison set**: This baseline vector is an average of all behaviour vectors in the comparison set.
3. **Create individual distance vectors**: For every individual, compare each raw behaviour vector value against the corresponding baseline vector value. The differences are stored in the corresponding components in a new distance vector for the individual.
4. **Score the population using sum of ranks**: Using sum of ranks, rank every distance vector in the population. The fitness score will also be combined at this stage.

Details of each step are discussed below.

2.5 Behaviour Vectors

Unlike the fitness score, diversity is open-ended, as it is measured relative to the current population or subset thereof. One can think of diversity as being a

Table 1. GP Parameters

PARAMETERS	
GENERATIONS	100
POPULATION SIZE	1000
CROSSOVER RATE	0.90
MUTATION RATE	0.10
CROSSOVER MAX DEPTH	17
MUTATION MAX DEPTH	17
CROSSOVER TRIES	10
MUTATION TRIES	1
TOURNAMENT SIZE	3
ELITISM SIZE	3
GROW MIN DEPTH	5
GROW MAX DEPTH	5
GROW RATE	0.50
HALF MIN DEPTH	2
HALF MAX DEPTH	6
HALF GROW RATE	0.50
TERMINAL RATE	0.10
NON-TERMINAL RATE	0.90

"distance measure" in an abstract behaviour space. The actual characteristics that might be used to define this space are numerous and varied, and may greatly impact the effectiveness of using diversity during evolution. We chose the following four rudimentary measurements for the behaviour space: (i) *wall impacts* – the number of times a predator hits a wall; (ii) *total time near prey* – total time in which prey is detected within the predator's sensing area (Fig. 1); (iii) *cells visited* – the number of unique arena cells visited by the predator; (iv) *average speed*. These measures were chosen because they represent different aspects of predator interaction with the environment, and do not overlap one other in drastic ways. They are also efficient to compute. Each individual in the population will have these measurements calculated during a simulation. The resulting values comprise its 4-dimensional *behaviour vector*:

$$< b_1^i, \ b_2^i, \ b_3^i, \ b_4^i >$$

Note that these metrics can use different scales of measurement. For example, wall impacts can range from 0 to 5000, while average speed will be a float value between 0.0 and 1.0. They will be normalized when they are converted to a score usable by evolution (Sect. 2.6).

Diversity distance is a measure of how different (far away, distant) an individual resides from other population members in 4D-behaviour space. To determine

Table 2. GP Language: Type C (conditional, boolean)

FUNCTION (C)	DESCRIPTION
Branch (C_1,C_2,C_3)	Returns C_2 if C_1 is true
	Returns C_3 otherwise
Ephemeral	Returns a random condition
	Condition is initialized on node creation
EqualTo (F_1,F_2)	Returns true if F_1 and F_2 are equal ($\pm$ 0.1)
GreaterThan (F_1,F_2)	Returns true of F_1 is greater than F_2
Get	Returns the value stored by "Set"
HittingWall	Returns true if against a wall
ProximityCheck	Returns true if a prey is sensed
SeesPrey	Returns true if a prey is seen
SeesWall	Returns true if a wall is seen
Set (C_1)	Returns C_1 and stores the value in memory
	Value will be unchanged until set again
ToCondition (F_1)	Returns true if F_1 is non-negative

the distance of an individual, one needs to compute its distance from a comparison set. In its simplest form, the comparison set can be the rest of the population. However, it can also be a subset of the population, as well as an archive of individuals from earlier populations. Once the comparison set is defined, a *baseline behaviour vector* is computed, which is a vector of average behaviour scores from the comparison set.

We explore the following variations of comparison sets:

- **All Neighbours (AN)**: Individuals are compared against the entire population.
- **All Neighbours with Archive (ANA)**: Individuals are compared against the entire population and an archive of the previous four generations' populations.
- **K-Nearest Neighbours (KNN)**: Individuals are compared against the K-nearest neighbours from the current population. We use K=32.
- **K-Nearest Neighbours with Archive (KNNA)**: Individuals are compared against the K-nearest neighbours (K=32) from the current population and the previous four generations' populations.

In each of the above strategies, the baseline vector is computed as the average behaviour value found in the defined comparison set:

$$< \bar{b}_1, \bar{b}_2, \bar{b}_3, \bar{b}_4 >$$

After a baseline vector is determined for the comparison set, an individual i's distance vector is determined:

$$< |b_1^i - \bar{b}_1|, |b_2^i - \bar{b}_2|, |b_3^i - \bar{b}_3|, |b_4^i - \bar{b}_4| >$$

Table 3. GP Language: Type F (float)

FUNCTION (F)	DESCRIPTION
Add (F_1,F_2)	Returns sum of F_1 and F_2
Average (F_1,F_2)	Returns average of F_1 and F_2
Branch (C_1,F_1,F_2)	Returns F_1 if C_1 is true
	Returns F_2 otherwise
BreakVectorX (V_1)	Returns X value from vector V_1
BreakVectorY (V_1)	Returns Y value from vector V_1
Cosine (F_1)	Returns cosine of F_1
CurrentTick (F)	Returns the current tick of the simulation
Divide (F_1,F_2)	Returns F_1 divided by F_2
	Dividing by zero produces zero
Dot (V_1,V_2)	Returns dot product of V_1 and V_2
Ephemeral	Returns a random float $[-1.0,+1.0]$
	Float is initialized on node creation
Get	Returns the value stored by "Set"
Invert (F_1)	Returns F_1 flipped
MaxXY $(F::=F_1F_2)$	Returns the greater value among F_1 and F_2
MinXY (F_1,F_2)	Returns the smaller value among F_1 and F_2
Multiply (F_1,F_2)	Returns F_1 multiplied by F_2
Set (F_1)	Returns F_1 and stores the value in memory.
	Value will be unchanged until set again
Sine (F_1)	Returns sine of F_1
StepsSinceLastCatch	Returns length of time since last catch
Subtract (F_1,F_2)	Returns F_1 subtracted by F_2
VectorLength (V_1)	Returns length of the vector V_1

The greater the difference between an individual's behaviour and the corresponding baseline behaviour, the more diverse that individual will be with respect to that behaviour.

2.6   Score the Population Using Sum of Ranks

The final step is to combine the population fitnesses and distance vectors into single scores for each individual, which will be used during evolution. As seen above, an individual's diversity is based upon the distance vector computed from the four raw behaviour measurements. High distance values are preferred.

We combine distance scores using the many-objective evaluation strategy called sum of ranks (or average rank), which has a successful track record in multi-objective optimization when four or more objectives are considered [2,5]. Sum of ranks allows us to combine multiple objective scores into one without

Table 4. GP Language: Type V (vector)

FUNCTION (V)	DESCRIPTION
Add (V_1,V_2)	Returns sum of V_1 and V_2
Branch (C_1,V_1,V_2)	Returns V_1 if C_1 is true
	Returns V_2 otherwise
Ephemeral	Returns a random vector ($[-1,1]$,$[-1,1]$)
	Vector is initialized on node creation
Get	Returns the value stored by "Set"
Inverse (V_1)	Returns the inverse (negated) V_1
MakeVector (F_1,F_2)	Returns vector (F_1,F_2)
Normalize (V_1)	Returns the unit vector of V_1
PredLocation	Returns predator's location vector
PredRotation	Returns predator's rotation vector
ProximityCheck	Returns location of a sensed prey
	Returns predator's location otherwise
ReplaceX (F_1,V_1)	Returns V_1 with X replaced by F_1
ReplaceY (F_1,V_1)	Returns V_1 with Y replaced by F_1
Rotate (F_1,V_1)	Rotates V_1 by F_1 degrees
Set (V_1)	Returns V_1 and stores it in memory
SightCheck	Returns location of a seen prey
	Returns predator's location otherwise
Subtract (V_1,V_2)	Returns V_1 subtracted by V_2
SwapXY (V_1)	Returns V_1 with swapped X and Y values

consideration of different measurement scales used among the scores. We will also use it to create a weighted sum of fitness and diversity during the final step in the process.

Consider Table 5 There are six individuals in the population, and two behaviour distance scores (*A* and *B*). The *ranks* column shows the integer rank of each separate objective; lower integers are assigned to preferred (higher distance) scores. If one sums these ranks, the result is the value in the *sum of ranks* column, where low values are preferred. In some applications, this score suffices as a measure to be used for fitness-based selection during evolution.

However, the raw sum of ranks score can be biased if one objective has a greater number of ranks than another. This is the case here, since *A* has five ranks in total, while *B* has three. By dividing the *rank* scores by the maximum rank for that respective objective, and summing these normalized values, the normalized sum of ranks is determined. Note how the rank ordering (integers in parentheses) of *sum of ranks* and *normalized sum of ranks* differ. Normalization has created a finer resolution of scores, by removing the bias caused by the disparity of total ranks between A and B.

Table 5. Sum of Ranks Example

INDIV.	Raw Fitness		Ranks		Sum of	Norm. Sum
	A	B	A	B	Ranks (Rank)	of Ranks (Rank)
1	10	50	2	1	3 (1)	0.73 (1)
2	2	30	3	2	5 (3)	1.27 (4)
3	1	50	4	1	5 (3)	1.13 (2)
4	0	30	5	2	7 (5)	1.67 (6)
5	0	50	5	1	6 (4)	1.33 (5)
6	16	9	1	3	4 (2)	1.20 (3)
Max rank =			5	3		

Once the normalized sum of ranks is determined for the diversity score, these values are normalized between 0.0 and 1.0. Likewise, the normalized ranks for fitness are calculated. The fitness and diversity scores are then combined for each individual i:

$$Final\_score^i = (w_1 Fitness^i) + (w_2 Diversity^i)$$

where $Fitness^i$ and $Diversity^i$ are normalized sum of ranks values, and w_1 and w_2 are appropriate weights. Our weights range from 0.0 to 1.0.

3   Experiments

Experiments uses a weighting to combine fitness and diversity score. This weighting is referred to in the format XF/YD, where X is the fitness weight and Y is the diversity weight. Every experiment has 30 runs.

3.1   Setting a Baseline

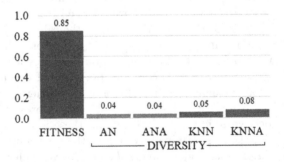

Fig. 2. Baseline Experiment (100F/0D) Final Scores. Note that diversity of solutions is measured, although it is not used during evolution.

To begin, we obtained baseline results that ignore diversity influences, by running GP using 100% fitness and 0% diversity (100F/0D). Figure 2 shows the fitness score of the baseline along with the diversity scores reported by each of the four diversity-based evaluation strategies. The baseline experiment demonstrates this convergence, as the low diversity scores show that the 30 evolved agents exhibit the same general raw behaviours.

3.2   Testing the Diversity Strategies

Four sets of experiments followed the baseline. Each set focused on a single diversity-based evaluation strategy (AN, ANA, KNN, or KNNA), wherein each experiment in the set used different weights. Specifically, each of the four sets explored five weighting combinations: 100F/0D (baseline), 75F/25D, 50F/50D, 25F/75D, and 0F/100D.

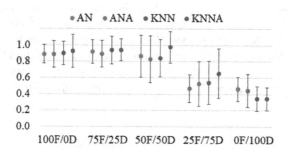

Fig. 3. Final Scores of All Strategies. The summed fitness and diversity scores (total score) achieved by each diversity strategy organized into the five weighting combinations.

A plot of combined fitness/diversity score distributions are shown in Fig. 3. Note that scores plotted for diversity-heavy weights (25F/75D, 0F/100D) are clearly weaker than the others, showing that fitness is not contributing significantly. Furthermore, the KNNA results (K-nearest neighbours with archive) are often superior to other strategies. Page constraints mean we will focus on the KNNA experiments henceforth. (See [*anon.*] for details on other experiments).

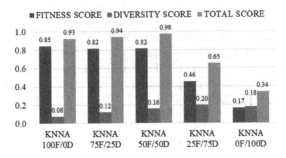

Fig. 4. KNNA Final Scores. Final scores of the best individual from each experiment. Scores are averaged over five test simulations during each of the 30 runs.

Figure 4 breaks down the KNNA experiment set to further analyse the five weighting combinations. As expected, a weighting shift from fitness to diversity causes the fitness score to decrease while the diversity score increases. Of note is that the fitness decreases very little to achieve a significant diversity increase. Specifically, KNNA 50F/50D traded 3% from the fitness score to gain 8% for the diversity score. Note that fitness and diversity scores are not directly comparable in terms of what they denote. As is shown below, the subjectivity of "interesting" emergent behaviours means that a small increase in diversity score may have a significant impact.

Table 6. Emergent Behaviour Descriptions

BEHAVIOUR	DESCRIPTION
WANDERING	Is moving without an obvious pattern
SCRAPING	Scrapes along the walls of the arena when moving
BOUNCING	Bounces off of walls upon impact
SCANNING	Scans up and down each row or column sequentially
CIRCLING	Moves in circular motions
SHAKING	Does not maintain a smooth rotation; field of view shakes
SPLITTING	Flips between two rotations every tick
CHASING	Chases a prey
POUNCING	Chases and immediately catches a prey
IDLING	Does not move
STALKING	Chases but intentionally avoids catching the prey
SLOWING	Gradually slows speed
CAMPING	Spends majority of time near one location
AVOIDING	Moves around prey
FLICKERING	Switches between moving and idling every other tick
HERDING	Chases prey against a wall and holds it there

An analysis of emergent behaviour of evolved predators is required in order to confirm that behaviours are truly more diverse. To do this, the best individual from every run of an experiment was observed in the simulation environment, and the exhibited emergent behaviours were recorded and tallied. The qualitative analysis process is subjective, since it must be done through human observation.

To mitigate biased decisions, simulations were watched in a random order, with the diversity strategy unknown to the observer until after recording. Table 6 lists our characterizations of observed emergent behaviours, which were collected during preliminary observations of evolved agents in the simulation.

Behaviours are labelled as either "primary" or "secondary". Primary behaviours are those that appear to be the most prominent during an entire simulation, and are always a type of locomotion-based behaviour, for example, wandering, scraping, and scanning. Secondary behaviours are observed to exist alongside the primary behaviour. This could include a predator that scrapes the walls but does so with shaky movement; scraping would be primary while shaking would be secondary in this example. A predator can also exhibit multiple locomotion-based behaviours, but only the most prominent will be marked as primary.

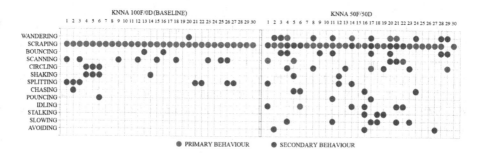

Fig. 5. Observed emergent behaviours - KNNA 100F/0D versus KNNA 50F/50D. Side-by-side comparison of the empirical analyses of the best solution from each of the 30 runs used in each experiment. Tallies indicate that the behaviour was observed at least once during that simulation.

Figure 5 shows a summary of emergent behaviours observed for the 100F/0D and KNNA 50F/50D runs. The baseline 100F/0D results are as expected: a single behaviour was overused and exploited. Due to the prey getting stuck on the walls for several turns at a time (caused by their random movement), the predator learned to simply scrape the environment walls in hopes of catching the unlucky prey. Some secondary behaviours arise on occasion, but they are infrequent and fleeting. On the other hand, almost every simulation from the KNNA 50F/50D experiment utilized scraping like the baseline, but a high diversity score is achieved by incorporating more secondary behaviours. More importantly, some new high-level emergent behaviours arose, for example:

1. **Sleeping To Scraping**: The predator sits idly and ignores any prey crossing its field of view. Once a prey crosses the predator's sensing radius, it chases the prey to a wall and begins scraping for the remainder of the simulation.
2. **Strolling To Scraping**: Similar to above, but the predator moves at an especially slow speed in a circular motion until hitting a wall. Then the predator begins scraping behaviour.
3. **Selective Bouncing**: The predator scrapes along three walls as normal, but bounces along the fourth. Sometimes during the bouncing, the predator pounces on nearby prey before returning to the wall. This is especially interesting since the pouncing behaviour requires the GP to evolve the correct vector mathematics to locate and navigate to detected prey, but scraping has no need for this technical ability.
4. **Passive Wall Scraping**: The predator scrapes the walls, but stops and waits for stuck prey to move out of the way before continuing forward. This occurs at seemingly random times, giving the impression that the predator is a "picky eater" and is waiting for a different catch.
5. **Scraping With Chasing**: The predator scrapes the walls as normal, but if a prey gets unstuck and leaves the wall moments before capture, the predator chases and catches the prey. After that, the predator turns around and returns to scraping.

In conclusion, we suspect that GP evolved solutions switch between different behaviours throughout the simulation in order to balance the fitness score with the diversity score. In most KNNA 50F/50D solutions, the predator efficiently scrapes the walls to catch as many prey as possible, but in order to maintain a high diversity score, alternate emergent behaviours appeared throughout the simulation, and most significantly, were not detrimental to the predator's hunting abilities.

3.3   Discussion

The primary goal of this research was to find a strategy that promotes diverse emergent behaviour, while minimally impacting the predator's prey capturing skill. The KNNA 50F/50D is the most successful candidate strategy in this regard. To support this claim, we apply a Mann-Whitney U test (alpha = 0.05, 30 data points per experiment) to the experimental fitness and diversity scores, comparing strategies with the 100F/0D vanilla GP. The intention is to highlight strategies in which solutions exhibit both statistically significant increases in diversity and negligible changes in fitness, when compared to 100F/0D. The following five strategies were flagged: AN 75F/25D, ANA 75F/25D, KNN 75F/25D, KNNA 75F/25D, and KNNA 50F/50D. However, we examined the emergent behaviours of the 75F/25D strategy solutions, and the observed emergent behaviours were unremarkable and too subtle to notice. On the other hand, the KNNA 50F/50D results clearly showed observable diverse emergent behaviours. We therefore designate the KNNA 50F/50D strategy as the most successful one.

4 Conclusion

This research explored strategies designed to encourage diverse behaviours while still maintaining an appreciable level of efficacy (fitness). The nature of fitness-based evaluation tends to exploit particular behaviours in monotonous and uninteresting ways. With the addition of diversity-based evaluation, this is remedied. K-Nearest Neighbours with Archive (KNNA) with an even 50/50 weighting between fitness and diversity scores achieved the best quantitative results. These results were confirmed by examining the emergent behaviours of solutions.

The use of many-objective sum of ranks was successful in combining fitness and the four raw behaviour measurements. Pareto multi-objective scoring, such as in [19], degenerates when four or more objectives are considered. Sum of ranks has proven success in other problem domains having 15 or more objectives [2,5]. Future work could consider more behaviours than the four we used, which would likely result in even richer varieties of emergent behaviours.

Another contribution is our evaluation of emergent behaviours of solutions, to confirm that high diversity scores equate to an increase in noticeable emergent behaviours. Admittedly, this is a very time-consuming and subjective analysis, and is prone to human bias. The emergent behaviours we identified, however, are not so straight-forward to measure, because they are inherently "unexpected" and hence difficult to program in advance for identification purposes. A user survey to identify generic verses interesting agent behaviour would mitigate human bias.

There are other directions for future work. Other intelligent agent problems such as herding and food gathering could be studied. A 3D environment with a physics simulation could be considered, as the types of emergent behaviour would be much more complex. The automatic identification of emergent behaviour of agents, perhaps via machine learning as in [9], is worth considering.

Acknowledgements. This research was supported by NSERC Discovery Grant RGPIN-2016-03653.

References

1. Alhejali, A.M., Lucas, S.M.: Evolving diverse ms. pac-man playing agents using genetic programming. In: 2010 UK Workshop on Computational Intelligence (UKCI), pp. 1–6. IEEE (2010)
2. Bentley, P.J., Wakefield, J.P.: Finding acceptable solutions in the pareto-optimal range using multiobjective genetic algorithms. In: Chawdhry, P.K., Roy, R., Pant, R.K. (eds.) Soft Computing in Engineering Design and Manufacturing, pp. 231–240. Springer London, London (1998). https://doi.org/10.1007/978-1-4471-0427-8_25
3. Bullen, T., Katchabaw, M.: Using genetic algorithms to evolve character behaviours in modern video games. In: Proceedings of the GAMEON-NA (2008)
4. Chatzilygeroudis, K.I., Cully, A., Vassiliades, V., Mouret, J.: Quality-diversity optimization: a novel branch of stochastic optimization. CoRR (2020). https://arxiv.org/abs/2012.04322

5. Corne, D.W., Knowles, J.D.: Techniques for highly multiobjective optimisation: some nondominated points are better than others. In: Proceedings of the 9th Annual Conference on Genetic and Evolutionary Computation, pp. 773–780 (2007)

6. Cowan, T.: Strategies for Evolving Diverse and Effective Behaviours in Pursuit Domains. Master's thesis, Brock University (2021)

7. Dockhorn, A., Kruse, R.: Combining cooperative and adversarial coevolution in the context of pac-man. In: 2017 IEEE Conference on Computational Intelligence and Games (CIG), pp. 60–67. IEEE (2017)

8. Gomes, J., Mariano, P., Christensen, A.L.: Devising effective novelty search algorithms: a comprehensive empirical study. In: Proceedings of the 2015 Annual Conference on Genetic and Evolutionary Computation, pp. 943–950 (2015)

9. Joseph, M.: Emergent Behaviour in Game AI: A Genetic Programming and CNN-based Approach to Intelligent Agent Design. Master's thesis, Brock University (2023)

10. Koza, J.R., Koza, J.R.: Genetic programming: on the programming of computers by means of natural selection, vol. 1. MIT press (1992)

11. Lehman, J., Stanley, K.O.: Abandoning objectives: evolution through the search for novelty alone. Evol. Comput. **19**(2), 189–223 (2011)

12. Lehman, J., Stanley, K.O.: Evolving a diversity of virtual creatures through novelty search and local competition. In: Proceedings of the 13th Annual Conference on Genetic and Evolutionary Computation, pp. 211–218 (2011)

13. Lehman, J., Stanley, K.O.: Novelty search and the problem with objectives. In: Riolo, R., Vladislavleva, E., Moore, J.H. (eds.) Genetic Programming Theory and Practice IX, pp. 37–56. Springer New York, New York, NY (2011). https://doi.org/10.1007/978-1-4614-1770-5_3

14. Lehman, J., Stanley, K.O., et al.: Exploiting open-endedness to solve problems through the search for novelty. In: Proceedings of the 11th International Conference on Artificial Life (ALIFE XI), pp. 329–336 (2008)

15. Liberatore, F., Mora, A.M., Castillo, P.A., Merelo, J.J.: Comparing heterogeneous and homogeneous flocking strategies for the ghost team in the game of ms. pac-man. IEEE Trans. Comput. Intell. AI Games **8**(3), 278–287 (2015)

16. Luke, S.: ECJ: A java-based evolutionary computation research system. https://cs.gmu.edu/~eclab/projects/ecj/. [Accessed 13 Oct 2021]

17. Montana, D.: Strongly typed genetic programming. Evol. Comput. **3**(2), 199–230 (1995)

18. Mouret, J.B., Doncieux, S.: Encouraging behavioral diversity in evolutionary robotics: an empirical study. Evol. Comput. **20**(1), 91–133 (2012)

19. Mouret, J.-B.: Novelty-Based Multiobjectivization. In: Doncieux, S., Bredèche, N., Mouret, J.-B. (eds.) New Horizons in Evolutionary Robotics, pp. 139–154. Springer Berlin Heidelberg, Berlin, Heidelberg (2011). https://doi.org/10.1007/978-3-642-18272-3_10

20. Pozzuoli, A., Ross, B.J.: Increasing features in map-elites using an age-layered population structure. In: IEEE Congress on Evolutionary Computation, CEC 2023, Chicago, IL, USA, July 1–5, 2023, pp. 1–8. IEEE (2023). https://doi.org/10.1109/CEC53210.2023.10254093

21. Rohlfshagen, P., Lucas, S.M.: Ms pac-man versus ghost team cec 2011 competition. In: 2011 IEEE Congress of Evolutionary Computation (CEC), pp. 70–77. IEEE (2011)

Using Evolution and Deep Learning to Generate Diverse Intelligent Agents

Marshall Joseph and Brian J. Ross[✉]

Department of Computer Science, Brock University, 1812 Sir Isaac Brock Way,
St. Catharines, ON L2S 3A1, Canada
bross@brocku.ca
http://www.cosc.brocku.ca/~bross/

Abstract. Emergent behaviour arises from the interactions between individual components of a system, rather than being explicitly programmed or designed. The evolution of interesting emergent behaviour in intelligent agents is important when evolving non-playable characters in video games. Here, we use genetic programming (GP) to evolve intelligent agents in a predator-prey simulation. A main goal is to evolve predator agents that exhibit interesting and diverse behaviours. First, we train a convolutional neural network (CNN) to recognize "generic" prey behaviour, as recorded by an image trace of a predator's movement. A training set for 6 generic behaviours was used to train the CNN. A training accuracy of 98% was obtained, and a validation performance of 90%. Experiments were then performed that merge the CNN with GP fitness. In one experiment, the CNN's classification values are used as a "diversity score" which, when weighted with the fitness score, allow both agent quality and diversity to be considered. In another experiment, we use the CNN classification score to encourage the evolution of one of the known classes of behaviours. Results were that this trained behaviour was indeed more frequently evolved, compared to GP runs using fitness alone. One conclusion is that machine learning techniques are a powerful tool for the automated generation of diverse, high-quality intelligent agents.

1 Introduction

In order for a player to be engaged in playing a video game, the game needs to be both challenging and interesting. "Challenging" implies that the player must develop skill to get a good score. On the other hand, "interesting" is not so easy to define. A very challenging game that the player cannot hope to win against will be decidedly uninteresting. However, one in which the computer opponent exercises the player's skill in unpredictable and innovative ways, would be an interesting one for the player.

The topic of computer games is a popular application area for AI research [22]. One much-studied topic is the automatic generation of computer opponents (aka non-playable characters, or NPCs). In pursuit (predator-prey) games, of which Pac-Man is an example, an intelligent agent is given the task to capture a set number of prey agents scattered across the environment [18]. Evolutionary algorithms are commonly used to evolve the predator's controller [14,15]. A fitness score tallies the number of prey the predator caught during a run and is used to determine who moves forward in the evolutionary process. A result, however, is that evolved predators often exhibit generic, monotonous behaviour, thanks to the effect of fitness pressure to obtain the best performing agent, which is not necessarily an interesting one.

Diverse behaviour helps agents perform under situations of uncertainty and creates more interesting computer opponents. Novelty search [11] rewards agents for discovering new and unexpected behaviours, regardless of their performance. Research in [5] combines fitness and novelty search to evolve predator agents, and a weighting of 50% resulted in good quality predators with assorted emergent behaviours that did not arise as frequently with fitness-only evolution. The work in [17] uses the MAP-Elites [13] quality-diversity algorithm to evolve an assortment of predator agents with varied emergent behaviours.

In the context of pursuit simulations, identifying and characterizing agent behaviour can be a difficult task due to the complexity of agent interaction with the environment. Predator agents often exhibit a wide range of behaviours and hunting strategies, such as ambushing or pursuit, which make it difficult to develop objective criteria for identifying and analyzing agent actions. Another problem is that it can be subjective and context-dependent, which can make it challenging to be consistent when labeling data. It can also be time consuming to manually label thousands of agents [5].

The goal of this research is to use deep learning to model the emergent behaviours of intelligent agents. By automating the recognition of agent behaviour, we can quickly distinguish common behaviours from unusual ones. This CNN will then be combined with genetic programming (GP) to assist in the evolution of high-quality, interesting predator agents. Fitness will use the number of prey captured, while the CNN will contribute a diversity score based on its evaluation of the behaviours seen. With this fitness-diversity scoring system, we have the ability to evolve agents that exhibit new emergent behaviours, while still encouraging prey-capturing skills. We can also use the CNN to help evolve specific known behaviours. More details of this research are in [6].

Section 2 presents the system design of the CNN and GP system. Section 3 discusses training and testing of the CNN model. Section 4 illustrates experiments that use the CNN and GP system together, to evolve diverse agents. Section 5 uses CNN and GP to evolve agents having a specific target behaviour. Section 6 gives final conclusions and suggestions for future work.

2   System Design

2.1   Pursuit Domain Simulation

A predator-prey simulation environment was implemented in Java. GP-evolved predators executed in the simulation are tasked to capture as many prey agents as possible within a fixed time limit. The number of prey captured will be the predator's fitness score. The system also records other details seen during a simulation, for example, a trace of the predator's movement. This recording will be used later to encode image traces of a predator's movements, which are required for processing by the CNN system for classifying predator behaviours.

The simulation maintains the following state information during its execution: (i) agent location (x, y coordinates); (ii) orientation (angle); (iii) GP tree and its evaluated (float) value, which updates the orientation value; and (iv) speed. The environment is a continuous, floating-point 200-by-200 unit arena in which the predator is free to move around. The advantage of a continuous environment over a discrete, integer-based grid is that it enables subtle changes in speed and direction, which will encourage interesting emergent behaviours. The predator agent always starts in the center of the environment and faces a random direction. At the start of the simulation, 25 prey agents are randomly placed inside the environment. After all agents are initialized, they then move in straight directions, with occasional random turns. When the predator is within 1 unit of a prey, the prey is considered captured.

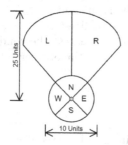

Fig. 1. Predator Perception

Figure 1 shows the perception fields of a predator agent who has the ability to seek and sense in every direction. A predator's ability to sense depends on the composition of its evolved controller (see Table 2). A full suite of sensors is not always guaranteed, as predator agents can differ from one another depending on their GP-evolved controller.

2.2 GP Architecture

Genetic programming (GP) is used to evolve predator agents [8,16]. We implement the GP system in ECJ 27 [12]. GP parameters used in experiments are in Table 1, and most are standard in the literature [8,16].

The GP language is shown in Table 2. The purpose of a GP tree is to perform predator sensing and movement in the simulation. Every clock tick of the simulator involves executing the GP tree, in order to determine speed and direction changes during that tick (see Sect. 2.1). *SetSpeed* saves the clamped floating

Table 1. GP Parameters

Parameter	Value
Generations	50
Number of Simulation Steps	5000
Population Size	500
Tournament Size	2
Maximum Tree Depth	15
Crossover Chance	90%
Mutation Chance	10%
Growth Algorithm	Half Builder
Growth Chance	50%
Initial Depth	2 to 6

point value as the next speed to use; the last execution of this function will be the one that is used. *MovesRemaining, MovesTaken, Orientation, PreyCaptured,* and *PreyRemaining* are terminals that return the current respective values from the environment. The functions prefixed with *Seek, Sense, TouchingWall,* and *GreaterThan* are all conditionals that test the condition and react accordingly. The remaining functions perform arithmetic, or return ephemeral constants or *pi.*

2.3 CNN Architecture

The CNN will take a trace image of a predator's movement during a simulation, and attempt to classify it according to its trained model of behaviours. The CNN uses a supervised approach for training, where each trace has been manually labelled based on one of six emergent behaviours. The behaviours were specifically chosen to be easy for manual identification.

The CNN architecture is shown in Table 3. The system is implemented using Python 3.7 [20] and TensorFlow 2.12 [1]. The architecture chosen was based on the work of [4], which performed a similar kind of image classification task to ours, but for L-systems images. The model was also inspired by VGGNet [19] and LeNet [10]. The CNN consists of 13 layers, each with different characteristics. The rescaling layer is the first layer. It takes an image input of (200, 200, 1), and normalizes all pixel values between 0 and 1. The following six layers consist of convolutional layers and max pooling layers. In each of the convolutional layers, a kernel size of (3, 3) is chosen alongside the ReLU activation function. The max pooling layers use a pool size of (2, 2). By combining these two layers and increasing the number of filters at each stage, various features can be extracted from the images. Layers 8, 10, and 12 are dropout layers which help to prevent overfitting by randomly dropping out some of the neurons. The first dropout layer uses a rate of 0.25, whereas the final two dropout layers use a rate of

Table 2. GP Language

GP Function	Description
SetSpeed (A)	Set predator speed to [0.5 <= A <= 5.0] and return A
MovesRemaining	The number of moves remaining in the current run
MovesTaken	The number of moves taken in the current run
Orientation	The current orientation of the predator
PreyCaptured	The number of prey captured in the current run
PreyRemaining	The number of prey remaining in the current run
SeekPreyLeft (A, B)	If prey in left vision radius return A, else return B
SeekPreyRight (A, B)	If prey in right vision radius return A, else return B
SeekWall (A, B)	If wall in vision radius return A, else return B
SensePreyNorth (A, B)	If prey in north sensing radius return A, else return B
SensePreyEast (A, B)	If prey in east sensing radius return A, else return B
SensePreyWest (A, B)	If prey in west sensing radius return A, else return B
SensePreySouth (A, B)	If prey in south sensing radius return A, else return B
TouchingWall (A, B)	If prey touching wall return A, else return B
GreaterThan (A, B, C, D)	If A < B return C, else return D
Add (A, B)	Return A + B
Subtract (A, B)	Return A − B
Multiply (A, B)	Return A · B
Divide (A, B)	If B ≠ 0 return A ÷ B, else return 0
Negative (A)	Return −A
Sin (A)	Return sin(A)
ASin (A)	Return arcsin(A)
Ephemeral	Random constant between [−1, 1]
Pi	The number π

0.50. The dense layers are fully connected and use batch normalization which allow for faster and more stable training of the network. The output layer uses a softmax activation function to produce a probability distribution over the six output classes.

2.4 Combined GP/CNN System

Figure 2 gives a high-level overview of the combined automatic agent generator system and its 3 key components. The GP component consists of the simulation itself as well as the GP controller used to control the predator agent in its environment. Once the simulation has run for 50 generations, a trace file is generated which depicts the best performing predator's movements. This is then converted to an image trace generator, for use by the CNN for classification. The CNN's classification score is then returned to the GP system, to be used for fitness/diversity evaluation.

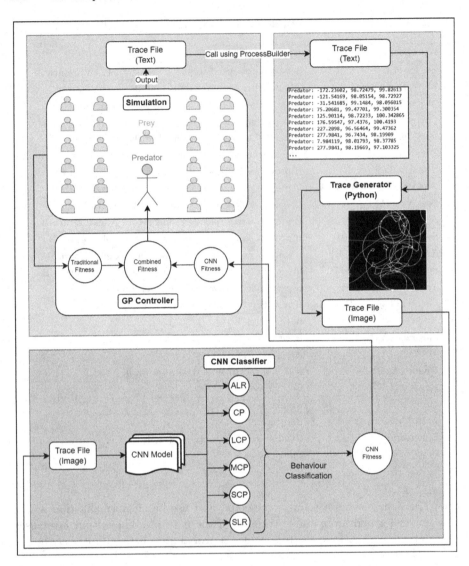

Fig. 2. System Architecture (Blue: GP Simulation, Green: Trace Generator, Purple: CNN Classifier) (Color figure online)

Table 3. CNN Architecture

#	Layer	Output Shape	Neurons	Activation Function	Note
1	Rescaling	(200, 200)	1		Input
2	Conv2D	(198, 198)	32	ReLU	Kernel Size = (3, 3)
3	MaxPooling2D	(99, 99)	32		Pool Size = (2, 2)
4	Conv2D	(97, 97)	64	ReLU	Kernel Size = (3, 3)
5	MaxPooling2D	(48, 48)	64		Pool Size = (2, 2)
6	Conv2D	(46, 46)	128	ReLU	Kernel Size = (3, 3)
7	MaxPooling2D	(23, 23)	128		Pool Size = (2, 2)
8	Dropout	(23, 23)	128		Rate = 0.25
9	Dense	(256)	256	ReLU	Batch Normalization
10	Dropout	(256)	256		Rate = 0.5
11	Dense	(128)	128	ReLU	Batch Normalization
12	Dropout	(128)	128		Rate = 0.5
13	Dense	(6)	6	Softmax	Output

Table 4. CNN Training Parameters

Parameter	Value
Validation Split	20%
Batch Size	32
Optimizer	Adam
Metrics	Accuracy, Loss
Loss Type	Sparse Categorical Cross-entropy
Epochs	20

3 CNN for Emergent Behaviour Classification

3.1 CNN Training

After many trial runs of GP and the pursuit simulator, we recognized six common emergent behaviours (Fig. 3) that were suitable for classification by the CNN. Classic pursuit is a commonly seen behaviour, which primarily involves a predator sensing and chasing a prey. The other behaviours are self-expanatory from their label descriptions.

To train the CNN, we required 2000 examples from each behaviour category for training and validation. In order to efficiently generate effective training sets for the categories, we took each trace instance of a behaviour, and performed sequences of vertical and horizontal image mirror flips, in which combinations of them are equivalent to 90-degree image rotations. Since CNN's are rotation-variant, each of the seven flipped/rotated image variants comprises a unique training example of that behaviour class.

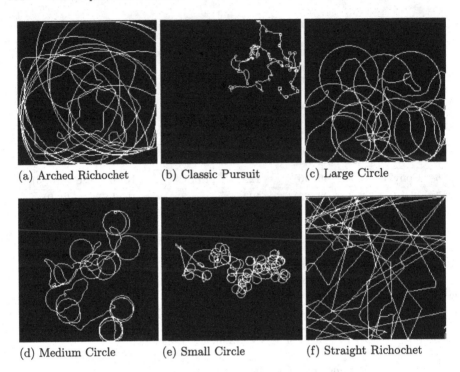

(a) Arched Richochet   (b) Classic Pursuit   (c) Large Circle

(d) Medium Circle   (e) Small Circle   (f) Straight Richochet

Fig. 3. Categories of Emergent Behaviours

The training parameters in Table 4 were used, and are common in the literature [2,3]. The validation split withholds 20% of the training examples to validate the model. The batch size chosen is 32. The adaptive moment estimation optimization algorithm, Adam [7], is used.

Figure 4 shows the results from training and validation. After training for 20 epochs, a training accuracy of 98% is achieved, indicating a strong fit of the model to the data set and negating any concerns of under-fitting, while a validation accuracy of 90% is attained. The training loss converges at 0.065 and the validation loss converges at 0.37.

To test the CNN model, 50 new trace images for each behaviour category were given to the CNN. Table 5 shows the testing accuracy of the CNN on each of the six categories. The behaviours are denoted as ALR (*arched line ricochet*), SLR (*straight line ricochet*), CP (*classic pursuit*), SCP (*small circle pursuit*), MCP (*medium circle pursuit*), and LCP (*large circle pursuit*). The first column denotes the labelled test behaviour, while rows denote the CNN classification for that labelled behaviour. The green diagonal shows the accuracy of correct classifications for each behaviour. The richochet (ALR, SLR) behaviours were the most challenging for the CNN, as there was often confusion between them, likely due to their similar trace characteristics. Classic pursuit (CP) and medium circle pursuit (MCP) were the easiest to identify.

Figure 5 visualizes the data in Table 5. For example, the third bar shows that almost all classic pursuit (CP) traces are correctly identified.

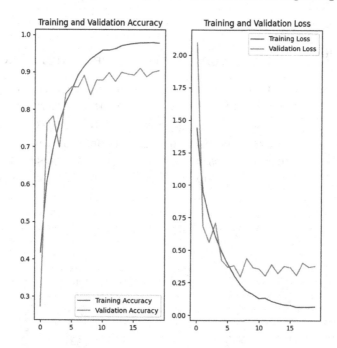

Fig. 4. CNN Training & Validation Results

4 Diverse Agent Generation

In this section, we combine GP and the trained CNN model in order to generate intelligent predator agents that are both fit and diverse. The fitness used by GP when generating training traces is kept, which is simply the normalized total of number of prey captured:

$$Fitness = (\# \ prey \ captured)/25 \tag{1}$$

The CNN classification (prediction) score is used for diversity:

$$Diversity = (Prediction - 16.67)/(100 - 16.67) \tag{2}$$

where *Prediction* represents the CNN's highest categorical prediction from the six categories, ranging from 16.67 to 100. This encourages the predator to combine as many of the behaviours as possible, ideally creating an even spread of each behaviour, rather than using a uniform known one. The overall combined fitness-diversity scores is an even balance of the above:

$$Combined \ Score = Fitness \ + \ Diversity \tag{3}$$

Table 6 summarizes fitness performance over the 30 runs per weighting. In the plots, lower fitness values are preferred. Note that fitness is weak when no fitness measure is used (0F100N), and it improves as fitness weights increase.

Table 5. Average Prediction per Category, CNN Testing. (50 examples per category)

Experiment / Class	ALR	SLR	CP	SCP	MCP	LCP
ALR	**35.1546**	30.8238	0.007	0.0196	0.0026	5.3338
SLR	32.2062	**68.9732**	0.007	0.015	0.0032	0.1524
CP	0.5358	0.0184	**97.3672**	0.4834	0.0028	0.0244
SCP	2.529	0.0184	0.9232	**83.0656**	0.8588	0.0258
MCP	0.3156	0.0548	1.6754	16.4	**98.9784**	5.135
LCP	29.2582	0.1112	0.0138	0.0134	0.1516	**89.327**

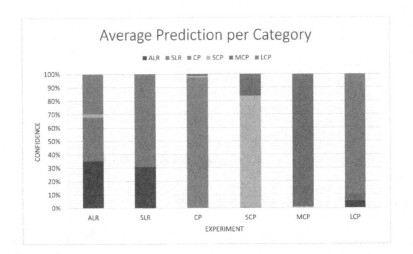

Fig. 5. Average Prediction per Category Histogram: CNN Testing

The "sweet spot" weighting is 75F25N, which performs better than 100F0N. This implies that the CNN is helping fitness quality by promoting genetic diversity in the population.

The final 3 columns in Table 6 are obtained by taking the best agent from each of the 30 runs per weighting, and re-executing it in the simulation 30 times, resulting in 900 traces per weighting category. The resulting averages give more insight into the fitness performance of best solutions, showing how many prey are caught on average, as well as the number of simulations in which more/less than 10 prey were caught.

Figure 6 summarizes the frequency in which different behaviours arise in the 900 traces. CP and SCP are the most common behaviours, while LCP and SLR the least common. Notably, the 75F25N runs show the best spread of evolved behaviours. We performed a statistical analysis, and found that changing weightings had a statistically significant effect on the CP and SLR behaviours ($p < 0.05$; see [6]).

Table 6. Fitness Analysis

	Avg. Mean Fitness	Avg. Best Fitness	Avg. Prey Caught	# Fit >10 Prey	# Unfit <=10 Prey
0F100N	0.721976	0.361333	4.846667	123	777
25F75N	0.530902	0.117333	8.437778	333	567
50F50N	0.352421	0.022667	12.96889	679	221
75F25N	0.245768	0.005333	14.16111	713	187
100F0N	0.261294	0.010667	14.41	693	207

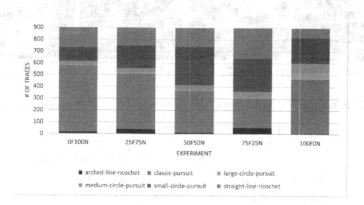

Fig. 6. Behaviour Analysis - Unique. Total of 900 traces per experiment.

Figure 7 shows examples of evolved behaviours. The most diverse trace is (b), which has the lowest score, and hence is closest to 16.67%. Others have higher scores, but do not show exclusive features of the labelled categories. Some, such as (d) and (e), exhibit characteristics of known behaviours – medium circles and straight-line richochet. Trace (a) shows a combination of straight line ricochet, small circles, and classic pursuit.

5   Targeted Behaviour Generation

Here, we combine the GP and CNN, but this time use the CNN prediction score to direct evolution towards one behaviour class – straight line ricochet (SLR). This behaviour was the least common of the six when creating training data (Sect. 3.1), and it was also uncommon when evolving diverse predators (Fig. 6). Hence our combined score is:

$$Combined\ Score = Fitness\ +\ (SLR\ Score/100) \qquad (4)$$

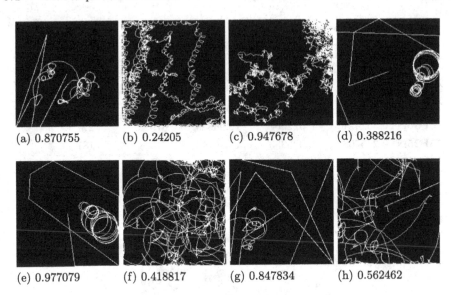

(a) 0.870755   (b) 0.24205   (c) 0.947678   (d) 0.388216

(e) 0.977079   (f) 0.418817   (g) 0.847834   (h) 0.562462

Fig. 7. Some Evolved Behaviours with Diversity Scores (low values preferred)

Table 7. Categorical Data - Targeted

	ALR	CP	LCP	MCP	SCP	SLR
25F75NUnique	40	468	13	35	193	151
75F25NUnique	55	248	9	50	283	255
25F75NTargeted	9	266	1	5	105	514
75F25NTargeted	44	366	14	30	109	337

We test both 25F75N and 75F25N. Each experiment is run 30 times, and the best solution from each run is simulated an additional 30 times, resulting in 900 traces per weight combination.

Table 7 shows the resulting behaviour frequencies. The unique rows refer to the results from Sect. 4. The most significant changes occur in the SCP and SLR columns, which show a significant decrease in SCP behaviours by approximately 50% and an increase in SLR behaviours by over 100%. The previous highest amount of SLR behaviours was in the 75F25N unique experiment which produced 255 SLR agents. In the 25F75N targeted experiment the number of SLR agents increased to 514 which is over 50% of the traces.

Figure 8 summarizes Table 7, including the corresponding weight combinations from the diversity experiment (from Fig. 6). The number of small circle pursuit behaviours has decreased significantly in the ricochet experiments, and the number of straight line ricochet behaviours has increased. In the 25F75N experiment over half of the traces are straight line ricochet which could be a reason why the fitness performance was better than in the 75F25N group. Since the

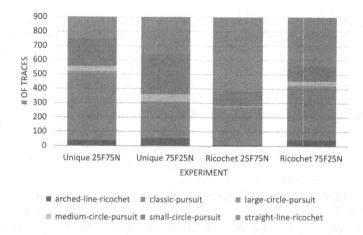

Fig. 8. Behaviour Analysis - Targeted. Total of 900 traces per experiment.

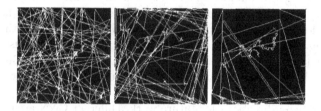

Fig. 9. Examples of Generated Straight Line Richochet Behaviours

straight line ricochet behaviour typically encourages the agent to travel across a large portion of the arena, it is possible that by targeting this behaviour specifically, the number of prey caught will increase. We found that the increase in SLR behaviours to be statistically significant ($p < 0.05$; see [6]).

Figure 9 shows some example SLR behaviours arising from evolved predators.

6   Conclusion

This research shows how deep learning can be used to model intelligent agent behaviour. One challenge while training the CNN was that it is time-consuming to collect and label thousands of trace images. Also, some categories are easy to confuse, for example, small and medium circle classes. We did study this subjective aspect of trace labelling, and found that the CNN's identification scores for ambiguous images somewhat matched our own: if we thought an image was 30% straight line richochet and 70% classic pursuit, the CNN often generated scores with similar values.

Our CNN design with softmax function must always generate a set of scores that sum to 1.0. Our assumption in Sect. 4 is that a diverse image does not have a strong score in any category. However, it is important to realize that *any greyscale image* given to this CNN will result in scores. For example, when a solid white image is supplied to it, it believes at a high certainty that it is an example of a classic pursuit. Therefore, one constraint on modelling with our CNN is that the images given to the CNN must be from the domain of *sensible* predator traces. Another approach is to either have a new category called "none of the others", or train another CNN to recognize sensible traces from nonsensical ones. [4] tried both approaches when using a CNN for L-system tree image classification. However, one pitfall with this approach is that the universe of nonsensical "not a real trace" images is far larger than the ones representing sensible predator traces. Thus CNN training becomes very difficult, because there is a non-exhaustive and unlimited number of negative training examples.

There are interesting differences between our approach to measuring and encouraging behaviour diversity with a CNN, and other research papers that use novelty search [5] and quality-diversity strategies such as MAP-Elites [17]. These other approaches require specific behavioural features to be measured, for example, counting the number of cells visited, the number of times an agent turns direction, or the average speed of an agent. Our CNN approach does not consider low-level measurements, but instead considers the entire high-level emergent behaviour as denoted by a trace image. One benefit of using low-level behaviour metrics is that there is no need for time-consuming CNN training. On the other hand, assessing emergent behaviours in those strategies requires significant human effort.

There are many directions for future work. Different CNN architectures can be explored. Feed-forward neural networks could be used, which might be trained on the text-based trace shown in Fig. 2. It would be interesting to use K-means clustering to identify broad classes of behaviour trace categories, perhaps in the behaviour space used by (for example) novelty search. Enhanced pursuit simulations can be considered, which would yield new forms of emergent behaviours. Applications such as food gathering [9], shepherding [21], and real video games such as Pac-Man [18] may be considered. It is also worth considering whether a CNN be trained to predict a predator's fitness from its behaviour trace.

Acknowledgements. This research was supported by NSERC Discovery Grant RGPIN-2016-03653.

References

1. Abadi, M.: TensorFlow: large-scale machine learning on heterogeneous systems (2015). https://www.tensorflow.org/, software available from tensorflow.org
2. Bengio, Y.: Practical recommendations for gradient-based training of deep architectures. In: Montavon, G., Orr, G.B., Müller, K.-R. (eds.) Neural Networks: Tricks of the Trade. LNCS, vol. 7700, pp. 437–478. Springer, Heidelberg (2012). https://doi.org/10.1007/978-3-642-35289-8_26
3. Bottou, L., Curtis, F.E., Nocedal, J.: Optimization methods for large-scale machine learning. SIAM Rev. **60**(2), 223–311 (2018). https://doi.org/10.1137/16M1080173

4. Chen, X.E., Ross, B.J.: Deep neural network guided evolution of l-system trees. In: 2021 IEEE Congress on Evolutionary Computation (CEC), pp. 2507–2514 (2021). https://doi.org/10.1109/CEC45853.2021.9504827
5. Cowan, T.: Strategies for Evolving Diverse and Effective Behaviours in Pursuit Domains. Master's thesis, Brock University (2021)
6. Joseph, M.: Emergent Behaviour in Game AI: A Genetic Programming and CNN-based Approach to Intelligent Agent Design. Master's thesis, Brock University (2023)
7. Kingma, D.P., Ba, J.: Adam: a method for stochastic optimization. arXiv (2017). https://doi.org/10.48550/arXiv.1412.6980
8. Koza, J.R.: Genetic programming - on the programming of computers by means of natural selection. MIT Press (1992)
9. Koza, J.R., Roughgarden, J., Rice, J.P.: Evolution of food-foraging strategies for the caribbean anolis lizard using genetic programming. Adapt. Behav. 1(2), 171–199 (1992). https://doi.org/10.1177/105971239200100203
10. LeCun, Y., Bottou, L., Bengio, Y., Haffner, P.: Gradient-based learning applied to document recognition. In: Proceedings of the IEEE, vol. 86, pp. 2278–2324 (1998). http://citeseerx.ist.psu.edu/viewdoc/summary?doi=10.1.1.42.7665
11. Lehman, J., Stanley, K.: Abandoning objectives: evolution through the search for novelty alone. Evol. Comput. 19, 189–223 (2011)
12. Luke, S.: ECJ evolutionary computation library (1998). http://cs.gmu.edu/~eclab/projects/ecj/ (Accessed 6 May 2022)
13. Mouret, J., Clune, J.: Illuminating search spaces by mapping elites. CoRR abs/arXiv: 1504.04909 (2015)
14. Panait, L., Luke, S.: Cooperative multi-agent learning: The state of the art. Auton. Agent. Multi-Agent Syst. 11, 387–434 (2005)
15. Parker, G., Parashkevov, I.: Cyclic genetic algorithm with conditional branching in a predator-prey scenario. In: 2005 IEEE International Conference on Systems, Man and Cybernetics, vol. 3, pp. 2923–2928 (2005)
16. Poli, R., Langdon, W., McPhee, N.: A Field Guide to Genetic Programming. Lulu Enterprises UK Ltd. (2008)
17. Pozzuoli, A., Ross, B.J.: Increasing features in map-elites using an age-layered population structure. In: IEEE Congress on Evolutionary Computation, CEC 2023, Chicago, IL, USA, 1–5 July 2023, pp. 1–8. IEEE (2023). https://doi.org/10.1109/CEC53210.2023.10254093
18. Rohlfshagen, P., Liu, J., Perez-Liebana, D., Lucas, S.M.: Pac-man conquers academia: two decades of research using a classic arcade game. IEEE Trans. Games 10(3), 233–256 (2018)
19. Simonyan, K., Zisserman, A.: Very deep convolutional networks for large-scale image recognition. arXiv: 1409.1556 (2014)
20. Van Rossum, G., Drake Jr, F.L.: Python reference manual. Centrum voor Wiskunde en Informatica Amsterdam (1995)
21. Werner, G.M., Dyer, M.G.: Evolution of herding behavior in artificial animals. In: From Animals to Animats 2: Proceedings of the Second International Conference on Simulation of Adaptive Behavior (SAB92). The MIT Press (Apr 1993). https://doi.org/10.7551/mitpress/3116.003.0053
22. Yannakakis, G.N., Togelius, J.: Artificial Intelligence and Games. Springer (2018). https://gameaibook.org

Vision Transformers for Computer Go

Amani Sagri[1], Tristan Cazenave[1], Jérôme Arjonilla[1(✉)],
and Abdallah Saffidine[2]

[1] LAMSADE, Université Paris Dauphine-PSL, CNRS, Paris, France
jerome.arjonilla@hotmail.fr
[2] The University of New South Wales, Sydney, Australia

Abstract. Motivated by transformers' success in diverse fields like language understanding and image analysis, our investigation explores their potential in the game of Go. Specifically, we focus on analyzing Transformers in Vision. Through a comprehensive examination of factors like prediction accuracy, win rates, memory, speed, size, and learning rate, we underscore the significant impact transformers can make in the game of Go. Notably, our findings reveal that transformers outperform the previous state-of-the-art models, demonstrating superior performance metrics. This comparative study was conducted against conventional Residual Networks.

Keywords: Computer Go · Transformer · Games

1 Introduction

Due to a huge game tree complexity, the game of Go has been an important source of work in the perfect information setting. In 2007, search algorithms have been able to increase drastically the performance of computer Go programs [11,12,16,20]. In 2016, a groundbreaking achievement occurred when AlphaGo became the first program to defeat a skilled professional Go player [26]. Currently, the level of play of such algorithms is far superior to those of any human player [26–28].

Over the years, various significant advances have been made to improve performance in the game of Go [3,31–34]. Many of these innovations find their roots in other domains, notably in computer vision, where the recognition and interpretation of the Go board's image serve as fundamental inputs. Algorithms such as ResNet [3,18] and MobileNet [5,7,19] have demonstrated exceptional performance by harnessing groundbreaking developments in computer vision. However, it is worth noting that one remarkable advancement in the realm of computer vision remains relatively untapped for Computer Go: *transformers* [30].

Transformers represent a groundbreaking leap in deep learning, reshaping how various tasks in natural language processing (NLP), computer vision, and beyond are approached. Initially developed for NLP tasks, transformers

© The Author(s), under exclusive license to Springer Nature Switzerland AG 2024
S. Smith et al. (Eds.): EvoApplications 2024, LNCS 14635, pp. 376–388, 2024.
https://doi.org/10.1007/978-3-031-56855-8_23

introduce a departure from conventional sequential methods by employing self-attention mechanisms. These mechanisms simultaneously capture intricate interdependencies among all elements in a sequence. This ability to understand nuanced relationships over long distances, without relying on recurrent or convolutional structures, has propelled transformers to the forefront of AI research. Notably, transformers have not only advanced language understanding, exemplified by models like BERT [13], but have also expanded their utility to image analysis, as seen in Vision Transformers (ViTs) [15] and other transformer-based models. EfficientFormer [21], a transformer-based model, achieves high performance and matches MobileNet's speed on mobile devices, proving that well-designed transformers can deliver low latency in computer vision tasks.

In this paper, we propose to analyze the impact of using Transformer methods in the game of Go. To do this, we use the EfficientFormer architecture. Our study analyses were done in comparison with other state-of-the-art vision architectures in Go such as Residual Networks on a wide range of criteria including prediction accuracy, win rates, memory, speed, architecture size, and even learning rate. We tune the learning rate and the size of the network for each network and we find that EfficientFormer improves better than Residual Network with longer training times. Both the policy accuracy, the Mean Squared Error (MSE), and the Mean Absolute Error (MAE) are better with longer training time. Other important properties of the tested networks are their latency and their memory use. To take into account the latency in the performance of the networks, we make them play using the same Monte Carlo Tree Search search time at every move and record their winning rates. We observe that EfficientFormer of size '19' with a learning rate of 0.0005 with 1,000 epochs of 100,000 states and a batch of 32 is better than Residual Networks on CPU and on GPU with the same number of epochs.

In Sect. 2, we present Computer Go, while Sect. 3 introduces the various algorithms and network architectures employed in this paper. Our results are detailed in Sect. 4, and the final section provides a comprehensive summary of our work along with insights into future avenues.

2   Computer Go

The game of Go is a turn-taking strategic board game of perfect information, played by two players. One player adds black stones to a vacant intersection of the board and the opponent adds white stones. After being placed, a player's stones cannot move. A group of contiguous stones is removed if and only if the opponent surrounds the group on all orthogonally adjacent points. The players aim at capturing the most territory and the game ends when no player wishes to move any further. There exist multiple rules for scoring. We have used the Chinese rule in our experiments: the winner of the game is defined by the number of stones that a player has on the board, plus the number of empty intersections surrounded by that player's stones and komi (bonus added to the second player as compensation for playing second).

Even though the rules are relatively simple, the game of Go is known as an extremely complex one in comparison to other board games such as Chess. On the standard board of size 19×19, the number of legal positions has been estimated to be 2.1×10^{170}. Algorithms based on Monte Carlo Tree Search (MCTS) [1] have been achieving excellent performance in the game of Go for many years. Combining deep reinforcement learning and MCTS as introduced in the *AlphaGo* series programs [26–28] has been widely applied. The neural network takes an image of the board as input and produces two outputs: a probability distribution over moves (policy head) and a scalar of score prediction (value head) (see Fig. 1).

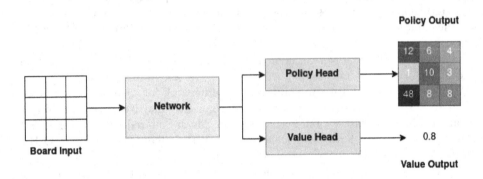

Fig. 1. AlphaZero network architecture

3   Network Architectures

3.1   Residual Network

Residual Networks are the standard networks for games [3,28]. They are used in combination with MCTS to evaluate the leaves of the search tree and to give a prior on the possible moves. To speed up the computation of the evaluation and of the prior the networks are usually run on a batch of states [6].

The employed residual layer in image classification integrates the input of the layer with its output, incorporating two convolutional layers before the addition. ReLU layers are applied following the first convolutional layer and after the addition. Figure 2 illustrates the structure of the residual layer. We will experiment with this kind of residual layer for our Go networks.

3.2   Transformer

Transformers are advanced neural network architectures that leverage the concept of self-attention to process and understand complex sequences of data, such

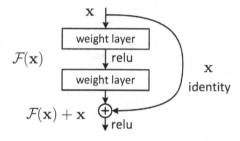

Fig. 2. The residual block.

as language. Self-attention allows a transformer model to analyze different elements within a sequence and determine their relative importance in relation to one another. By calculating attention scores based on the similarity of these elements, the model can dynamically weigh their significance and understand how they interrelate. Transformers employ multiple self-attention mechanisms (multihead self-attention) operating in parallel, enabling them to capture intricate patterns, dependencies, and contextual nuances across the entire input sequence.

Transformer was originally proposed as a sequence-to-sequence model [29] for machine translation. Later works show that Transformer-based pre-trained models (PTMs) [23] can achieve state-of-the-art performances on various tasks. In addition to language-related applications, Transformer has also been adopted in computer vision [2,15,22], audio processing [10,14,17] and even other disciplines, such as chemistry [25] and life sciences [24]. In natural Language Processing, the mechanism of attention of the Transformers tried to capture the relationships between different words of the text to be analyzed, in Computer Vision the Vision Transformers try instead to capture the relationships between different portions of an image.

3.3 Efficient Former

The EfficientFormer model is a big step forward in making transformer architectures work better for tasks that need real-time results, especially on devices with not much computing power. Adding a dimension-consistent plan allows the model to easily switch between different ways of organizing its parts, like in 4D and 3D setups. This way of thinking helps the EfficientFormer model to make the time it takes for predictions much shorter. By focusing on making predictions happen fast, a set of EfficientFormer models emerges, each achieving a careful equilibrium between performance and latency. This change in approach reaches its peak with models like EfficientFormer-l1, which impressively demonstrates outstanding top-1 accuracy on benchmarks like ImageNet-1K. At the same time, it manages to keep inference latency remarkably low on mobile devices, aligning closely with the efficiency of optimized versions of MobileNet.

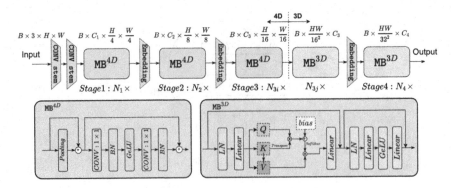

Fig. 3. Overview of EfficientFormer architecture [21].

In Fig. 3, we illustrate the network architecture of the EfficientFormer. The network begins its operations with a convolution stem. This part usually carries out a set of actions to pull out basic features from the input data. These actions typically involve using convolutional layers, pooling layers, and normalization layers. Following that, there is a set of MetaBlocks (MB). There are two types of MetaBlocks: MB^{4D} and MB^{3D}. The layers that make up each block are illustrated in Fig. 3. In between these blocks, we have the embedding layers. These layers break down the input into patches of a fixed size and assign each patch to a high-dimensional vector representation. These patch representations are then sent into the transformer blocks (within MB^{3D}) for further processing.

EfficientFormer is available in various sizes, denoted as 11, 13, 17, and 19. Each size is linked to a tuple of information where the first information is the width and the second information is the depth. The width is a list designing different dimensionalities (number of channels) of the feature vectors processed by different layers and blocks within the neural network. The width represents the number of blocks in different levels of the EfficientFormer architecture. The sizes tested throughout the paper are the following:

- '11' : ([48, 96],[3, 4])
- '13' : ([64, 128],[4, 6])
- '17' : ([96, 192],[6, 8])
- '19'; ([128,256],[8,10])

In order to clarify the notations used for transformers, we are going to explain in detail the parameters of the first example '11'. The entire structure comprises 7 MetaBlocks (MBs): 3 MBs^{4D} and 4 MBs^{3D}. The count of these blocks is determined by examining the second list [3, 4]. Let's denote the set of the 3 MBs^{4D} as X and the set of the 4 MBs^{3D} as Y. The number 48 from the first list represents the number of channels in all layers of X. While 96 is the number of channels in all layers of Y. Between the set X and the set Y, we have an embedding layer. This embedding layer allows the transition of the number of channels from 48 in X to 96 in Y.

3.4   Adaptation for the Game of Go

In this paper, we took the same work on EfficientFormer model of Li *et al.* [21] and adapt the transformer mechanism to the Go game prediction. This necessitated modifying the final layers, which were originally designed for tasks like classification or segmentation, and instead, replacing them with layers tailored for policy and value (*i.e.*, the probability of winning the game) prediction.

This modification transformed the tasks into a dual output setting, combining multiclass classification and regression functionalities. The value head uses Global Average Pooling followed by two Dense layers [4, 8]. The policy head uses a 1×1 convolution to a single plane that defines the convolutional policy [8].

Another significant adjustment involved the downsampling and embedding layers commonly used in image classification tasks to detect features by reducing the image size before feeding it into the transformer. However, in the context of Go, the input board's dimensions were fixed at 19×19, and it was imperative to preserve this size throughout the training process to avoid losing critical information. Therefore, to retain the richness of the board data, the height and width of the board were maintained during training, ensuring that no valuable details were lost in the process. This tailored architectural approach played a pivotal role in optimizing the models for Go game prediction.

4   Experimental Results

4.1   Dataset

The data used for training comes from the Katago Go program self played games [31]. KataGo is one of the strongest open-source Go bots available online. There are 1,000,000 different games in total in the training set. Input data consists of 31 planes with a size of 19×19 each, encompassing information such as the color to play, ladders, the current state on two planes, and two previous states on four planes. The output targets are the policy (a vector of size 361 with 1.0 for the move played, 0.0 for the other moves), and the value (close to 1.0 if White wins, close to 0.0 if Black wins).

4.2   Experimental Information

To compare the different network architectures we trained them on multiple epochs. One epoch uses 100,000 states randomly selected from the Katago dataset with two labels: a one-hot encoding of the Katago move and an evaluation between 0 and 1 by Katago of the winrate for White. The training is done with Adam and cosine annealing [9] without restarts. Cosine annealing leads to better convergence by modifying the learning rate of Adam.

In the next tables, we denote Residual(X, Y), the Residual Network of X blocks of Y planes and we denote Efficient(lX), the lX architecture of EfficientFormer. Among the different metrics used, we compute the accuracy, mean squared error (MSE), mean absolute error (MAE), and when possible the winning rate against an opponent. The winning rate is more informative than the other because it combines the impact of improving policy and value network. It also takes into account the latency of the networks. Accuracy measures the closeness of strategies between the policy network and Katago data. In the following experiment, we present the result on the validation set.

4.3   Training and Playing

Our research paper focuses on the study of transformers and their impact on several performance criteria, as well as speed and memory. In our experiments, we first analyze the best hyperparameters obtained on performance for models of varying sizes (11/13/15/19) and with varying lengths of training (100/500/1000 epochs). Our results on Efficient are compared against Residual Network, which is the current state of the art in the field. After that, we delve into a comprehensive analysis with a specific focus on assessing memory consumption and processing speed for each model.

In Table 1, we analyze the impact of the learning rate on the Accuracy, MSE, and MAE for multiple Residual networks across 1000 epochs in comparison to Efficient(19). We will then use the best learning rate in Table 2 where we analyze the Accuracy, MSE, and MAE for both Residual and Transformer networks across 500 epochs. We found that Efficient(19) performs better than the other Efficient Former and Residual networks. Additionally, Fig. 4 provides a visual comparison of the training curves for Accuracy, MAE, and MSE between Efficient (19) and Residual (20,256) over 1000 epochs.

In Tables 3 and 4, an analysis of winning rates among multiple algorithms compared to Efficient (19) over 500 and 1000 epochs is presented. It is noteworthy that Efficient (19) exhibits superiority over the majority of algorithms, demonstrating a remarkable performance trend. An exception is observed in the case of the 500-epoch scenario 3, where Efficient (17) achieves comparable performance on CPU, and Residual (20,256) attains equivalent performance on GPU. The longer training time and the better GPU are in favor of the transformer network.

In Table 5, we conduct an analysis of latency and peak memory utilization on GPU and CPU for the various network architectures under examination. The experiments on GPU were carried on a *RTX 2080 Ti* with 11 Gb of Memory and *Epyc* server for the CPU. On GPU, it is noteworthy that both Residual Networks and Transformers exhibit similar latency characteristics, however, Transformers incur a memory usage approximately three times greater than that of Residual Networks. On CPU (Table 5), for smaller network, we observe that Efficient architecture achieve a lower evaluation rate per second than Residual. Nevertheless, this trend shifts as network sizes increase, with the Efficient architecture ultimately surpassing Residual Networks in evaluation per second.

Lastly, Table 6 charts the evolution of GPU latency concerning batch size variation for the different network configurations. Large batch sizes are relevant to self-play in Alpha Zero style [28,31]. Smaller batch sizes are relevant to normal play with batch parallel MCTS [6]. This analysis sheds light on how network performance scales with batch size changes. The Residual Networks use relatively more playouts since they parallelize better with current GPU hardware and software.

Table 1. Comparison of large network architectures for 1000 epochs of 100,000 states per epoch.

Network	Size	Learning Rate	Batch	Accuracy	MSE	MAE
Residual	20,256	0.0002	64	52.13%	0.0496	0.1510
	20,256	0.0001	64	53.30%	**0.0458**	0.1456
	20,256	0.00005	64	**53.82%**	0.0465	**0.1434**
	20,256	0.00002	64	53.28%	0.0471	0.1483
	20,256	0.00001	64	51.23%	0.0516	0.1595
	40,256	0.0002	64	45.68%	0.0717	0.1990
	40,256	0.0001	64	51.88%	0.0513	0.1599
	40,256	0.00005	64	52.64%	**0.0475**	**0.1493**
	40,256	0.00002	64	**52.90%**	0.0486	0.1497
	40,256	0.00001	64	51.10%	0.0551	0.1625
Efficient	19	0.0005	32	**56.22%**	**0.0360**	**0.1240**

Table 2. Comparison of networks for 500 epochs of 100,000 states per epoch and a batch size of 64. The Accuracy, MSE and MAE were computed on a set of 50,000 states sampled from 50,000 games that were never seen during training.

Network	Size	Learning Rate	Accuracy	MSE	MAE
Residual	10,128	0.0002	49.12%	0.0534	0.1649
	20,128	0.0002	50.29%	0.0516	0.1618
	20,256	0.00005	**52.50%**	**0.0476**	**0.1518**
	40,256	0.00005	51.27%	0.0499	0.1586
Efficient	11	0.002	49.35%	0.0553	0.1659
	13	0.002	51.28%	0.0484	0.1519
	17	0.002	53.01%	0.0440	0.1422
	19	0.001	**54.29%**	**0.0405**	**0.1351**

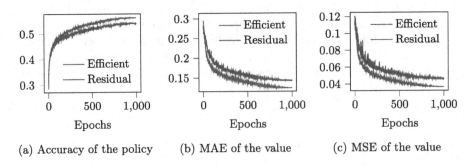

(a) Accuracy of the policy   (b) MAE of the value   (c) MSE of the value

Fig. 4. Efficient(l9) versus Residual(20,256) comparisons on Accuracy, MAE and MSE over 1,000 epochs of 100,000 states. All values are computed with the test set that contains games that were never seen in the training set.

Table 3. Comparison of networks for 500 epochs of 100,000 states per epoch and a batch size of 64. The winrate WinCPU is the result of 800 randomized matches on CPU against Efficient(l9) with 10 s of CPU per move for both sides. The GPU winrate is calculated by using the same RTX 2080 Ti GPU time for both networks with a batch of 64. The learning rate is fixed to 0.0005 for the Residual network and 0.001 for the Efficient network.

Network	Size	vs Efficient(l9)	
		WinCPU	WinGPU
Residual	10,128	33.5%	20.4%
	20,128	31.6%	25.8%
	20,256	30.6%	46.4%
	40,256	8.9%	29.7%
Efficient	l1	11.6%	8.1%
	l3	31.0%	19.4%
	l7	50.4%	38.3%

Table 4. Comparison of the winning rate for multiple networks trained during 1,000 epochs of 100,000 states per epoch. The winrate WinCPU is the result of 400 randomized matches on CPU with 10 s of CPU per move for both sides. The GPU winrate is calculated by using the same A6000 GPU time for both networks with a batch of 32 for the inference.

Network	Size	vs Efficient(l9)	
		WinCPU	WinGPU
Residual	20,256	30.5%	39.0%
	40,256	15.0%	33.0%

Table 5. Latency and number of evaluations per second on CPU/GPU for different architectures and networks of different sizes and peak memory on GPU. The CPU used is *Epyc* server. The GPU used is a RTX 2080 Ti GPU with 11 Go. The latency and the peak memory on the GPU are measured using a batch of 64 states. They are averaged over 100 calls to predict after a warmup of 100 previous calls. The latency is the average time in seconds to make a forward pass.

Network	Size	CPU		GPU		
		Latency	Evals/s	Latency	Evals/s	Peak Memory
Residual	10,128	0.043	23.07	0.0890	719	436,656,640
	20,128	0.082	12.24	0.0943	679	350,025,728
	20,256	0.304	3.29	0.1185	540	452,578,816
	40,256	0.455	2.20	0.1580	405	529,187,072
Efficient	11	0.065	15.27	0.0958	668	1,101,474,048
	13	0.074	13.52	0.1106	579	1,148,030,976
	17	0.092	10.90	0.1307	490	1,159,418,368
	19	0.159	6.30	0.1700	376	1,179,129,088

Table 6. Evolution of the A6000 GPU latency with the size of the batch. The latency and the peak memory are the median values of 7 runs. Each run is the average over 100 forwards after a warmup of 100 forwards.

Network	Size	Batch	GPU Latency	Evals/s on GPU	Peak Memory
Residual	20,256	32	0.111	288	253,801,472
	20,256	64	0.126	508	548,938,240
	20,256	128	0.159	805	800,936,192
	20,256	256	0.227	1,128	1,566,134,528
	20,256	512	0.368	1,391	2,954,716,416
	20,256	1,024	0.667	1,535	4,793,448,960
Efficient	19	32	0.128	250	589,454,592
	19	64	0.168	381	1,141,404,672
	19	128	0.224	571	2,297,159,168
	19	256	0.346	740	4,359,236,608
	19	512	0.583	878	8,672,660,992
	19	1,024	1.062	964	17,121,701,376

5   Conclusion

This paper investigates the impact of the Vision Transformer architecture on the game of Go. Building upon the proven success of the Transformer architecture across diverse domains, our investigation seeks to explore its potential in the context of Go. Our analysis traverses a multitude of critical dimensions, ranging from prediction accuracy and win rates to memory utilization, processing speed, and learning rates. Through this examination, we underscore the significant role that Transformers can play in enhancing performance in the game of Go.

Significantly, our findings highlight the benefits of the EfficientFormer architecture, showcasing remarkable performance enhancements on both CPU and GPU platforms. Notably surpassing the benchmarks set by the previous state-of-the-art algorithms, this superiority becomes particularly pronounced in the context of larger networks, underscoring the scalability and efficiency of the EfficientFormer.

In addition, it is essential to emphasise that the impact and adaptability of the EfficientFormer architecture goes beyond the boundaries of the game of Go, extending its applicability to a wide range of games and domains. This versatility positions the EfficientFormer as a promising candidate for pushing the boundaries of artificial intelligence not only in strategic board games but also in various other complex decision-making scenarios.

References

1. Browne, C., et al.: A survey of Monte Carlo tree search methods. IEEE Trans. Comput. Intell. AI in Games **4**(1), 1–43 (2012)
2. Carion, N., Massa, F., Synnaeve, G., Usunier, N., Kirillov, A., Zagoruyko, S.: End-to-end object detection with transformers. In: Vedaldi, A., Bischof, H., Brox, T., Frahm, J.-M. (eds.) Computer Vision – ECCV 2020: 16th European Conference, Glasgow, UK, August 23–28, 2020, Proceedings, Part I, pp. 213–229. Springer International Publishing, Cham (2020). https://doi.org/10.1007/978-3-030-58452-8_13
3. Cazenave, T.: Residual networks for computer go. IEEE Trans. Games **10**(1), 107–110 (2018)
4. Cazenave, T.: Spatial average pooling for computer go. In: Computer Games: 7th Workshop, CGW 2018, IJCAI 2018, Stockholm, Sweden, July 13, 2018, Revised Selected Papers 7, pp. 119–126. Springer (2019)
5. Cazenave, T.: Improving model and search for computer go. In: 2021 IEEE Conference on Games (CoG), pp. 1–8. IEEE (2021)
6. Cazenave, T.: Batch monte carlo tree search. In: Browne, C., Kishimoto, A., Schaeffer, J. (eds.) Computers and Games: International Conference, CG 2022, Virtual Event, November 22–24, 2022, Revised Selected Papers, pp. 146–162. Springer Nature Switzerland, Cham (2023). https://doi.org/10.1007/978-3-031-34017-8_13
7. Cazenave, T.: Mobile networks for computer go. IEEE Trans. Games **14**(1), 76–84 (2022)
8. Cazenave, T., et al.: Polygames: improved zero learning. J. Int. Comput. Games Assoc. **42**(4), 244–256 (2020)

9. Cazenave, T., Sentuc, J., Videau, M.: Cosine annealing, mixnet and swish activation for computer go. In: Browne, C., Kishimoto, A., Schaeffer, J. (eds.) Advances in Computer Games: 17th International Conference, ACG 2021, Virtual Event, November 23–25, 2021, Revised Selected Papers, pp. 53–60. Springer International Publishing, Cham (2022). https://doi.org/10.1007/978-3-031-11488-5_5

10. Chen, X., Wu, Y., Wang, Z., Liu, S., Li, J.: Developing real-time streaming transformer transducer for speech recognition on large-scale dataset. In: ICASSP 2021–2021 IEEE International Conference on Acoustics, Speech and Signal Processing (ICASSP), pp. 5904–5908. IEEE (2021)

11. Coulom, R.: Efficient selectivity and backup operators in monte-carlo tree search. In: van den Herik, H.J., Ciancarini, P., Donkers, H.H.L.M.J. (eds.) CG 2006. LNCS, vol. 4630, pp. 72–83. Springer, Heidelberg (2007). https://doi.org/10.1007/978-3-540-75538-8_7

12. Coulom, R.: Computing elo ratings of move patterns in the game of Go. ICGA J. **30**(4), 198–208 (2007)

13. Devlin, J., Chang, M.W., Lee, K., Toutanova, K.: BERT: pre-training of deep bidirectional transformers for language understanding. In: Proceedings of the 2019 Conference of the North American Chapter of the Association for Computational Linguistics: Human Language Technologies, pp. 4171–4186 (2019)

14. Dong, L., Xu, S., Xu, B.: Speech-transformer: a no-recurrence sequence-to-sequence model for speech recognition. In: 2018 IEEE International Conference on Acoustics, Speech and Signal Processing (icassp), pp. 5884–5888. IEEE (2018)

15. Dosovitskiy, A.,et al.: An image is worth 16x16 words: Transformers for image recognition at scale. arXiv preprint arXiv:2010.11929 (2020)

16. Gelly, S., Silver, D.: Monte-Carlo tree search and rapid action value estimation in computer Go. Artif. Intell. **175**(11), 1856–1875 (2011)

17. Gulati, A., et al.: Conformer: convolution-augmented transformer for speech recognition. arXiv preprint arXiv:2005.08100 (2020)

18. He, K., Zhang, X., Ren, S., Sun, J.: Deep residual learning for image recognition. In: Proceedings of the IEEE Conference on Computer Vision and Pattern Recognition, pp. 770–778 (2016)

19. Howard, A.G., et al.: Mobilenets: efficient convolutional neural networks for mobile vision applications. arXiv preprint arXiv:1704.04861 (2017)

20. Kocsis, L., Szepesvári, C.: Bandit based monte-carlo planning. In: Fürnkranz, J., Scheffer, T., Spiliopoulou, M. (eds.) Machine Learning: ECML 2006, pp. 282–293. Springer Berlin Heidelberg, Berlin, Heidelberg (2006). https://doi.org/10.1007/11871842_29

21. Li, Y., et al.: Efficientformer: vision transformers at mobilenet speed. Adv. Neural. Inf. Process. Syst. **35**, 12934–12949 (2022)

22. Parmar, N., et al.: Image transformer. In: International Conference on Machine Learning, pp. 4055–4064. PMLR (2018)

23. Qiu, X., Sun, T., Xu, Y., Shao, Y., Dai, N., Huang, X.: Pre-trained models for natural language processing: a survey. SCIENCE CHINA Technol. Sci. **63**(10), 1872–1897 (2020)

24. Rives, A., et al.: Biological structure and function emerge from scaling unsupervised learning to 250 million protein sequences. Proc. Natl. Acad. Sci. **118**(15), e2016239118 (2021)

25. Schwaller, P., et al.: Molecular transformer: a model for uncertainty-calibrated chemical reaction prediction. ACS Cent. Sci. **5**(9), 1572–1583 (2019)

26. Silver, D.: Mastering the game of go with deep neural networks and tree search. Nature **529**, 484–489 (2016)

27. Silver, D., et al.: Mastering chess and shogi by self-play with a general reinforcement learning algorithm. CoRR **abs/1712.01815** (2017), http://arxiv.org/abs/1712.01815

28. Silver, D., et al.: A general reinforcement learning algorithm that masters chess, shogi, and go through self-play. Science **362**(6419), 1140–1144 (2018)

29. Sutskever, I., Vinyals, O., Le, Q.V.: Sequence to sequence learning with neural networks. In: Advances in Neural Information Processing Systems, vol. 27 (2014)

30. Vaswani, A., et al.: Attention is all you need. In: 31st International Conference on Neural Information Processing Systems (NIPS 2017), pp. 6000–6010 (2017)

31. Wu, D.J.: Accelerating self-play learning in go. arXiv preprint arXiv:1902.10565 (2019)

32. Wu, I.C., Wu, T.R., Liu, A.J., Guei, H., Wei, T.: On strength adjustment for MCTS-based programs. In: Proceedings of the AAAI Conference on Artificial Intelligence. vol. 33, pp. 1222–1229 (2019)

33. Wu, T.R., Wei, T.H., Wu, I.C.: Accelerating and improving AlphaZero using population based training. In: Proceedings of the AAAI Conference on Artificial Intelligence. vol. 34, pp. 1046–1053 (2020)

34. Wu, T.R., et al.: Multilabeled value networks for computer go. IEEE Trans. Games **10**(4), 378–389 (2018)

Surrogate-Assisted Evolutionary Optimisation

Integrating Bayesian and Evolutionary Approaches for Multi-objective Optimisation

Tinkle Chugh$^{(\boxtimes)}$ ⓘ and Alex Evans

Department of Computer Science, University of Exeter, Exeter, UK
{t.chugh,aje220}@exeter.ac.uk

Abstract. Both Multi-Objective Evolutionary Algorithms (MOEAs) and Multi-Objective Bayesian Optimisation (MOBO) are designed to address challenges posed by multi-objective optimisation problems. MOBO offers the distinct advantage of managing computationally or financially expensive evaluations by constructing Bayesian models based on the dataset. MOBO employs an acquisition function to strike a balance between convergence and diversity, facilitating the selection of an appropriate decision vector. MOEAs, similarly focused on achieving convergence and diversity, employ a selection criterion. This paper contributes to the field of multi-objective optimisation by constructing Bayesian models on the selection criterion of decomposition-based MOEAs within the framework of MOBO. The modelling process incorporates both mono and multi-surrogate approaches. The findings underscore the efficacy of MOEA selection criteria in the MOBO context, particularly when adopting the multi-surrogate approach. Evaluation results on both real-world and benchmark problems demonstrate the superiority of the multi-surrogate approach over its mono-surrogate counterpart for a given selection criterion. This study emphasises the significance of bridging the gap between these two optimisation fields and leveraging their respective strengths.

Keywords: Bayesian optimisation · Gaussian processes · Pareto optimality · Evolutionary computation · Many-objective optimisation

1 Introduction

Real-world optimisation problems may involve more than conflicting objectives to be optimised. Such problems are known as multi-objective optimisation problems (MOPs) and are defined as:

$$\text{minimise } \mathbf{f} = (f_1(\mathbf{x}), \ldots, f_m(\mathbf{x})) \qquad \text{subject to} \quad \mathbf{x} \in S,$$

where $m \geq 2$ is also the number of objectives and S is the (nonempty) feasible space and is a subset of the decision space $\mathbb{R}^n$ and consists of decision vectors $\mathbf{x} = (x_1, \ldots, x_n)^T$. There is no single solution to such problems due to trade-offs between objectives but a set of so-called Pareto optimal solutions. Multi-objective optimisation methods such as multi-objective evolutionary

S. Smith et al. (Eds.): EvoApplications 2024, LNCS 14635, pp. 391–406, 2024.
https://doi.org/10.1007/978-3-031-56855-8_24

algorithms (MOEAs) provide mechanisms to consider the potential trade-offs between objectives and provide users with an approximated set of Pareto optimal solutions.

In some real-world MOPs, the objective functions involve computationally intensive assessments. Such problems typically involve black-box evaluators (or simulators) lacking closed-form expressions for the objective functions. To mitigate the computational burden and find an approximate set of Pareto optimal solutions with minimal function evaluations, Bayesian optimisation (BO) comes into play. These approaches rely on a Bayesian model acting as a surrogate or metamodel for the objective functions. They identify promising decision vectors through the optimisation of an acquisition function. The Bayesian model frequently takes the form of a Gaussian process because it offers a meaningful quantification of uncertainty, which is subsequently utilised in optimising the acquisition function. The acquisition function strikes a balance between exploration and exploitation, guiding the search process. In multi-objective BO, several works exist [6,10,23,26] on developing efficient acquisition functions.

Multi-objective Bayesian optimisation (BO) typically involves two distinct approaches to constructing a Bayesian model. In the first approach, separate models are created for each objective function (computational complexity is at most $O(mN^3)$, where N is the size of the data set), and an acquisition function that utilises these models is then employed to identify promising decision vectors. This approach is known as the multi-surrogate approach. An example of a multi-objective BO employing the Expected Hypervolume Improvement (EHVI) [11,12,28] belongs to the multi-surrogate approach. The second approach develops a single Bayesian model (computational complexity of $O(N^3)$) after aggregating the objective functions using a scalarising function. This approach is referred to as the mono-surrogate approach. A well-known algorithm that falls into the mono-surrogate category is ParEGO [15,21], where the weighted Tchebycheff [27] function was used as the scalarising function. The second approach effectively reduces the number of objectives from m to one. Additionally, a single-objective acquisition function can be used in the mono-surrogate approach. The mono-surrogate approach assumes the distribution of the scalarising function is Gaussian. In a recent work [3,4], the author showed that the distribution of the weighted Tchebycheff in the multi-surrogate approach is not Gaussian and empirically showed that the multi-surrogate is better than the mono-surrogate approach.

Decomposition-based MOEAs such as MOEA/D [29] and RVEA [2] decompose a multi-objective optimisation problem into a number of single objective optimisation problems using a uniformly distributed set of weight vectors [5]. A scalarising function such as weighted sum, weighted Tchebycheff, Penalty boundary intersection [29] and angle penalised distance [2] is then used as the selection criterion to select a solution for each weight vector. The scalarising function attempts to balance convergence and diversity when selecting a solution, which is similar to the role of the acquisition function in multi-objective Bayesian optimisation. Inspired by the wide use of decomposition-based MOEAs, we utilise their potential in multi-objective BO by building Bayesian surrogate models on the scalarising function. We embed these models in both mono- and multi-surrogate approaches in the context of multi-objective BO. To summarise, the contributions of the current work are as follows:

1. Leveraging the Capabilities of Decomposition-Based Multi-Objective Evolutionary Algorithms (MOEAs) in Multi-Objective Bayesian Optimization.
2. Examination of Mono- and Multi-Surrogate Approaches through the Construction of Bayesian Models on the Selection Criterion of Decomposition-Based MOEAs.
3. Validation of the Efficacy of the Multi-Surrogate Approach through Empirical Testing on Real-World and Benchmark Multi-objective Optimisation Problems.

The rest of the paper is structured as follows. Section 2 provides a comprehensive background on multi-objective Bayesian Optimization and offers an overview of decomposition-based MOEAs. Section 3 provides the main methodology explaining how the selection criterion in decomposition-based MOEAs can be effectively combined in multi-objective Bayesian optimisation. Section 4 provides the results from the implementation of the proposed methodology, and Sect. 5 concludes the paper with the future research directions.

2 Background

2.1 Multi-objective Bayesian Optimisation

The steps of a multi-objective BO algorithm are outlined in Algorithm 1. It starts with an initial data set, $D = \{(\mathbf{x}_i, \mathbf{f}(\mathbf{x}_i))\}_{i=1}^{N}$ of size N. This data set can be obtained through random sampling or other techniques such as design of experiment [18]. A scalarised function, G is then used to aggregate objective functions, i.e. $\{g(\mathbf{x}_i) = G(\mathbf{f}(\mathbf{x}_i), \mathbf{w}, \boldsymbol{\zeta})\}_{i=1}^{N}$, where $\mathbf{w}$ is the weight vector and $\boldsymbol{\zeta}$ is the vector of parameters for a given scalarising function. The existing data is then used to build one Gaussian process model on $\{(\mathbf{x}_i, g_i(\mathbf{x}_i))\}_{i=1}^{N}$ (mono-surrogate approach) or m surrogate models on $\{(\mathbf{x}_i, f_1(\mathbf{x}_i))\}_{i=1}^{N}, \ldots, \{\mathbf{x}_i, f_m(\mathbf{x}_i))\}_{i=1}^{N}$ (multi-surrogate approach).

Algorithm 1. Bayesian optimisation

Input: Data Set $D = \{(\mathbf{x}_i, \mathbf{f}_i(\mathbf{x}_i))\}_{i=1}^{N}$
Output: Evaluated solutions

1: **while** Termination criterion is not met **do**
2:     Train the $\mathcal{GP}$ models on the data set
3:     Optimise the acquisition function i.e. $\mathbf{x}^* \leftarrow \operatorname{argmax}_{\mathbf{x}} \alpha(x)$
4:     Evaluate $\mathbf{x}^*$ and add to the data set
5: **end while**

Gaussian process ($\mathcal{GP}$) models are non-parametric and are defined by a multivariate normal distribution with a mean and covariance function (or kernel). The kernel captures the correlation between different data points. The parameters of the kernel, such as length scale, amplitude and noise, can be estimated by maximising the marginal likelihood function [22]. An acquisition function, $\alpha(\mathbf{x})$, such as expected improvement (a widely used and efficient acquisition function) is then optimised with the models and existing data set using an appropriate optimiser. The EI is defined as:

$$\alpha_{EI}(\mathbf{x}) = \int_{-\infty}^{g'(\mathbf{x})} (g'(\mathbf{x}) - g(\mathbf{x}))dg,$$

where $g'(\mathbf{x})$ is the lowest scalarised function value. In the mono-surrogate approach, the EI has a closed-form expression as the posterior predictive distribution on the scalarising function is Gaussian and is derived as:

$$\alpha_{EI}(\mathbf{x}) = (g'(\mathbf{x}) - \mu(\mathbf{x}))\Phi\left(\frac{g'(\mathbf{x}) - \mu(\mathbf{x})}{\sigma(\mathbf{x})}\right) + \sigma(\mathbf{x})\phi\left(\frac{g'(\mathbf{x}) - \mu(\mathbf{x})}{\sigma(\mathbf{x})}\right), \quad (1)$$

where $\mu(\mathbf{x})$ and $\sigma(\mathbf{x})$ are the posterior mean and standard deviation from $\mathcal{GP}$, $\Phi(\cdot)$ and $\phi(\cdot)$ are cumulative and probability distribution functions of standard normal distribution, respectively. In the multi-surrogate approach, EI may not have a closed-form expression and can be estimated by approximation such as Monte Carlo. This work uses EI as the acquisition function in both mono3- and multi-surrogate approaches.

2.2   Decomposition-Based MOEAs

Decomposition-based MOEAs [7,14,17,25,29] have gained widespread popularity, particularly for their efficacy in addressing optimisation problems characterised by a large number of objectives, often referred to as many objectives. These algorithms can effectively explore the trade-offs and dependencies between objectives by decomposing the problem into a number of subproblems. Two fundamental components that define the success of these MOEAs are decomposition utilising weight vectors and the selection criteria.

The selection criterion in a decomposition-based MOEA plays a pivotal role in determining the quality of solutions and the convergence behaviour of the algorithm. As highlighted in the introduction, various scalarising functions, including weighted sum, weighted Tchebycheff, and penalty boundary intersection, can be employed to choose solutions for a given subproblem effectively. The selection criterion becomes instrumental in navigating the algorithm toward Pareto optimal solutions, embodying a critical aspect of the algorithm's overall performance.

2.3   Scalarising Functions

Weighted sum (WS) [13,19] is considered one of the simplest and most straightforward scalarising functions, valued for its ease of implementation. It is defined as:

$$g(\mathbf{x}) = \sum_{i=1}^{m} w_i f_i(\mathbf{x}), \tag{2}$$

where w_i is the ith component of the weight vector, $\mathbf{w}$.

Weighted Tchebycheff (TCH) [1,27] is derived from the L_p metric with $p = \infty$. In multi-objective BO, the function (with the augmentation term) was first used in ParEGO [15](mono-surrogate approach). It is defined as:

$$g(\mathbf{x}) = \max_{i=1,\dots,M} [w_i | f_i(\mathbf{x}) - z_i^* |], \tag{3}$$

where z_i^* is the ith component of the ideal objective vector, $\mathbf{z}^*$. In this work, we normalise the objective function values between [0,1] and therefore, the ideal objective vector is a vector of zeros.

Penalty boundary intersection (PBI) [5,29] handles convergence and diversity and was first applied in MOEA/D [29]. It is defined as

$$g(\mathbf{x}) = d_1 + \beta d_2, \tag{4}$$

where $d_1 = \left| \mathbf{f}(\mathbf{x}) \cdot \frac{\mathbf{w}}{\|\mathbf{w}\|} \right|$ and $d_2 = \left\| \mathbf{f}(\mathbf{x}) - d_1 \frac{\mathbf{w}}{\|\mathbf{w}\|} \right\|$, $\| \cdot \|$ represents the norm. The performance of the scalarising function can be influenced by the parameter β.

Angle Penalised distance (APD) [2] uses the angle, θ, between the objective vector and the weight vector and balances convergence and diversity. It is defined as:

$$g(\mathbf{x}) = (1 + P(\theta)) \cdot \|\mathbf{f}(\mathbf{x})'\|, \quad \text{where } P(\theta) = m \left(\frac{FE}{maxFE} \right)^{\delta} \frac{\theta}{\gamma}, \tag{5}$$

where FE and $maxFE$ are the number and maximum number of evaluations, respectively. γ is the minimum of all angles between a weight vector and other weight vectors, and δ is a user-defined parameter. $\mathbf{f}(\mathbf{x})'$ is the translated objective vector: $\mathbf{f}(\mathbf{x})' = \mathbf{f}(\mathbf{x}) - \mathbf{z}^*$. As can be seen in the equation; the function depends on two parameters, $maxFE$ and δ. Contour plots of all four scalarising functions for a weight vector $= [0.5, 0.5]$, $\beta = 5$, $\delta = 1$, $FE = 1$ and $maxFE = 10$ are shown in Fig. 1.

3   Multi-objective BO with MOEAs Selection Criterion

We first create a data set D with Latin hypercube sampling and generate a set of uniformly distributed weight vectors [5]. We randomly select a weight vector, $\mathbf{w}$ and estimate $g(\mathbf{x})$ using scalarising function, $G(\mathbf{f}(\mathbf{x}), \mathbf{w}, \zeta)$. For the mono-surrogate approach, we build a single Gaussian process model on the scalarising function and in the multi-surrogate approach, we build m independent models for each objective function. Given the data set and model(s), we optimise the EI to find the next decision vector. This process is continued until a termination criterion is met. The mono and multi-surrogate approaches in multi-objective Bayesian optimisation are outlined in Algorithm 2.

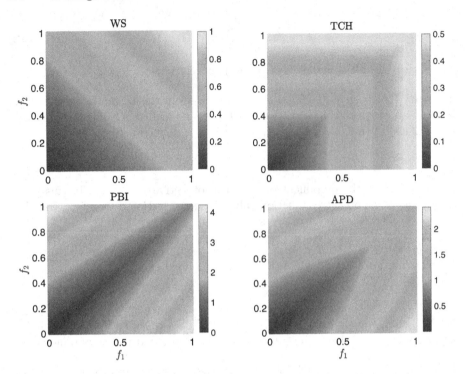

Fig. 1. Contour plots of four scalarising functions, weighted sum (WS), weighted Tchebycheff (TCH), penalty boundary intersection (PBI) and angle penalised distance (APD)

3.1 Estimation of Acquisition Function

In the mono-surrogate approach, the EI has a closed-form expression (Eq. 1) as the model is directly built on $g(\mathbf{x})$. However, the EI may not have a closed-form expression in the multi-surrogate approach except when using the weighted sum in the multi-surrogate approach. In the multi-surrogate approach when using the weighted sum, the distribution of the scalarising function after building independent $\mathcal{GP}$ models on the objectives is also Gaussian:

$$g(\mathbf{x}) = \sum_{i=1}^{m} w_i f_i(\mathbf{x}), \quad \text{where } f_i(\mathbf{x}) \sim \mathcal{N}(\mu_i(\mathbf{x}), \sigma_i(\mathbf{x})^2) \text{ OR}$$

$$g(\mathbf{x}) \sim \mathcal{N}\left(\sum_{i=1}^{m} w_i \mu_i(\mathbf{x}), \sum_{i=1}^{m} w_i^2 \sigma_i(\mathbf{x})^2 \right).$$

The weighted Tchebycheff scalarising function can be approximated in the form of a Gumbel distribution [3]. However, this does not provide a closed-form expression of EI. Therefore, we use the Monte-Carlo approximating the EI for other scalarising functions in the multi-surrogate approach.

Algorithm 2. Mono- and multi-surrogate approaches in multi-objective BO

Input: Data Set $D = \{(\mathbf{x}_i, \mathbf{f}_i(\mathbf{x}_i))\}_{i=1}^{N}$; A set of uniformly distributed weight vectors

Output: Evaluated solutions

1: **while** Termination criterion is not met **do**
2:     Select a random weight vector $\mathbf{w}$ from the set of weight vector
3:     Estimate $\mathbf{g}(\mathbf{x}) = G(\mathbf{f}(\mathbf{x}), \mathbf{w}, \zeta)$
4:     Train $\mathcal{GP}$ model on $\mathbf{g}(\mathbf{x})$ or m models on $f_1(\mathbf{x}), \ldots, f_m(\mathbf{x})$
5:     Optimise the acquisition function i.e. $\mathbf{x}^* \leftarrow \text{argmax}_{\mathbf{x}}\, \alpha(\mathbf{x})$
6:     Evaluate $\mathbf{x}^*$ and add to the data set
7: **end while**

3.2 Demonstration

We provide a demonstration of the differences between mono- and multi-surrogate approaches on a bi-objective and one decision variable optimisation problem. The data set, underlying objective space, the underlying scalarising function space and the Pareto front of the problem are shown in Fig. 2 (row 1). The second and third rows show the mono-surrogate approach, involving building of a $\mathcal{GP}$ model on scalarising functions and the landscape of EI as the acquisition function.

Figure 3 displays the multi-surrogate approach, where independent $\mathcal{GP}$ models are built on the objectives (row 1). The second and third rows in Fig. 3 illustrate the Monte Carlo approximation of the scalarising functions and the EI after building independent $\mathcal{GP}$ models on the objectives.

Two observations emerge from this demonstration: (1) Predictive models on the scalarising functions in the multi-surrogate approach more accurately replicate the underlying landscape of the scalarising function compared to the mono-surrogate approach. (2) The landscape of the acquisition function and the locations of maxima differ between both approaches for a given scalarising function. Upon evaluating the maxima with the underlying objectives, the solution obtained after one iteration is shown in Fig. 4. Notably, solutions obtained through both approaches with the same scalarising functions differ. The multi-surrogate approach, particularly when using weighted sum, weighted Tchebycheff, and angle-penalized distance, approximates proximity to the Pareto front after a single iteration. It is important to note that these results reflect outcomes after one iteration and further iterations will be necessary for both approaches to converge towards the Pareto front.

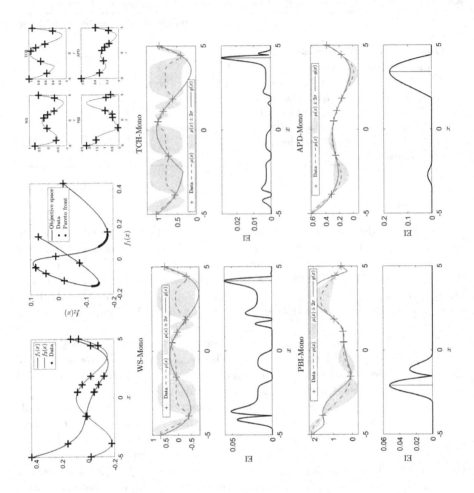

Fig. 2. Data set and underlying objective space, Pareto front and underlying scalarising function landscape (row 1). $\mathcal{GP}$ model on scalarising functions (mono-surrogate) with posterior means and standard deviations and landscape of EI. (rows 2 and 3).

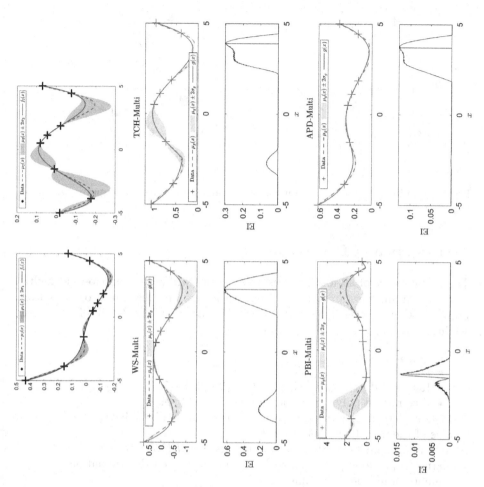

Fig. 3. Independent $\mathcal{GP}$ models on the objective functions (row 1). The rows 2–3 show the approximation of scalarising functions (multi-surrogate) and EI with Monte-Carlo approximations.

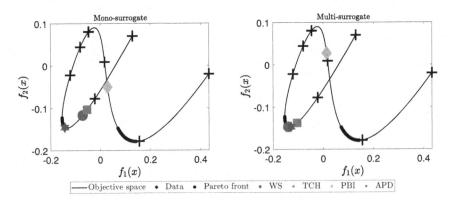

Fig. 4. Results after one iteration on mono- and multi-surrogate approaches from Figs. 2 and 3.

4   Results and Discussion

The performance of all four scalarising functions within both mono- and multi-surrogate approaches are evaluated on benchmark and real-world optimisation problems with 2–9 objectives and 2–7 decision variables. The details of numerical settings with parameters are as follows:

- Problems:
  - DTLZ [9]: $(m, d) = (2, 5)$
  - Real-world problems [24]: $(m, d) = (2 - 9, 2 - 7)$
  - Free radical polymerisation [20]: $(m, d) = (3, 4)$.
  Further details of real-world problems are provided in Table 1.
- Size of the initial data set: $10 \times d$
- Maximum number of function evaluations, $maxFUN = 30 \times d$
- Kernel: Squared exponential with automatic relevant determination
- Number of independent runs: 11
- Parameters, ζ in scalarising functions: $\beta = 5$, $\delta = 1$
- Number of weight vectors (with Simplex-Lattice design method [5]): $100(m = 2); 105(m = 3); 120(m = 4); 126(m = 5); 132(m = 6); 174(m = 9)$
- Performance indicator: Hypervolume ratio (Hypervolume of solutions divided by the hypervolume of Pareto front). The approximated Pareto fronts of real-world problems are provided on the GitHub repository, https://github.com/ryojitanabe/reproblems, and for FRP, one run of NSGA-II [8] was used to approximate the Pareto front.
- Optimiser to maximise marginal likelihood in $\mathcal{GP}$: BFGS with 10-restarts
- Optimiser to maximise the EI: Genetic algorithm

Figures 5 and 6 illustrate the hypervolume ratio with the number of expensive function evaluations for different problems. As can be observed, the multi-surrogate approach performed better than their mono-surrogate counterparts across the majority of the problems. Figure 7 provides a visual representation of

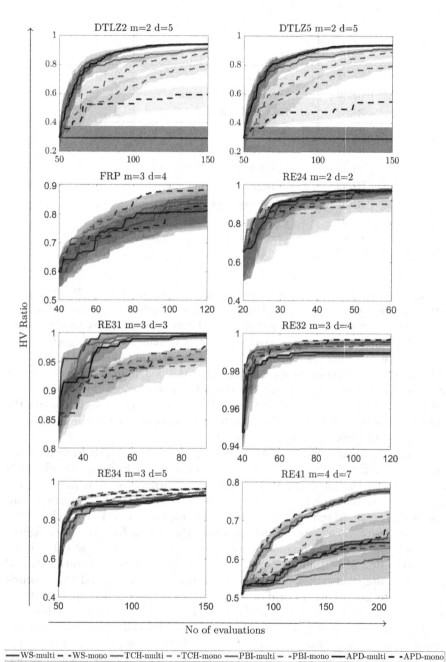

Fig. 5. Hypervolume ratio with the number of expensive function evaluations.

Table 1. Real-world problems and their dimensionality in objective and decision spaces. For more details, see [20, 24]

Abbreviation	Problem	m	d
RE24	Hatch cover design	2	2
RE31	Two bar truss design	3	3
RE32	Welded beam design	3	4
RE33	Disc brake design	3	4
RE34	Vehicle crashworthiness design	3	5
RE37	Rocket injector design	3	4
RE41	Car side impact design	4	7
RE42	Conceptual marine design	4	6
RE61	Water resource planning	6	3
RE91	Car cab design	9	7
FRP	Free radical polymerisation	3	4

the proportion where an algorithm was either the best or statistically equivalent to the best algorithm. A statistical analysis was conducted to rigorously compare the performance of the four scalarising functions in each approach and collectively. An approach was considered the best if it achieved the largest median hypervolume value among all runs after the maximum number of function evaluations. To establish statistical significance, a one-sided, paired Wilcoxon signed-rank test [16] was employed, with the Holm-Bonferroni correction to adjust for multiple comparisons. This analysis aimed to robustly assess and compare the performance of the different scalarising functions and approaches in solving the given multi-objective optimisation problems.

In both the mono- and multi-surrogate approaches, the Tchebycheff scalarising function demonstrated superior performance, emphasising its relevance, particularly in the context of the ParEGO algorithm as a mono-surrogate approach. However, upon considering all eight combinations (comprising four scalarising functions and two approaches), Tchebycheff and APD in the multi-surrogate approach consistently outperformed the others. Notably, the weighted sum exhibited good performance by either being the best or statistically equivalent to the best algorithm in certain problems within both mono-surrogate and multi-surrogate approaches. To verify these findings, we checked the approximated Pareto fronts of real-world problems, which are available from the GitHub repository. These Pareto fronts exhibit convex shapes, validating the efficacy of the weighted sum approach. These intriguing results call for a more in-depth investigation to understand the specific characteristics of problems (as evident in [14] for the decomposition-based MOEAs) and their correlation with the performance of scalarising functions in the context of multi-objective BO.

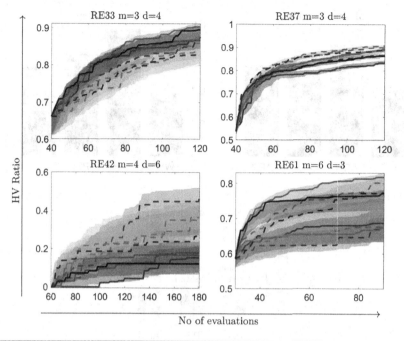

Fig. 6. Hypervolume ratio with the number of expensive function evaluations.

5   Conclusions

This work employed the selection criteria of decomposition-based multi-objective evolutionary algorithms within the context of multi-objective Bayesian optimisation. Four distinct selection criteria, functioning as scalarising functions, were utilised in both mono- and multi-surrogate approaches: weighted sum, weighted Tchebycheff, penalty boundary intersection, and angle-penalized distance. Notably, the multi-surrogate approach outperformed the mono-surrogate approach overall. For certain selection criteria, weighted Tchebycheff, penalty boundary intersection, and angle-penalized distance in the multi-surrogate approach, closed-form expressions were not available. Consequently, the Monte Carlo approximation was employed for estimating the acquisition function. Among the scalarising functions, weighted Tchebycheff and angle-penalized distance demonstrated superior performance.

It is noteworthy that two scalarising functions, penalty boundary intersection and angle-penalized distance, are reliant on specific parameters that can influence their performance. Future research will include sensitivity analysis and the automatic learning of these parameters when integrating them into multi-objective Bayesian optimisation. The future work will also explore the performance of various scalarising functions based on the shape and properties of the

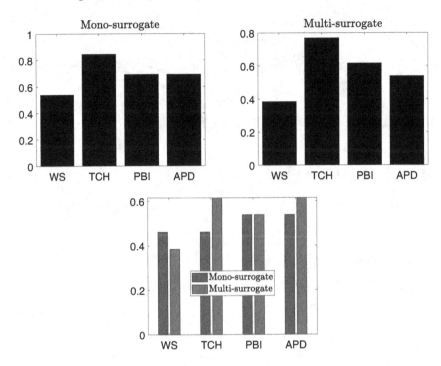

Fig. 7. Performance of different scalarising functions. The bar heights correspond to the proportion of time a scalariser (mono, multi or both) was best or statistically equivalent to the best.

multi-objective optimisation problem, including aspects such as convexity and characteristics of Pareto fronts, such as disconnected and degenerated fronts. Additionally, we acknowledge that the multi-surrogate approach is computationally more expensive than the mono-surrogate approach, and while this computational time may be negligible for solving expensive multi-objective optimisation problems, future work will address strategies to alleviate the computation cost associated with the multi-surrogate approach.

References

1. Bowman, V.J., et al.: On the relationship of the Tchebycheff norm and the efficient frontier of multiple-criteria objectives. In: Thiriez, H., Zionts, S. (eds.) Multiple Criteria Decision Making: Proceedings of a Conference Jouy-en-Josas, France May 21–23, 1975, pp. 76–86. Springer Berlin Heidelberg, Berlin, Heidelberg (1976). https://doi.org/10.1007/978-3-642-87563-2_5
2. Cheng, R., Jin, Y., Olhofer, M., Sendhoff, B.: A reference vector guided evolutionary algorithm for many-objective optimization. IEEE Trans. Evol. Comput. **20**, 773–791 (2016)

3. Chugh, T.: Mono-surrogate vs multi-surrogate in multi-objective Bayesian optimisation. In: Proceedings of the Genetic and Evolutionary Computation Conference Companion, pp. 2143–2151. ACM (2022)

4. Chugh, T.: R-MBO: A multi-surrogate approach for preference incorporation in multi-objective Bayesian optimisation. In: Proceedings of the Genetic and Evolutionary Computation Conference Companion, pp. 1817–182. ACM (2022)

5. Das, I., Dennis, J.E.: Normal-boundary intersection: a new method for generating the pareto surface in nonlinear multicriteria optimization problems. SIAM J. Optim. **8**(3), 631–657 (1998)

6. Daulton, S., Balandat, M., Bakshy, E.: Differentiable expected hypervolume improvement for parallel multi-objective bayesian optimization. In: Larochelle, H., Ranzato, M., Hadsell, R., Balcan, M.F., Lin, H. (eds.) Advances in Neural Information Processing Systems. vol. 33, pp. 9851–9864. Curran Associates, Inc. (2020)

7. Deb, K., Jain, H.: An evolutionary many-objective optimization algorithm using reference-point-based nondominated sorting approach, part I: solving problems with box constraints. IEEE Trans. Evol. Comput. **18**, 577–601 (2014)

8. Deb, K., Prarap, A., Agarwal, S., Meyarivan, T.: A fast and elitist multiobjective genetic algorithm: NSGA-II. IEEE Trans. Evol. Comput. **6**, 182–197 (2002)

9. Deb, K., Thiele, L., Laumanns, M., Zitzler, E.: Scalable test problems for evolutionary multiobjective optimization. In: Abraham, A., Jain, L., Goldberg, R. (eds.) Evolutionary Multiobjective Optimization, pp. 105–145. Springer-Verlag, London (2005). https://doi.org/10.1007/1-84628-137-7_6

10. Emmerich, M., Beume, N., Naujoks, B.: An EMO algorithm using the hypervolume measure as selection criterion. In: Coello Coello, C.A., Hernández Aguirre, A., Zitzler, E. (eds.) Evolutionary Multi-Criterion Optimization, pp. 62–76. Springer Berlin Heidelberg, Berlin, Heidelberg (2005). https://doi.org/10.1007/978-3-540-31880-4_5

11. Emmerich, M., Deutz, A., Klinkenberg, J.: Hypervolume-based expected improvement: Monotonicity properties and exact computation. In: Proceedings of the IEEE Congress on Evolutionary Computation, pp. 2147–2154. IEEE (2011)

12. Emmerich, M., Giannakoglou, K., Naujoks, B.: Single- and multiobjective evolutionary optimization assisted by Gaussian random field metamodels. IEEE Trans. Evol. Comput. **10**, 421–439 (2006)

13. Gass, S., Saaty, T.: The computational algorithm for the parametric objective function. Naval Res. Logist. Quart. **2**(1–2), 39–45 (1955)

14. Ishibuchi, H., Setoguchi, Y., Masuda, H., Nojima, Y.: Performance of decomposition-based many-objective algorithms strongly depends on pareto front shapes. IEEE Trans. Evol. Comput. **21**(2), 169–190 (2017)

15. Knowles, J.: ParEGO: a hybrid algorithm with on-line landscape approximation for expensive multiobjective optimization problems. IEEE Trans. Evol. Comput. **10**, 50–66 (2006)

16. Knowles, J. D., Thiele, L., Zitzler, E.: A tutorial on the performance assessment of stochastic multiobjective optimizers. Tech. rep. (2006)

17. Majumdar, A., Chugh, T., Hakanen, J., Miettinen, K.: Probabilistic selection approaches in decomposition-based evolutionary algorithms for offline data-driven multiobjective optimization. IEEE Trans. Evol. Comput. **26**, 1182–1191 (2022)

18. Mckay, M., Beckman, R., Conover, W.: A comparison of three methods for selecting values of input variables in the analysis of output from a computer code. Technometrics **42**, 55–61 (2000)

19. Miettinen, K.: Nonlinear multiobjective optimization. Kluwer, Boston, MA (1999)

20. Mogilicharla, A., Chugh, T., Majumder, S., Mitra, K.: Multi-objective optimization of bulk vinyl acetate polymerization with branching. Mater. Manuf. Processes **29**, 210–217 (2014)
21. Rahat, A.A.M., Everson, R.M., Fieldsend, J.E.: Alternative infill strategies for expensive multi-objective optimisation. In: Proceedings of the Genetic and Evolutionary Computation Conference, pp. 873–880. GECCO '17, ACM, New York, NY, USA (2017)
22. Rasmussen, C.E., Williams, C.K.I.: Gaussian processes for machine learning. The MIT Press (2006)
23. Shahriari, B., Swersky, K., Wang, Z., Adams, R.P., de Freitas, N.: Taking the human out of the loop: A review of Bayesian optimization. Proc. IEEE **104**(1), 148–175 (2016)
24. Tanabe, R., Ishibuchi, H.: An easy-to-use real-world multi-objective optimization problem suite. Appl. Soft Comput. **89**, 106078 (2020)
25. Trivedi, A., Srinivasan, D., Sanyal, K., Ghosh, A.: A survey of multiobjective evolutionary algorithms based on decomposition. IEEE Trans. Evol. Comput. **21**(3), 440–462 (2017)
26. Wagner, T., Emmerich, M., Deutz, A., Ponweiser, W.: On expected-improvement criteria for model-based multi-objective optimization. In: Schaefer, R., Cotta, C., Kołodziej, J., Rudolph, G. (eds.) Parallel Problem Solving from Nature, PPSN XI, pp. 718–727. Springer Berlin Heidelberg, Berlin, Heidelberg (2010). https://doi.org/10.1007/978-3-642-15844-5_72
27. Wierzbicki, A.P.: The use of reference objectives in multiobjective optimization. In: Fandel, G., Gal, T. (eds.) Multiple Criteria Decision Making Theory and Application, pp. 468–486. Springer Berlin Heidelberg, Berlin, Heidelberg (1980). https://doi.org/10.1007/978-3-642-48782-8_32
28. Yang, K., Emmerich, M., Deutz, A., Bäck, T.: Multi-objective Bayesian global optimization using expected hypervolume improvement gradient. Swarm Evol. Comput. **44**, 945–956 (Feb 2019)
29. Zhang, Q., Li, H.: MOEA/D: a multiobjective evolutionary algorithm based on decomposition. IEEE Trans. Evol. Comput. **11**, 712–731 (2007)

Author Index

Printed in the United States
by Baker & Taylor Publisher Services